Arid Land Resources and their Management

Arid Land Resources and their Management

Jordan's Desert Margin

Foreword by

HRH Crown Prince Hassan bin Talal and HRH The Duke of Kent

Edited by

Roderic W. Dutton, John I. Clarke and Anwar Battikhi

KEGAN PAUL INTERNATIONAL
London and New York

First published in 1998 by
Kegan Paul International Limited
UK: P.O. Box 256, London WC1B 3SW, England
Tel: (0171) 580 5511 Fax: (0171)436 0899
E-mail: books@keganpau.demon.co.uk
Internet: http://www.demon.co.uk/keganpaul/
USA: 562 West 113th Street, New York, NY 10025, USA
Tel: (212) 666 1000 Fax: (212) 316 3100

Distributed by

John Wiley & Sons Ltd
Southern Cross Trading Estate
1 Oldlands Way, Bognor Regis,
West Sussex, PO22 9SA, England
Tel: (01243) 779 777 Fax: (01243) 820 250

Columbia University Press
562 West 113th Street
New York, NY 10025, USA
Tel: (212) 666 1000 Fax: (212) 316 3100

Printed in Great Britain by
TJ International Ltd. Padstow

ISBN 0 7103 0612 1

British Library Cataloguing in Publication Data

Arid land resources and their management: Jordan's desert margin
1. Arid regions – Jordan 2. Land use – Jordan 3. Jordan – Economic Conditions
1. Dutton, Roderic, 1942 – II. Clarke, John I. (John Innes), 1929 –
III. Batticki, Anwar Munir
333.7'36'095695

ISBN 0710306121

Library of Congress Cataloguing-in-Publication Data

Arid Land resources and their management: Jordan's desert margin / Editors, Roderic W. Dutton, Jon I. Clarke, Anwar Batticki: Foreword by HRH Crown Prince Hassan bin Talal and HRH The Duke of Kent.
p. cm.
Includes bibliographical references.
ISBN 0-7103-0612-1
1. Arid regions—Jordan. I. Dutton, Roderic. II. Clarke, John Innes.
III. Batticki, Anwar.
GB618.81.J6A75 1988
333.73'6'95697—dc21 97-48326
CIP

Patrons

HRH The Crown Prince El-Hassan of Jordan
HRH The Duke of Kent

Counterpart Organizations

Higher Council for Science and Technology, Jordan
and
Royal Geographical Society
(with the Institute of British Geographers), UK
with
Centre for Overseas Research and Development
University of Durham, UK

Programme Directors

Mohammed Shahbaz
Higher Council for Science and Technology

Roderic W. Dutton
Centre for Overseas Research and Development

H.H. Sharifa Zein bint Nasser
Finance and Public Relations, Higher Council for Science and Technology

The Jordan Badia Research and Development Programme acknowledges the financial and in-kind contributions made by the following organizations

Major financial contributors, Jordan and UK

Ministry of Finance, Jordan
Higher Council for Science and Technology, Jordan
Department for International Development, UK
Department of the Environment, UK

Major in-kind contributors

Jordan Armed Forces
Royal Jordanian Air Force
Public Security
Royal Arab League/Border Guards
Ministry of Agriculture
Ministry of Water and Irrigation
Ministry of Municipal and Rural Affairs and Environment
Ministry of Public Works and Housing
Ministry of Health
Department of Statistics
Department of Meteorology
Natural Resources Authority
Royal Jordanian Geographic Centre
NASA (Pathfinder Program), USA
Jordan News Agency (Petra)
Jordan Television (JTV)
Gold Star Company
The Jordan Cement Factories Co. Ltd.
Arab Potash Company
Jordan Phosphate Mines Co. Ltd.
Izzideen Naseef

Corporate Benefactors

Royal Jordanian
Land Rover

Smiths Industries
Gernot Langes Swarovski

Sponsors

Barclays Private Bank
British Council
British Embassy, Amman
Foreign and Commonwealth Office
Greencard Charitable Trust
Hunting Technical Services
Iraq Petroleum Company
Johnston Pipes
Lloyds Bank Charitable Trust
Scott Wilson Kirkpatrick
Siebe plc
Sir Alexander Gibb and Partners
Walter Guinness Charitable Trust
Whinney Murray & Co.
Royal Botanic Gardens, Kew
Royal Society
Tropical Agricultural Association

Steering Committee Members

(Aug 1994 to present unless otherwise stated)

Dr. Hani El Mulki, co-Chairman for HCST, Aug 1994-Mar 1997

Dr. Khalid El-Shuraydeh, deputized for Dr.El Mulki

Dr. Usama El Khaldi, co-Chairman for HCST, Mar 1997-

Dr. John Hemming, co-Chairman for RGS-IBG

Dr. Bassam Snober, Member and Chairman of the Administration Committee, BRDP (Dean of Agriculture, University of Jordan), Aug 1994-July 1997

Dr. Abdulla Al Musa, Member and Chairman of the Administration Committee, BRDP (Dean of Agriculture, University of Jordan), July 1997-

Prof. John I. Clarke, Member and Chairman of the RGS-IBG Planning Group

Brigadier Salim Khalifeh, Member (Director General Royal Jordanian Geographic Centre)

Prof. Mahmud Duwayri, Member (Director General National Centre for Agricultural Research and Technology Transfer), Aug 1994-May 1995

Mr. Nabil Katkhuda, deputized for Prof. Duwayri

Dr. Awni Taimeh, Member (Director General National Centre for Agricultural Research and Technology Transfer), June 1995-

Dr. Fakhruddin Daghistani, Member (Director General Natural Resources Authority), Aug 1994-Mar 1996

Khalid Al-Shiab, Member (Director General Natural Resources Authority), Mar 1996-

Administration Committee Members

(Aug 1994 to present unless otherwise stated)

Dr. Bassam Snobar, Chairman (Dean of Agriculture University of Jordan), Aug 1994-July 1997

Dr. Abdulla Al Musa, Chairman (Dean of Agriculture University of Jordan), July 1997-

H.H. Sharifa Zein bint Nasser, Finance and Public Relations Director

Dr. Zahir Rawajfeh, Member (Dean of Agriculture JUST University), Aug 1994-Jan 1996

Dr. Abdur-Rahman Al-Fataftah, Programme Director, Aug 1994-Jan 1995, Member (Dean Faculty of Science, Yarmuk University)
Mohammed Shahbaz, Programme Director, Jan 1995-
Dr. Roderic Dutton, co-Director (Durham University, UK)
Dr. Wajeh Awais, Member (Dean Faculty of Science, Yarmuk University)
Dr. Salman Al-Bdour, Member (Representing al Al-Bayt University)
Ma'arouf Al-Bakheet (Jordan Armed Forces)
Hassan Al-Mehasen, Member (Jordan Armed Forces), Aug 1994-Mar 1995
Hamdan A. Smairan, Member (Jordan Armed Forces), July 1996-
Walid Kreshan, Member (Jordan Armed Forces), Mar 1995-July 1996
Rouhe Al-Kaied, Member (Ministry of Interior), Aug 1995-Feb 1997
Saaed Qawqazeh, Member (Ministry of Interior), Feb 1997-
Kriem Ashoshan, Member (Public Security)
Dr. Saied Khatari, Member (Dean of Agriculture, Mut'ah University), Aug 1995-June 1997
Dr. Khaled Al Tarawneh, Member (Dean of Agriculture, Mut'ah University, June 1997-
Dr. Marwan Muala, Member (Dean of Agriculture, JUST University), Jan 1996-
Shaykh Sualem Al Sufian, Member (Local Community), Sep 1995-Feb 1996
Mr. Zaied Al-Bnaian, Member (Local Community), May 1996-Feb 1997
Falha Al-Rshoud, Member (Local Community), Mar 1997-Sep 1997
Shaykh Rakad Al-Shabib (Local Community), Oct 1997-

CONTENTS

ACRONYMS

ACSAD	Arab Center for the Studies of Arid Zones and Dry Land
ANOVA	Analysis of variance
ASFR	Age specific fertility rate
AWS	Automatic weather station
AWSA	Amman Water and Sewerage Authority
BRDP	(Jordan) Badia Research and Development Programme
CHC	Comprehensive health centre
EC	Electrical conductivity
HCST	Higher Council for Science and Technology
IMR	Infant mortality rate
IPC	Iraq Petroleum Company
JPFHS	Jordan Population and Family Health Survey
JTV	Jordan Television
JUST	Jordan University for Science and Technology
MCH	Maternity and child health
NASA	National Aeronautics and Space Agency
NCARTT	National Centre for Agricultural Research and Technology Transfer
NERC	Natural Environment Research Council
NGO	Non government organization
NRA	Natural Resources Authority
PC	Personal computer
PHC	Primary health centre
PPR	Peste des petits ruminants
RGS-IBG	Royal Geographical Society (with the Institute for British Geographers)
RJGC	Royal Jordanian Geographic Centre
RRA	Rapid rural appraisal
TDS	Total dissolved solids
TFR	Total fertility rate
UJ	University of Jordan
UNDP	United Nations Development Programme
UNFPA	United Nations Fund for Population Activities
USAID	United States Agency for International Development
USSCS	United States Soil Conservation Service
WAJ	Water Authority of Jordan
XRD	X-ray defraction
XRF	X-ray fluorescence

برنامج بحث وتطوير البادية الاردنية

JORDAN BADIA RESEARCH AND DEVELOPMENT PROGRAMME

The wonderful world of the desert is revealed in all its glories and challenges in the Jordanian Badia. With its dunes and pavements, Bedouin and cultivators, its wadis and oases, oryxes and gazelles, it stretches from the Fertile Crescent into the heart of the Arabian peninsula. The Badia covers well over four-fifths of the country and although we cannot be precise about its area, because there is no defined linear boundary and the level of desertification oscillates from year to year, the desert is clearly the prevalent ecosystem.

With a rapidly growing population, arising from a combination of natural increase and immigration, Jordan is faced with increasing population pressure, especially in its already densely settled areas, which tend to be separate from the Badia. Faced with such uneven human occupation, Jordan naturally seeks to integrate the Badia more closely into its living space; to stimulate greater awareness of its value, and to see how it can contribute to the economy and society in a sustainable way without damaging the fragile desert environments.

The Jordan Badia Research and Development Programme has been an excellent collaborative venture between Jordan's Higher Council for Science and Technology and the Royal Geographical Society (with the Institute of British Geographers) of London. It has brought together numerous Jordanian and British scientists, as well as those of other nations, to study a wide variety of aspects of the physical and human environment with a view towards sustainable development. Based upon the Field centre at Safawi, the study area covers an extensive part of the northern Badia, which forms a pilot project for the whole of the Jordanian Badia. We feel sure that Safawi will gain further recognition as an important regional centre for desert research, of increasing value to other countries in the region.

We are happy to welcome this first volume of scientific papers, synthesizing much of the work so far. It will complement the existing Working Papers, and will soon be joined by other more specialized volumes on different research themes.

HRH Crown Prince El Hassan bin Talal
Patron

HRH The Duke of Kent
Patron

ROYAL JORDANIAN الملكية الأردنية

OFFICIAL CARRIER TO THE PROGRAMME — الناقل الرسمي للبرنامج

برعاية
صاحب السمو الملكي
الامير الحسن ولي العهد
صاحب السمو الملكي
دوق كنت

Patrons
HRH Crown Prince El-Hassan
HRH The Duke of Kent

المجلس الاعلى للعلوم والتكنولوجيا
ص.ب: ٩٢٥٩٦٧
عمان – الأردن
هاتف: ٨٤٠٤٠١ (٦-٩٦٢)
فاكس: ٨٤٠٥٨٩ (٦-٩٦٢)

The Higher Council for Science and Technology (HCST)
P. O. Box: 925967
Amman - Jordan
Tel: (962-6) 840401
Fax: (962-6) 840589

الجمعية الجغرافية الملكية
Royal Geographical Society (RGS)
1 Kensington Gore
London SW7 2AR
Tel: (071) 589 5466
Fax: (071) 823 7200

CORD
مركز البحوث والتنمية لما وراء البحار
Centre for Overseas Research and Development (CORD)
University of Durham,
Durham, UK, DH1 3LE
Tel/Fax: (091) 374 2495

PREFACE

The arid lands, or Badia, of north-east Jordan have been the subject of many forces of change throughout the course of the twentieth century, particularly since the end of the First World War. National boundaries drawn in the 1920s separated Jordan's Badia from similar areas in Syria, Iraq and Saudi Arabia and meant that the people living within these new boundaries became Jordanian and had to develop a set of relationships with the new Jordanian government in Amman which now provides basic services to the population, including education, health care, electricity and water. Then in the 1930s the Iraq Petroleum Company (IPC) built a pipeline from Kirkuk to Haifa through northern Jordan, creating pumping stations known as H4 and H5 around which grew the settlements of Ruwayshid and Safawi. Today, although some pastoralists retain their traditional lifestyle of full mobility with their sheep and goats, most are only partially mobile for parts of the year or have adopted a fully settled way of life in a cluster of new villages between Safawi and Mafraq, or further afield. Vehicles have replaced camels for most forms of transport (except desert police patrols), but a growing number of sheep and goats overgraze the natural vegetation and are dependent on imported grain-based concentrates for survival during much of the year. Irrigated farming has been introduced, dependent on deep bores, high pressure pumps and sophisticated, piped irrigation systems. Large volumes of good quality water are pumped from the area to Amman for domestic consumption. In addition, a network of paved roads has been built, including the international highway linking Amman with Baghdad. International links are also increased by some members of the population travelling to find work overseas, while others commute to Mafraq and Amman.

These many changes to the population, the resource base and the management of the Badia's natural resources are the subject of the Jordan Badia Research and Development Programme, based at Safawi in a field study area comprising the western half of Jordan's pan-handle. The Programme, initiated by Crown Prince El-Hassan, is under the aegis of the Higher Council for Science and Technology in Amman and the Royal Geographical Society in London. Some of the main research findings from the first phase of the Programme, 1992-6, are brought together in this volume which is an edited selection from about 50 field reports made by Jordanian, British and other researchers during this period.

The two overview chapters (1 and 2) respectively define the changing relationship between the people and their environment and describe the natural resources upon which the people in the Badia depend. Both chapters set the findings in the study area into a wider Jordanian Badia context. Subsequent chapters, which focus more directly on the study area, divide into groups. Chapters 3 to 10 cover aspects of the hydrological cycle from precipitation to water demand while Chapters 11 to 13 look at aspects of the relationship between rainfall and irrigation and geomorphological processes. A checklist of the flora (Ch. 14), including rain dependent grazing plants, then precedes an introduction to the changing pastoral system for the management of sheep and goats (Ch. 15). A final group of papers describes demographic change (Chs. 16 and 17), provides an introduction to aspects of social structure (Chs. 18 and 19) and discusses the provision of health and education services (Chs. 20 to 22). The volume concludes with a selection of abstracts of papers by Jordanian authors in Arabic and English.

Collectively the papers establish a baseline for the natural and human resources of the study area. More baseline studies are in hand in phase two of the Programme, 1997-2000, and together they will facilitate the monitoring and evaluation of change in future years.

The Programme, long-term, is also deeply concerned with questions of economic and social productivity. The aim of the research is to provide decision makers, in both the national government and local households, the information they require for sustainable management of Badia resources by government agencies and local people on behalf of the population and the nation as a whole.

Such a large collaborative research programme has benefited from the close involvement of numerous people, from our keenly interested Patrons who have supported us wholeheartedly to the youngest undergraduates who have set foot in the Badia for the first time. The long lists of committee members and authors give only an indication of the numerous scholars, administrators, researchers and officials who have contributed to the success of the Programme, and to whom thanks must be given. To them must be added, in particular, the people of the Badia without whom this work could not be undertaken. To all who have made contributions to the Programme our grateful thanks are extended.

Roderic W. Dutton
John I. Clarke
Durham, June 1997

Overviews

CHAPTER 1

Population, Environment and Development

RODERIC W. DUTTON
(University of Durham)

1. CONCEPTS AND DEFINITIONS

There is an ever-changing dynamic balance between the population of a given locality and the environment in which that population lives; each impacts on the other and each responds to alterations in the other. By environment is here implied that combination of natural resources to which the population under consideration (the local community) has direct access. The dynamic balance between the people and their directly accessible natural resources may be referred to as the inner system, which is spatially defined and contained. In previous eras of low world population, simple technologies and limited demand on resources, the inner system was not significantly perturbed by the wider context, or outer system, within which it operated. The fact that it maintained itself over long time periods is also proof that it was sustainable. Change was slow, allowing the impact of change to be monitored and self-correcting mechanisms put in place to restore the dynamic balance. The outer system is comprised of the actions of the political and economic decision-making communities of people that are external to and beyond the control of the local community. These external communities become an integral part of the overall system as soon as they make decisions which affect the local community under consideration and the natural resources to which it has direct access, and therefore the impact of each upon the other.

It is generally true to say that the inhabitants of arid lands are not primary engineers of change or development. Typically they have evolved a relatively unchanging way of life that is in harmony with the scarce resources that are available to them. They perceive a need for only a few physical possessions. Frequently they are (or traditionally were) also mobile pastoralists, moving to follow the seasons or the occurrence of the unpredictable and localized rainfall that is characteristic of arid lands and which generates the pastures that are essential for the well-being of the pastoralists' flocks and herds.

Where people in arid lands are truly dependent on their environment for life, they tend to treat it with great respect. The local community evolves a deep and practical sense of responsibility for the environment in which it lives, based on an awareness of its real long-term self interest. Local custom and law ensure, for example, that pastures are not over-grazed, trees are not felled and sources of water are not fouled.

However, during the twentieth century the growing power of urban communities (typically in regions richer in natural resources) has increasingly impinged upon both the local communities and the physical environments of arid lands. Ideas and technologies developed in the urban centres have created new demands for arid land resources (mainly for the benefit of the urban populations), or generated new services and products which have become integrated into the lifestyle of the arid land population. This has altered the relationship of the people to their lands, and encouraged a desire for a lifestyle (including better services and more physical possessions) which cannot be supported by local natural resources. It tends also to reduce their sense of responsibility for their local physical environment, as their relative dependence on it diminishes, frequently leading to the depletion and degradation of those resources upon which the people previously depended. This matters, because it creates instability in both a physical sense (for example, flash floods, increased erosion, squandering of the already scarce resources because of a loss of the traditional sense of responsibility for them) and a political sense (disaffected people alienated from the central national authorities).

But in practice is there anything that can be done to improve the situation? In Jordan this is the objective of the Jordan Badia Research and Development Programme (the Badia Programme or BRDP). Research is being undertaken in order to gain a better understanding of the natural resource base of the Programme study area and the traditional and evolving usage of these resources, and thus to suggest actions which will lead to the protection of the resources and to their sustainable usage for the long-term benefit of the local population. The Programme took shape in the months following the Gulf War, after earlier discussions between Jordan's Higher Council for Science and Technology (HCST) and the UK's Royal Geographical Society (RGS) at the request of HRH Crown Prince El-Hassan of Jordan, in his role as Chairman of HCST. Thus the Programme is currently a joint Jordan-UK enterprise under the aegis of HCST in Jordan and the RGS in London with HRH Crown Prince El-Hassan and HRH the Duke of Kent as patrons. In the UK, the Centre for Overseas Research and Development (CORD), University of Durham, acts on behalf of the RGS.

During December 1991 discussions between colleagues from Jordan and the UK defined the aims and objectives, and the broad outlines of the research work that would be required to achieve them. It was also agreed that the focus of the fieldwork would be the area around the small town of Safawi some 150km east of Amman. Some of the buildings at the old oil-pumping station known as H5 at Safawi were made available to the Programme, and they have been under renovation ever since. Some fieldwork started in 1992 and 1993, including a baseline human science survey undertaken in January-February 1993. The Programme was formally inaugurated in April 1994 when a framework of research activities for the first main phase of the Programme was finalized (BRDP 1994). In the autumn of 1994 two long-term researchers from the UK took up residence at Safawi to study aspects of livestock management and production. They became the nucleus of what has become a large Jordan-UK livestock research team (Ch. 15). Other work initiated includes climate (Ch. 3) and hydrology (Chs. 4-8), earth sciences (Chs. 11-13), plant taxonomy (Ch. 14), biodiversity, water management (Chs. 9-10), demography (Chs. 16 and 17), aspects of development anthropology and gender studies (Chs. 18 and 19) and research into the key education and health services (Chs. 20-22). A long-term programme framework was agreed in May 1995 (Shahbaz and Dutton 1995a), and a three-year second phase framework (1996-8) was approved in December 1995 (Shahbaz and Dutton 1995b). At the same meeting it was agreed that the core Jordan-UK activities should be complemented by regional and international contributions to the research programme.

The Badia Programme area comprises some 11,210km^2 of the western half of the Jordan panhandle that extends north-east towards the boundary with Iraq, and is equal to around 14 per cent of the total land mass of Jordan (Fig. 1). The area has international boundaries with Syria to the north and Saudi Arabia to the south. The eastern border is coincident with the western border of the Hammad Basin project, centred at Ruwayshid. It also approximately follows the eastern edge of the basalt outcrops which cover much of the area's land surface. In the north-west the border of the Badia Programme is roughly coincident with the Azraq basin but the western border to the programme area then arches eastwards and runs to the Saudi border east of Azraq. However, for certain aspects of our research, data are being collected from further afield.

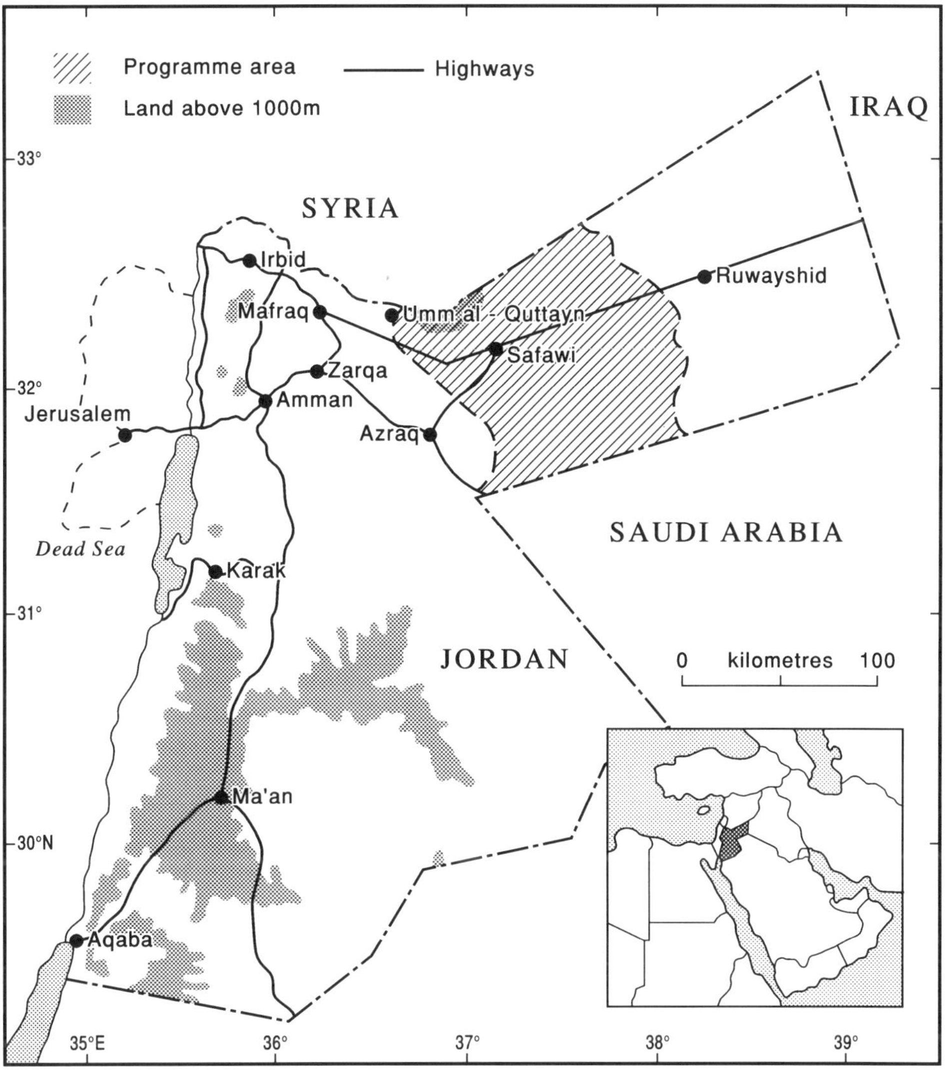

Fig. 1. The location of the Badia Research and Development Programme.

2. THE ENVIRONMENT OF THE STUDY AREA

The Badia Programme is located within a very large area of semi-arid and arid lands which extend to the west, north and east to the so-called 'fertile crescent' of terrain which is relatively well provided with water from rainfall and springs (to the west), rainfall (to the north) and the rivers Euphrates and Tigris (to the east). These well-watered lands gave rise to some of the first agricultural communities known in the world, and thence to urban centres including Jericho, Damascus, Mari, Uruk and others. In the Badia lived the Bedouin basing their livelihood on livestock and trade. There were of course no telecommunications, no roads, no motorized transport and no international boundaries.

2.1. Terrain and Soils

Tertiary-quaternary continental basalt flows and tuffs cover almost the entire Programme area. They form distinct strata dating from 23 to less than 0.5 million years ago for some exposed rocks that originated in lava flows centred on the Jabal al-Arab. Significant mineral resources have been identified including zeolite, basalt, tuff, porcellanite, bentonite and others (Ch. 2). Their economic value remains to be confirmed but some of them could become a significant focus for future development. Some small gasfields and oilfields have also been identified.

Topographically the area is not complex. The highest ground, 1,100-1,200m, occurs to the north-west, south of the Jabal al-Arab in Syria. The other relatively high ground is formed by the north-south chain of volcanic cones towards the eastern edge of the study area. Between the two, from the north, the weakly defined wadi Rajil descends towards the Azraq basin which is the low point of the area. Much of the land surface is composed of gently undulating low hills with elevation differences of 25-30m within short distances of each other, and gentle concavo-convex slopes.

In the basalt fields the dominant feature is large basalt boulders overlying poorly weathered basaltic soils, but towards the south-east small, angular chert fragments are more common. Near the Saudi boundary there is some sand. A combination of wind and water erosion has washed some of the soils into the low points which show up on satellite imagery as distinct off-white patches between the black boulder covered higher land. Some of the low points are basins, or near basins, locally known as *qa'a*, which collect water after the rains and tend to be covered with fine deposits of very limited permeability and which may be somewhat saline. Other low points have a small but distinct gradient so that water flows through them leaving coarser grained non-saline deposits that are more permeable, and are locally known as *marab*.

A better understanding of the characteristics of the basalt boulder surfaces (Ch. 13), of surface water infiltration (Ch. 12) and of soil capping characteristics (Ch. 11) under the impact of rainfall and irrigation will materially assist with hydrological modelling and water management.

2.2. Climate and Water Availability

On this high steppe land (mean elevation 800m) far from marine influence the climate is continental with big diurnal and seasonal temperature fluctuations. In winter, mean minimal temperatures may drop to 2-9° C (Meteorological Department 1988). Snowfall and sub-zero temperatures are known, and the chill factor can be quite severe when cold air sweeps down from the north-west. In the summer absolute maxima may exceed 46° C, and mean maxima 35-38° C. Summer air has low humidity, and the potential evapotranspiration is estimated to be over thirty times the mean annual rainfall. Occasional very strong winds create dust-storms. Though generally pleasant in the spring and autumn, in winter and summer it is a harsh climatic regime, and one that creates conditions that can become very unpleasant for all forms of life.

The mean rainfall is very low (Ch. 3), and there is a strong sub-regional variation from up to 250mm in the north-west on the high land along the boundary with Syria, to less than 50mm in the south. The variation from west to east is less pronounced, meaning that most of the precipitation is in the northern half of the study area. Most of the rain falls between December and March but otherwise the temporal and spatial variation is highly variable, as is the intensity of individual showers and storms. Some catchment areas are quite large and the vegetation sparse, so runoff and spate wadi flow are more frequently generated than might otherwise be expected. However, the gently undulating terrain gives rise to many small basins (the *qa'a* and *marab* described above) in which sediments have collected and which receive an effective rainfall far higher than the nominal 50-100mm. And as the rain falls in the cool months when there is little evaporation the production of vegetation can be quite high in some basins. Extensive pools form in some *qa'a*. Soil moisture is also increased by the presence of the basalt boulders which inhibit evaporation and from which dew trickles into the soil.

Three aquifer complexes underlie the study area: upper, middle and lower (Ch. 4). The upper aquifer defines the Azraq groundwater basin whose main natural outlets are the springs at Azraq, including the Druze springs fed from the north and the springs at Azraq Shisham fed from the south. In the mid-1950s the discharge from the Druze and Shisham springs was measured at $400m^3$/hr and $2600m^3$/hr respectively. These discharges gave rise to the Azraq oasis, and a biodiversity rich wetland which received international recognition.

2.3. Plants and Animals

No systematic plant taxonomic or biomass studies are known from the period prior to this century. Therefore it is difficult to make a comparison between people-plant interactions then and now. In the spring of 1992 after only 10 days fieldwork the total list of recorded plant species was increased from 150 to 223, including several taxa newly recorded for Jordan and one possibly new to science (Ch. 14). Recent studies have also identified over 100 species of birds and evidence of some larger mammals. To some extent the animals are protected by paucity of resources from external intervention, and also by the very rugged terrain which makes travel by vehicle very difficult and slow.

3. CHANGING POPULATION-ENVIRONMENT INTERACTIONS DURING THE TWENTIETH CENTURY

The environmental changes that have occurred have been entirely or largely beyond the control of the local population. They are mainly a result of political and economic decisions made by people living in relatively distant urban centres within the region or by political decisions and technological developments made by people in countries beyond the region, but whose impact has altered the physical environment of the study area and the ways in which local natural resources can be exploited. The changes themselves and the responses by the local population to the changes have increased pressure on the environment and its resources. The archaeological record suggests that the relationship between the population and the environment of the study area has been far from static. However, it is here postulated that change during the twentieth century has been of a different order; much faster, much more varied, more threatening to the environment and more problematic for sustainable development, as is indicated below.

3.1. Population Growth Nationally and in the Study Area

Jordan has one of the highest rates of natural population increase in the world. From data published in 1990 it is estimated that the population could double in as little as 17 years (Chs. 16 and 17). In addition, Jordan has received large numbers of migrants, particularly as Palestinians fled east in movements associated with the creation and expansion of Israel, and most recently as a consequence of the Gulf War when an estimated 300,000 Palestinians fled Kuwait and settled mainly in Amman. There are also an estimated 250,000 migrant labourers. Thus the population of Jordan rose from 680,000 in 1952 to 4.2 million by the census of December 1994. National surveys of 1976, 1983 and 1990 indicate that Jordan has a high but falling total fertility rate (TFR) with a big differential between Amman and rural areas. In the latter the TFR was much higher, associated with the lower educational levels of women.

Amman and its satellite towns of Zarqa and Suweilah have grown most rapidly as a result of a combination of natural increase, rural-urban migration, migrant labour and Palestinian settlement. Greater Amman contains perhaps half of the population of the whole country and is two-thirds Palestinian. In surface area, Greater Amman has also grown rapidly. It is mostly a low density city with new suburbs eating into that small proportion of Jordan's land which receives adequate rainfall for non-irrigated agricultural production. Its primary concerns, therefore, are with its own growth problems and with the lands to the west - the Jordan Valley, the West Bank and Israel. The population in general has very little interest in or knowledge of the large and very sparsely populated lands to the east of Amman, that 80 per cent of Jordan's territory which comprises the Badia. Hence in recent years a new interest has been created, envisaging the Badia as Jordan's 'strategic reserve'. The Badia Programme is a consequence of this interest. One of its aims, in addition to research and development, is to make the population of Greater Amman and western Jordan more aware of Badia issues and more sympathetic to them.

The national picture of rapid population growth is reflected also in the study area. The baseline demographic survey conducted by the Badia Programme in January/February 1993 showed that a total of 15,655 people were grouped into 2,398 households. Some 48.5 per cent were less than 15 years of age compared with 44 per cent for Jordan as a whole (Ch. 16). A remarkable one in six of the population was under one year of age due to high fertility rates, and only a small number of people survive to old age. The crude birth rate was a high 48.3 per thousand compared with 36.1 for Jordan as a whole and 39.0 for rural areas. The crude death rate was 11.6 per thousand, including a high rate of infant mortality. There was, therefore, a very high rate of natural increase in the programme area of 3.7 per cent per annum which would lead to a doubling of the population during 1993-2013 (Ch. 17). Compared with both the 1976 and 1990 national surveys, the area has a very high TFR. However, in many respects the fertility in 1993 resembled that of Jordan as a whole in 1976. This indicates a 20 year time-lag in the area, though the youngest cohort of women (15-19 years) were conforming to the nation-wide trends towards later marriage and consequently had fewer children than would be predicted by assumptions of a 20 year time-lag. It seems, therefore, that the time-lag is being reduced in recent years perhaps as education and health services are becoming more widespread in the Badia and as modern communications bring town and country closer together. From work in the southern Badia, Findlay and Maani (Ch. 16) suggest that as opportunities for waged employment in the urban secondary and tertiary sectors expand families may see it in their interests to have fewer children for whom they can afford a full education in order to enter the urban labour market. As far as these services are concerned, the present study area may be in transition. On the one hand, the baseline demographic survey revealed that of the 4,650 children of school age (6-14 years) 22 per cent of the boys and 34 per cent of the girls were reportedly not at school (BRDP 1994). On the other hand, Brandenburg (Ch. 20) found a lot of dissatisfaction expressed about the quality and relevance of the education being delivered, and the difficulty of getting to school, which indicates that many parents are now aware that a good education is important to their children's future. If this is the case, rural and urban areas and interdependencies will come still closer together. For labour, the Badia will indeed become a strategic resource but opposite to the way envisaged when the term was first coined - people will find work in the capital rather than work moving to the Badia. Already in this supposedly pastoral area, of the 2,856 people who reported a major activity, 41 per cent said they followed a profession, including government service. Many of these work outside the area and remit money back to their families, and much of the money is spent on goods and services which originate from outside the study area. The physical environment in which the people are living is losing its economic relevance to them.

3.2. Territory, International Boundaries and Land Tenure

The Bedouin of the study area were pastoralists and traders. They were few in number, and with only a limited dependence on material goods and technologies. At the beginning of the century the Bedouin concepts of territory were expressed by the term *dirah*, the area throughout which a group migrated, generally including pastures and some cultivated zones. The extent and degree of control over a *dirah* depended on the number of people and their alliances, the personality of their leader, the number and type of livestock owned, the vagaries of the weather and the corresponding strength or weakness of neighbouring groups. The boundaries of a *dirah* were therefore always in flux, as were loyalties to groups external to the *dirah*. The land and the resources within the *dirah* were those on which the group depended for life, and for the conservation of which they had total responsibility, except in some cultivated zones.

Major changes to the tribe-*dirah* relationship came with the ending of the Ottoman empire and the settlement which followed the First World War. Britain secured a corridor of territory linking Jordan with Iraq and separating the Arabian Peninsula from Syria, and thus an air route from Amman to Baghdad and India (Bocco and Tell 1995) and a route for bringing crude oil to the Mediterranean. Thus was Jordan's panhandle created (of course without any reference to the inhabitants of the Badia), with international boundaries dividing Jordan's Badia from those of Syria, Iraq and Saudi Arabia. Along with the new concept of the nation state which they implied, they literally and metaphorically cut across the *dirah*. The Rwala tribe, for example, practised a transhumance which was cut by all of Jordan's new boundaries with its eastern neighbours.

As the years passed and the political and social organization and the economies of the countries on either side of the new boundaries diverged, so they acquired an ever greater significance. This is perhaps at an extreme today due to the complete collapse of the economy of Iraq as a result of the trade embargo following the Gulf War. The situation in Iraq contrasts with the great wealth of Saudi Arabia and the intermediate position of Syria. This has of course created great opportunities for unofficial cross-border trade, which partly explains the growth of Safawi and Ruwayshid towns. And as these opportunities have grown so have the efforts of the government to control that trade, through the employment of more local people, either to keep out illegal goods or to secure customs duties. Thus the attempts to control the trade (as much as the trade itself) have provided ever greater income to the people of the Badia, and further reduced their dependence on local environmental resources and their interest in their conservation.

It is sometimes said that the pastoralists, and others in the study area, ignore the international boundaries. Although this may have been true in the years immediately following their creation, it is now much truer to say that the people evade the restrictions imposed by the boundaries (whenever and however they are able to) while turning the economic, financial, social and political differences between the countries on either side to their advantage. As far as sheep are concerned, the pastoralists tend to raise their animals where food is least costly and most plentiful and the quality of the support services is cheapest and best, and sell them and their by-products where the prices are highest. In recent years the north-east Badia has been a relatively good place to rear sheep; heavily subsidized concentrate feeds can be supplemented by grazing in the early spring months, and the sheep are only one border away from the high market prices of Saudi Arabia. Sheep crossing Jordan from Syria to Saudi Arabia have to pay a JD5/head tariff when they enter Jordan. People in the north-east also profit from the movement of other goods across the borders and invest some of this profit in maintaining larger sheep herds. These factors have helped increase the number of sheep in the area. The sheep overgraze and degrade the rangelands, reducing the production of plant biomass and increasing erosion. Thus the boundaries, imposed by external forces, together with the feed subsidies have had an impact on the relationship between the population and their environment to the detriment of the latter. One predicted consequence of the removal of the feed subsidy in 1996 is a reduction in the number of sheep and therefore less grazing pressure.

The existence of the boundaries created the need for the government in Amman to have the ability to defend their territory, to seek to define which tribes were Jordanian, and to ensure the allegiance of those tribes to the government (Tell 1994). One method to achieve each of these three interlinked objectives was the creation of the Desert Patrol Force (DPF) during the 1930s. Members of the Bedouin tribes, as many as one or two people from each family, were paid a regular salary as soldiers in the DPF to protect their *dirah* against outside incursions. Also a policy of paying subventions to local sheikhs was enacted by Glubb Pasha. In part the sheikhs and the members of the DPF were being paid to recognize themselves as being of Jordanian nationality at the same time as the nation took greater authority for and control over the resources of the *dirah*. In addition to the DPF, army and air bases have been established in the Badia providing more military and civilian employment. But this not only secured Jordan's Badia, it also fundamentally altered the relationship between government and tribes, and between the tribes and their environment. As tribal dependence on their environmental resources lessened (as income from military employment grew), so their sense of management responsibility for them diminished.

Concepts of land tenure, as expressed in the relationship between the tribe and the *dirah*, were also much modified by the creation of the nation state with its international boundaries during the 1920s. Two decades later, at the end of the British mandate, the

Jordanian government declared that all uncultivated land belonged to the state. In the Badia programme area the government in practice did little or nothing with this land. It certainly did not manage the land and so it was impossible to interpret what the declaration meant in practical terms. It simply further weakened the authority which the tribes had over their traditional *dirah*, and their sense of responsibility for its resources, without any effective management system being created in its place. From early in the mandate period greater stability in the Mafraq area led to pastoralists acquiring land and settling to become farmers (Jaradat *et al.* 1993). This land was formally registered but the lands further east have never been registered and so although individuals have laid claim to portions of it by ploughing, ownership is not recognized in law. In practice there is a state of confusion between traditional tenure (the *dirah*), and state and private ownership. For most of the land in the programme area no one has the right to keep other people's livestock off the land, and so nobody has the authority or responsibility for managing the range and protecting it from degradation.

3.3. Mobility and Settlement

Encouraging people to settle was another means of defining them as Jordanian. While people lived in tents moving at irregular intervals between all four countries of the region their loyalties were equally mobile. They also remained truly independent so long as they did not rely on subventions, employment or services being provided by Jordan, or the other governments. Accepting subventions and employment compromised their independence, though usually not enough from the viewpoint of the government which at different times has sought to encourage the Bedouin into permanent settlement. In practice it was also difficult for the government to deliver or the Bedouin to receive services while they retained a fully mobile lifestyle - piped water is usually delivered to permanent houses, while schools and clinics are normally associated with villages or towns. In consequence, as the twentieth century has proceeded, successive generations of pastoralists have become more interested in village life and the comforts and services that it provides. The number of villages and hamlets has grown rapidly (including ten new ones discovered by the Programme's 1993 census) to the point where today the overwhelming majority of the Bedouin live in stone or concrete houses in one of 34 villages that are all located in the north-western corner of the study area (Ch.17, Fig. 1) where there is relatively good rainfall and groundwater and also easy road access to jobs and services in Mafraq (the Governorate town) and Amman. The Bedouin are also in transition from being pastoralists to becoming catch-crop farmers and civil servants. True, some of them depart from their villages with their sheep and return for months at a time to their mobile tent-dwelling former existence (Ch. 15), but the average number of sheep in a herd owned by a villager is much less than in a real pastoralist's herd and many people have given up sheep keeping altogether. Thus the trend towards settlement (by a mix of government inducement and local desire) is reducing the detrimental pressure on the rangelands. However, it is also associated with new environmental pressures. Domestic consumption of water is far greater, the protective layer of basalt boulders is being cleared to create new fields, and some 27 farms are irrigating horticultural crops using groundwater from the Azraq basin which already has a negative water balance (Ch. 10).

3.4. Technology and Communications

As has been said, a major reason for the creation of Jordan's panhandle was to provide an uninterrupted corridor of land under British influence to bring Iraqi oil from Kirkuk to the Mediterranean. The Kirkuk-Haifa pipeline opened in 1934 including two pumping stations in Jordan's eastern Badia, stations known as H4 and H5. These desert pumping stations were very large affairs and created a lot of local as well as expatriate employment to maintain and protect the pumps and the pipeline. In turn this gave rise to the two townships of Ruwayshid (H4) and Safawi (H5), and inevitably the pipeline was serviced by a road which gradually increased in quality until today it is an important international highway (Fig. 1) linking Amman and Aqaba in Jordan with Baghdad and the whole of Iraq. Safawi, the main town in the Badia Programme area, suffered in 1948 when the pipeline closed but subsequently grew as a town servicing the main road, particularly in the 1980s during the Iran-Iraq war. The economy of Safawi is only marginally dependent on the environment in which it is situated. It is the creation of ideas, ambitions, economies and technologies that have nothing to do with the population of the area. Yet it has become part of their environment. Some people live in it and others respond to it as a source of goods and services which are supplied from outside the area. It has also created some employment, but imposed an extra burden on scarce local resources, particularly water.

The oil industry's road and the security forces brought motorized vehicles and telecommunications to the study area. Motor vehicles are an example of a sophisticated technology developed without any involvement of the people of the study area but which since their arrival have had direct impacts on the people and the environment and on the relationship of each with the other. At the same time they allow other populations to impinge on the local environment and draw on its resources. Telecommunications have allowed the spread of ideas from other parts of Jordan and from other countries in the world into the study area. The combination of new technologies and new ideas and a growing external demand for resources is changing the external assessment of the value of and demand for some of the resources in the study area, notably water.

The people of the Badia have made a progressive switch from camels to trucks and pick-ups as the primary means of transport for themselves and for the food and water needed by their sheep and goats. Today it is common to see a herd of sheep in a very remote location attended by a feed truck and a water bowser. Larger herds can be maintained further from water points than was previously possible, and the herds are no longer compelled to move to the feed because the feed can be brought to the herd. Indeed one reason for siting the herd is relative ease of access from the main road. Bowsers may travel great distances between water sources and the herds. The concentration of destructive tyre marks of the trucks and bowsers is highly visible. Sometimes the water sources are boreholes drilled for this purpose by the government for the Bedouin. Thus the pastoralists, using external finance and technologies, have extended their environmental impact into aquifers which were previously unknown to them and quite unreachable by their simple technologies. The use of vehicles of course accounts for the great reduction in the number of camels in the region, but the vehicles, used also for trading, mark a radical departure from the traditional Bedouin disdain for material goods. Vehicles are expensive to buy and service, and they can only be maintained by increased levels of trading activity made possible by the vehicles travelling to distant towns and by salaried employment. The vehicles bring back other material goods, even television sets which require easy access to the towns to maintain the televisions and the small generators that provide electricity for them. The growth in the number of vehicles around the 34 settlements in the north-west of the study area has also forced the government to satisfy an ever-growing demand for secondary roads in this area with strong direct and indirect environmental impacts.

Another visible expression of the environmental impact of a combination of external technologies, finance and ideas is the network of lines cleared through the basalt boulders to permit seismic surveys in what has so far proved to be a fruitless search for oil and gas. Other tracks were created by the DPF. It is possible also in the future that some of the minerals listed in the sub-section on terrain and soils, above, will be commercially exploited causing pollution and leaving permanent scars on the landscape. The extent to which the local population will be consulted about these activities, or derive direct benefit from them, is yet to be seen.

3.5. Water Requirements and Water Engineering

As the population of Amman has grown so has its demand for water. Today this has become an international issue of great significance to the peace talks and to negotiations between Jordan, Syria, Palestine and Israel. Jordan is extremely short of water, and the population growth and political strength of Amman ensure that decisions are made to pump water into the capital from wherever it can be found. This includes the upper basalt aquifers of the Azraq basin, which is roughly coincident with the Badia Programme field area. Abstraction rates have increased rapidly since 1980 reaching an estimated total, for all purposes, from the basin of 43 million m^3 in 1993 (Ch. 4). The biggest single extraction point is the Amman Water and Sewerage Authority well-field near Azraq which pumps approximately 15-20 million m^3 of drinking water annually to Amman.

In addition to demand from the capital, greater total abstractions take place in the agricultural areas east of Azraq and north of the Mafraq-Safawi road. The first boreholes were drilled during the 1940s, and a recent inventory has located over 600 boreholes in the immediate area of Azraq. Current abstraction rates are considerably in excess of recharge, and 'mining' of groundwater is occurring. One consequence has been a steady decline in the springs which fed the Azraq oasis during the past 40 years until the last spring completely dried up in 1992 (Ch. 4).

The situation is made worse by the series of small dams that the Syrian authorities have built on the southern flanks of Jabal al-Arab thereby damming wadis which previously flowed into the study area, and finally into the Azraq oasis.

A complicating factor is the shift from a pastoral lifestyle to settled communities in which people live in fixed houses with water piped to them by the local authority. Although this is a small proportion of the total used it is a highly visible service, and as the water is in short supply it invites criticism of the water authority and of the policy which takes local water to the capital.

According to the Badia Programme baseline census of 1993, over 80 per cent of households (those in the villages) obtained their water piped by the water authority (Chs. 10 and 16). In the country as a whole the proportion was 91 per cent and in Amman city 98 per cent. The predicted rapid increase in population in the Badia, the likely increase in proportion of homes with piped water and the increased usage of water that a piped system engenders, all suggest that domestic water demand and consumption in the study area will rise rapidly in the coming decades.

Therefore a combination of decisions taken in urban centres (Amman and Damascus) and the development from outside the region of technologies for drilling, pumping and dam building, which are applied locally without any reference to the people of the study area, have greatly decreased the water locally available at the same time that demand for it is growing as a result of a growth in population, a change in lifestyle and a growing local awareness of the power of new technologies to make available to the local population that part of the environment (groundwater) that simple technologies previously could not reach.

One response by the people of the study area has been to demand water collecting structures for themselves. The Water Authority of Jordan therefore has built a dam in the wadi Rajil and designed a number of other small dams and pools and other structures for collecting water. The armed forces have also taken it upon themselves to use heavy equipment to clean old dams of silt so that they can be re-used. Technically the new structures are well enough designed. However, insufficient attention has been given to their economic return or their environmental impact. It is now estimated that the dam in the wadi Rajil will only collect water in very exceptional years, such as 1994/5, so even its social return is doubtful.

Another local response has been to drill boreholes to raise water for irrigation. Thus 27 irrigated farms have been created in the settled north-west of the programme area (Ch. 10). However, many of the boreholes east of Azraq have been drilled by people not from the area. Wealthy citizens from the capital have sunk speculative wells here but water level and quality have both declined and the enterprises are not economic. This is another example of outside forces, beyond the control of the local population, having an adverse environmental impact. More recently, farmers from Zarqa have started to rent tribal land in Wadi Sirhan, sink wells and grow water melons and tomatoes. During the season the wells may be operating 24 hours per day. Not only is the water table threatened but also prior bulldozing of the natural vegetation (relatively dense in this area) is destroying biodiversity in the name of unsustainable short-term economic gain by people outside the study area.

3.6. Livestock and Agriculture

As the interdependencies between the government and the people in the Badia Programme area have increased so the government has responded to their stated needs for services to support their livestock. One of the main services has been the provision of subsidized grain-based concentrates to sheep and goats at the rate of 0.5kg/head/day. Other grain, at well below international prices, has also been available. It is generally believed that the people of the Badia, including the study area, have responded to the subsidies by increasing their number of sheep. Feed from rangeland plants has become a minor part of the total annual feed intake, and therefore efforts to increase or even maintain rangeland productivity were becoming not worth the trouble. However, in another example of ideas being imposed from outside the study area, the government in accordance with the structural adjustment programme and international trade agreements phased out the grain subsidies in 1996, so in future the private sector will take responsibility for feed grain imports. This will have an impact on the people, their sheep and their environment - possibly reducing the number of sheep and therefore giving better protection to the rangeland plants.

Sheep will still continue to be reared if they are profitable. Today the market price for fat lambs is set by Saudi Arabia, and there is thus a general movement of Awassi lambs and sheep from Jordan (and Iraq and Syria) for slaughter in Saudi Arabia. Thus the oil wealth of Saudi is affecting the number of sheep in the study area and their consumption of rangeland plants.

In theory there is a free veterinary service but in practice this has either not been available to the people of the study area or is held in such low repute that people make little or no use of it. Some of the major diseases present in the sheep in the Badia have only very recently been identified and their wide distribution realized, including foot-and-mouth, peste de petits ruminants (PPR) and blue tongue. They have caused relatively little complaint because they do not manifest themselves in large numbers of deaths, but they are debilitating diseases which severely reduce female fertility and the growth of lambs and make the animals susceptible to parasites and other disease organisms. As the animals do not die, they continue to eat the rangeland plants causing environmental loss without equivalent concomitant economic gain.

In recent decades there has been a steady expansion in agriculture in the west of Jordan and a parallel expansion in towns, villages and roads as the population has grown. Agriculture has created more crop residues to feed to sheep and other livestock, but their numbers have also grown. The combination of these animals and the more restricted access to pastoralists' sheep from the east due to road networks and urban by-laws have limited the freedom of these sheep to feed on crop residues in the summer months. Indeed, there has been something of a reverse trend in sheep movements. Animals normally based in the west (or even in the dry south of the country) are taken by truck to graze in the north-eastern Badia if heavy winter rains generate good grazing there. Hence further environmental loss without economic gain to the local population.

3.7. Summary

The basic changes that have taken place in population-environment relations in the study area this century are due to the increasing involvement from outside the study area, and the growing strength of politico-financial decision makers and of new technologies. Some of the decision makers are Jordanian, based in Amman, others are in those Middle East countries that adjoin the north-east Badia, and others are from countries and international organizations further afield. The new technologies are mainly Western in origin. Normally they are simply the instruments of the decision makers, though sometimes they are employed by members of the local population themselves. The decisions taken and the technologies used have had direct effects on both the local population and the environment, and they have had indirect effects by allowing or encouraging interactions of each with the other.

The above changes have had two primary results. First, the decision makers have expressed a political and economic interest in the environment and its resources. Second, the local population has become relatively less dependent on those resources as it has come to rely on material goods and services that originate from outside the study area, and which are paid for by employment unrelated to local environmental resources. A consequence has been the loss of a sense of responsibility by the local population for the environment and its resources because they are only partially dependent on them and because the decision makers have taken responsibility for the environment from them. Unfortunately the primary concerns of the decision makers have been for the well-being of the urban centres, and of other countries and international organizations where they live and by which they are employed. Thus their tendency is to exploit the natural resources for their own ends and have little regard for the local, long-term environmental or developmental consequences.

In these circumstances the local population has grown as have the demands on the natural resources by the local population and outsiders, and the environment and the development potential for local people have suffered in ways which the Badia Programme is revealing.

4. DEVELOPMENT AND POPULATION-ENVIRONMENT DYNAMICS IN THE BADIA PROGRAMME

4.1. Research to Development: Aims and Approach

How can the research aspect of the Badia Programme do anything to improve the above situation, given the fact that it has neither the mandate nor the financial resources to be directly involved in change? To begin with, the Programme is seeking to provide relevant and accurate information and focus as strongly on human factors (through the disciplines of demography, socio-economics and development anthropology) as on technical factors, remembering that in most cases the long-term management of change will be in the hands of people of the study area. In addition, the Programme is involving the local people and the external decision makers, as far as possible, in the research process - particularly to agree the important issues, and at the stages of translating findings into conclusions, and conclusions into recommendations. It is also of great importance that recommendations must be economically as well as socially and environmentally acceptable. Finally, it must be made clear with whom responsibility for development and environmental protection lies; as far as possible, in local hands and to the benefit of local people.

As was stated in the original Badia Programme Framework Document (RGS 1992) the aim of the programme, and therefore of the research that is being undertaken, is:

> " the sustainable development of the desertified Badia environment and the improvement of the standard of living of the inhabitants... under management systems which conserve the natural resources so that production levels will be sustainable in the long term."

In a revision of the Framework Document (Shahbaz and Dutton 1995a) agreed by the Steering Committee in May 1995 the development aim was stated as:

> " Development should make the best sustainable use of the scarce local Badia resources to the benefit of the people of the region."

It is essential that the research must be genuine scientific research, with results and conclusions that carry conviction with those involved in subsequent development planning and project implementation. Those involved in development include, on the one hand, the local population of the study area, and, on the other hand, providers of funds, planners and other government and private sector decision makers.

Concerning the local population, in many cases the long-term management of change, as noted above, will depend upon the people who are directly affected by the change. Therefore they will need to be convinced that change is in their own individual and collective interests, and they will have to be involved in the whole process of change, from the beginning. One aspect of research will be to identify people who can act as disinterested spokespersons for definable interest groups within the study area and to create a framework for an on-going dialogue between them and officials representing governmental and other organizations who are in a position to contribute development funds and appropriate development expertise. Only through such dialogue, it is argued, will appropriate development decisions be reached and devolve responsibility as far as possible on people from the area to manage change in their own interests, and in those of the environmental resources upon which beneficial change will depend.

Concerning the decision makers, the Badia Programme stresses the importance of working with other organizations in Jordan so that they can make their special contributions to the Programme's understanding of Badia issues and to consequent beneficial development, and bring resources (human and financial) into the area. In this way the Programme is maintaining on-going contact with a range of governmental and other organizations which have decision-making power and influence in Jordan.

4.2. Types of Development

Three distinct, though interrelated, types of development have been locally identified: beneficial economic change, conservation of the environment, and improved delivery of services.

The potential that exists for these types of development, within the likely range of financial and other constraints, is still being evaluated. One possible topic of each type of development is outlined below, noting how each hopes to achieve a beneficial effect on the population or on population-environment interactions.

4.3. Improving Livestock Productivity

Awassi sheep afford the main potential for beneficial economic change in the livestock sector. In the study area there are no cattle and only a few camels. Awassi sheep comprise about 80 per cent of the total small ruminant stock. Research is being conducted by people employed by the Programme, by the Jordan Co-operative Organization (JCO) and by the Jordan University of Science and Technology (JUST). The overall team incorporates established researchers and post-graduate research students in fields spanning veterinary science, sheep production, milk production and the use of milk and wool by-products, livestock marketing, demography and pastoral livestock management systems. The research therefore has major technical and human science components. Inevitably it involves daily contact with the owners of the sheep who therefore are fully aware of the aims of the research and contribute to it. By collecting information from control flocks and from flocks vaccinated against diseases, coupled with the production, management and marketing information mentioned above, it is becoming possible to quantify the production and economic gain due to vaccinations in ways which should convince the livestock owners and the decision makers alike. Work is also focusing on nutrition, seeking an economically and more environmentally sound balance of feedstuffs from rangeland plants, concentrate feeds and plant/animal by-products. The previous heavy subsidy on concentrate feeds, provided per head of adult stock, encouraged the maintenance of large numbers of animals without regard to their productivity. They then devoured any rangeland plants generated by winter rains before they reached their full biomass potential and before they set seed. The subsidy was removed in mid-1996. A working hypothesis (not an assumption) is that a new generation of pastoralists confident of the health and greatly enhanced productivity of their animals will keep fewer of them, thereby spending less money on maintenance whilst maintaining total production and increasing profit margins. Fewer sheep will make less demand on the range (and require less labour) particularly if there is a more rational use of plant/animal by-products in the diet and if lambs are taken to be fattened in feedlots. We hope therefore that by collecting and demonstrating the required information, the pastoralists will modify their production systems, supported by better structured external support, because it is to their own advantage to do so. At the same time they will be improving the environment by making less destructive and unproductive use of the rangeland plants. From the autumn of 1996, following a workshop in September to discuss the Programme findings, formal channels of direct communication with Ministry of Agriculture are being established. Also the Badia Programme livestock activities entered a new phase, with a focus on delivering extension information based on a combination of basic knowledge and innovations. Radical change became essential when the grain subsidy was ended.

4.4. Conservation of Water

Water is the main conservation issue. The Programme has already demonstrated that the total volume of water 'mined' to date from the upper aquifer is between 40 and 205 million m^3. This wide variation between lower and higher estimates depends on which value for recharge is assumed (which indicates a future research requirement). The age of the water is also uncertain and a project is now in hand to determine how recent or ancient it is; is it 'fossil' water or is it being replenished? Nevertheless, it is certain that the springs that fed the Azraq oasis have stopped flowing and that the local water balance is in deficit. However, a high proportion of the water is used by people not from the study area, some taken by the Amman Water and Sewerage Authority to Greater Amman and the rest pumped to irrigate farms east of Azraq some of which are owned by people from the capital area. It is also generally believed that water is used locally far in excess of the volumes authorized by licence. Therefore the problems are complex and serious, and beyond the control of the local community. The government needs to intervene, firstly to decide upon a water management strategy for the upper aquifer. If a 'safe yield' strategy is agreed so that the overall development of the resource is sustainable, then total abstraction must be halved (Ch. 4). This will require a ban on new boreholes, stopping illegal abstractions and setting limits on the abstractions from licensed wells. It will also require an effective system of enforcement. Thought will have to be given also to prioritizing water usage so that after people's drinking and sanitary needs have been met, the remaining water is used in ways which add most value to the economy. The development task of the Programme is to ensure that the authorities have all the information they need to formulate, enact and enforce an environmentally sound and economically beneficial water strategy.

4.5. The Demand for Improved Health and Education Services

As in all other countries, the delivery of the health and education services appears to be far from perfect. One problem from the government perspective is that the population is small and widely dispersed and therefore relatively costly to serve. In addition, there is a considerable reluctance by people from the urban centres in Jordan to run the clinics or staff the schools in the Badia.

Many discussions have been held with health workers and the recipients of healthcare in the study area (Chs. 20-2). All complain of shortage of facilities. Women and men complain that the treatment of women is remote, impersonal and perfunctory. Relationships between the medical staff and the patients are often not good enough to ensure the best diagnosis and treatment. Uptake of the services is limited by such things as a shortage of women doctors, insufficient opening hours in the small, primary centres, and shortage of public transport. The condition of children's teeth appears to be a particular cause for concern. Of course, the health service providers will have their reasons and excuses for inadequacies in the system of health provision, including pointing out that there is a shortage of money with which to improve facilities and services. However, an underlying problem appears to be a lack of effective communication between the local population and those staff in the health service who are sufficiently senior to be able to respond to their complaints and suggestions for improvement. The Badia Programme will have performed its role only if it manages to create a regular meeting ground where influential officials and community representatives can inform each other about constraints and requirements, and discuss ways - practical steps - by the which the present delivery of the health services can be improved and made both more sympathetic and more cost effective. The meetings, particularly the early ones, will need to be very carefully prepared so that the deliverers and receivers of healthcare see themselves as together fighting a common cause, and not as mutually uncomprehending, hostile adversaries.

As for education, the Programme census, as noted above, revealed that 56 per cent of the adults (over 16 years of age) had either no education or only limited reading and writing ability, and that of the children of school age (6-14 years), 22 per cent of the boys and 34 per cent of the girls were reportedly not at school. Clearly these statistics are indicators of educational need, but only when the data are fully analysed and backed up by a listing of current educational provision to the area and also by more detailed household based studies will it be possible to target future improvements to be made to the local educational inputs. However, we now know more about the inadequacies of the educational facilities, and we are also acquiring a better understanding of the reasons why boys and girls leave school early and why others never enrol at all. There are a number of technical reasons (such as transport, cost and lack of facilities) coupled with poorly motivated teachers and lack of parental support. The Badia Programme needs to identify and quantify the relative importance of the various factors contributing to poor education uptake, and then to stimulate discussion at various levels to find cost-effective ways to overcome the problems. It has already been suggested that community participation has a role to play, encouraging a greater flow of information between parents and teachers with a focus on communities solving their own problems through joint-participation in developing educational programmes (Ch. 20).

5. CONCLUSIONS

During the course of this century the relationship between the local population of the study area and their immediate physical environment has changed continuously and rapidly under the influence of a wide range of stimuli - political, financial, economic, technological and social. The origin of these stimuli has, for the most part, been outside the study area, amongst the decision-making and technology-developing communities in

Amman, in other parts of the Middle East, and in countries further afield. To a large extent the local population have been passive partners in the change, though seeking to turn change to at least short-term advantage where they have the capacity to do so. The changes have allowed people from outside the study area to exploit local resources for their own advantage (notably water) whilst the people living in the study area - the local population - have become far less dependent on their local environmental resources and have lost a lot of the management responsibility that they previously had for them. In consequence the environment has suffered, as the population and its demands on scarce resources have grown. The Badia Programme has begun the research task of trying to define and quantify the developmental and environmental potentials and problems of the area, and to suggest some ways in which they might be addressed in order to create a new harmony between the demand for development and the requirement for resource conservation.

ACKNOWLEDGEMENTS

I would like to acknowledge the roles of HCST in Jordan and RGS-IBG in the UK for creating the opportunity to establish the Jordan Badia Research and Development Programme.

REFERENCES

Bocco, R. and Tell, T. (1995) Frontières, tribus et état(s) en Jordanie orientale à l'époque du mandat. In R. Bocco, and C. Velud, (eds). *Tribus, tribalismes et états en Moyen-Orient*. pp.26-47.

BRDP (1994). Jordan Badia Research and Development Programme. Stage One: Safawi. BRDP.

Jaradat, A., Oglah, M. and Tutweiler, R. (1993) Characterizing dryland farming systems of northern Jordan, paper presented at the JUST/ICARDA seminar on 'characterization of the farming systems in Mafraq area', JUST.

Meteorological Department (1988) *Jordan Climatological Data Handbook*. The Hashemite Kingdom of Jordan, pp.104.

RGS (1992). Framework Document: the Jordan Badia Research and Development Programme 1992-1995 (Stage one: Safawi). Royal Geographical Society.

Shahbaz, M. and Dutton, R.W. (1995a) Research Framework 1995/96-1998/99. BRDP.

Shahbaz, M. and Dutton, R.W. (1995b) Priority Research Themes and Topics, 1996-98 and Beyond. BRDP.

Tell, T. (1994). Les origines sociales de la glasnost Jordanienne. In R. Bocco, and M. Djalili, (eds). *Moyen-Orient: Migrations, Démocratization, Médiations*. PUF, Genéve, pp.201-20.

CHAPTER 2

Geology, Geomorphology, Hydrology, Groundwater and Physical Resources

ROBERT J. ALLISON, DAVID L. HIGGITT,
ALASTAIR J. KIRK, JEFF WARBURTON
(University of Durham, UK)
AZM S. AL-HOMOUD
(Jordan University of Science and Technology)
BASSAM F. SUNNA
(formerly Ministry of Energy and Mineral Resources, Jordan)
KEVIN WHITE
(University of Reading, UK)

1. INTRODUCTION

The Badia encompasses a wide and significant part of the Hashemite Kingdom of Jordan. It covers an area of approximately 72,600km^2, which constitutes 81 per cent of the total area (89,400km^2) of Jordan (Fig. 1). The region is sub-divided into three geographical areas as follows:

- Northern Badia which comprises 35 per cent (25,930km^2) of the Badia total area;
- Middle Badia which comprises 13 per cent (9,634km^2) of the Badia total area;
- Southern Badia which comprises 51 per cent (37,096km^2) of the Badia total area.

The Badia region falls in an arid climatic zone. Rainfall is erratic both spatially and temporally, with a maximum of 200mm annually. Air temperature fluctuates widely from a daily mean minimum of 10° C, mean maximum of 24.5° C and a mean daily temperature of 17.5° C. Occasionally absolute minimum and maximum temperatures might reach -5° C and 46° C respectively.

Jordan's Badia holds numerous and rich natural resources including mineral deposits, surface and groundwater, tourist sites, renewable natural rangeland and cultivable land suitable for agriculture and livestock production. The area also has the potential for development of non-polluting renewable energy sources, namely solar and wind energy. As the Badia extends into the borders of neighbouring countries, there is the additional benefit of its being a junction for export-import activities at the regional level. The Badia is rich with mineral ores, which are mainly exported as raw materials. Mineral deposits include phosphate, white cement, pure limestone, dolomite limestone, quartz glass sand, clay, marble, feldspar, tripoli, silica, rock salts, tuff and raw building materials.

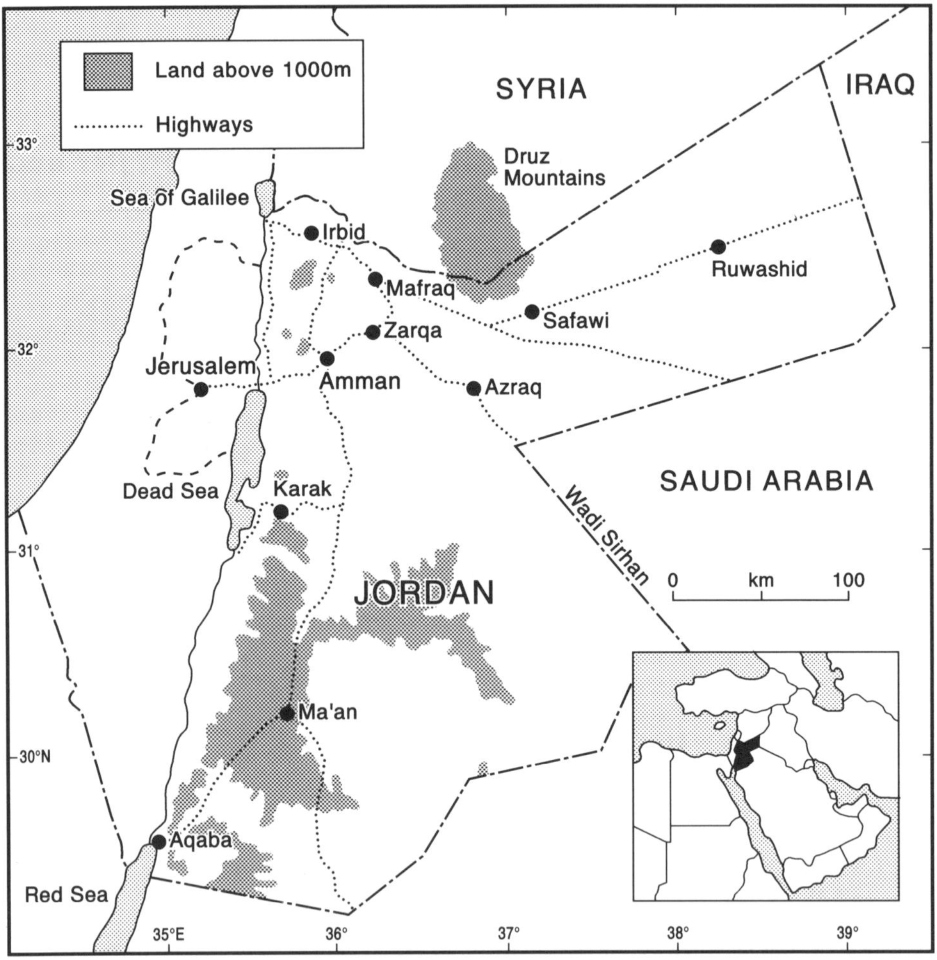

Fig. 1. Jordan land distribution.

Rangeland is the largest resource. Potential pasture land covers large parts of the Badia, although the vegetation cover is not dense and surface water mostly absent. Barley is the main field crop in dry farming. Forage, vegetables, fruit orchards and wheat are under irrigation. Most farm animals in the country (61%) are located in the Badia. Livestock production includes sheep, goats, cattle and camels. Poultry production includes broiler and egg-laying hen farms. Around 70 per cent of Jordan's animal products are produced in the Badia.

The total population of the Badia represents 5 per cent of the whole population of the country. Only 5 per cent of the Badia population are still nomadic. The rest are now settled. The majority of the Badia population are involved in agriculture, livestock production, the civil service, commerce, the armed forces, mining and prospecting industries. More than 85 per cent of the Badia population are accommodated within a well established basic infrastructure of roads, water, electricity and telephones. Roads serve more than 88 per cent of the Badia communities, while more than 72 per cent have post-offices equipped with telecommunication systems.

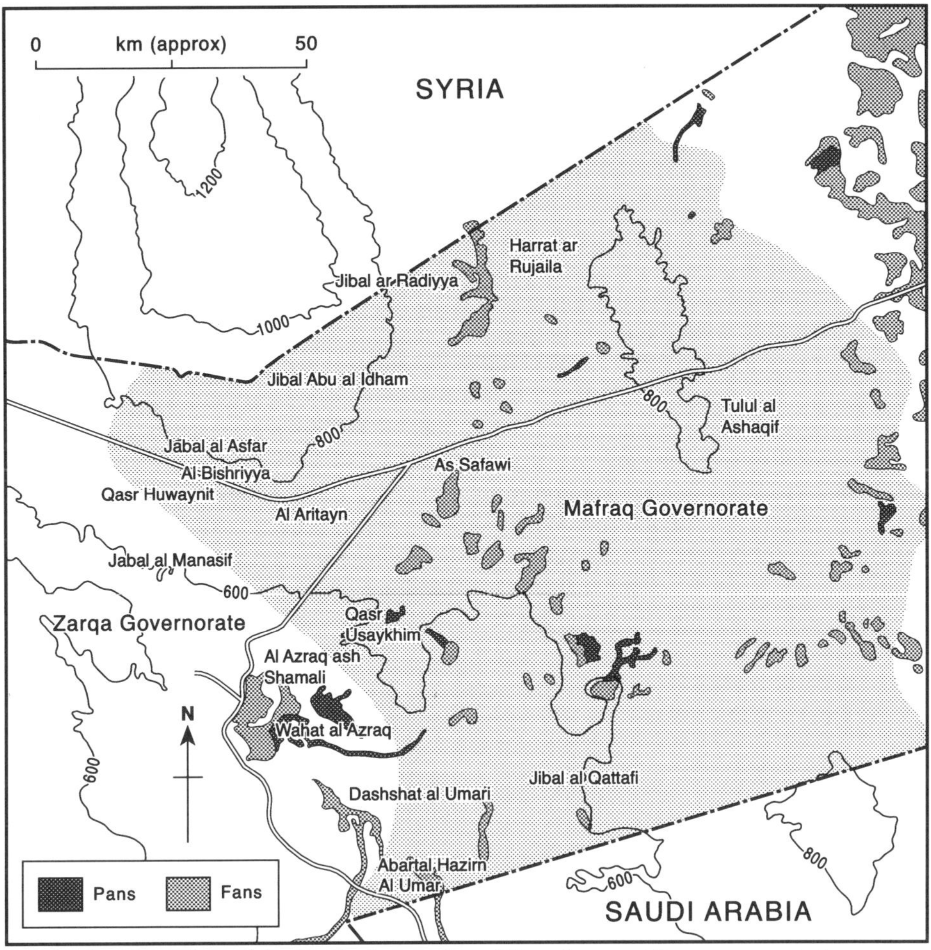

Fig. 2. Location of the Badia Research and Development Programme study area.

The general objective of the Jordan Badia Research and Development Programme, which involves collaboration between the Higher Council for Science and Technology (HCST) of Jordan and the Royal Geographical Society with the Institute of British Geographers (RGS-IBG) of the United Kingdom, is *the sustainable development of the desertified Badia environment and the improvement of the standard of living of its inhabitants.* The general objective is to be achieved through enhancing teamwork and self-reliance amongst the inhabitants of the region and through the introduction of appropriate technologies, to improve productivity and income from livestock and agriculture, under management systems which conserve the natural resources so that production levels will be maintained in the long term. Within the context of the general objective of the Programme, the geomorphology and physical resources research covers the topics of geology, mineral resources, geomorphology, water and soil. Nowhere are the constraints to development resulting from a lack of information and understanding about the physical environment more apparent than in the Badia region of the Hashemite Kingdom of Jordan (Fig. 1). Successful development of the Badia can only be achieved by reaching a comprehensive knowledge and understanding of the physical environment, including processes and landforms, natural resources, human effects and impacts on natural systems and the past and potential future influence of climate change.

The Programme Area, as defined for the purposes of the Jordan Badia Research and Development Programme, comprises some 11,210km^2 (Fig. 2). The area is bounded by Syria to the north and Saudi Arabia in the south. The east and west margins approximately follow the perimeter of extensive basalt outcrops, which cover a large part of the ground surface between the towns of Azraq and Ruwayshid. The region comprises around 14 per cent of the total land area of Jordan and is predominantly widely rolling, desert plain, covered with dark basalt rocks. The study area is centred on the town of Safawi, 156km north-east of Amman. Safawi, the largest centre of local population, grew in the 1930s around one of the pumping stations, H5, on the Baghdad to Haifa oil pipeline. It is one of the main towns on the Amman-Baghdad highway. The Programme Area includes a number of other recognized villages. Current estimates for the Programme Area show a population of around 15,000 people. Most of the region is governed locally from the Governorate of Mafraq, although a small sector in the south-west comes under the Governorate of Zarqa. Extensive tracts of land have no current use although parts of the region, particularly in the north, are utilized in an increasingly structured manner for cash-crop production, such as fruit and early salad vegetables. Uncontrolled grazing frequently causes deterioration to the quality of the land. There are few management practices or strategies for maintaining soil quality.

2. GEOLOGY

Sedimentary rocks cover most of Jordan. Some Precambrian plutonic and metamorphic rocks are exposed in the south-west of the country. The thickness of the sediments increases in a north-east direction, where progressively younger sediments are exposed (Bender 1974). Unmetamorphosed Cambrian to Silurian sedimentary rocks unconformably overlie the Precambrian. These Palaeozoic rocks consist mainly of clastics, with some thin carbonates. They dip gently north and north-east beneath the Cenozoic sequences (Fig. 3).

Following a combination of uplift, tilting and likely peneplanation, Triassic rocks were deposited. The pattern of north-west to south-east transgressions was repeated several times during the Mesozoic. Jurassic to Lower Cretaceous sequences are predominantly clastics and marine carbonates. These, together with the Palaeozoic sediments, form an important aquifer complex in Jordan. Widespread deposition of carbonates occurred during the Upper Cretaceous, comprising a variety of marls, limestone and dolomitic limestone. Carbonate deposition continued into the Tertiary and the sequence forms another major aquifer. Late Tertiary and early Quaternary basalt rocks are found in the north-east of Jordan (Ch. 13). They cover about one-seventh of the country. Minor basaltic intrusions are also found among the east escarpment of the Jordan Rift Valley, usually associated with faults.

The Tertiary-Quaternary continental basalt flows and tuffs cover approximately 11,000km^2, the majority of the Safawi area (Fig. 4). The resulting basalt plateau is between 50km and 170km wide from east to west and extends 180km from the Syrian border, east and south of the Azraq basin, towards Saudi Arabia. The basalts are part of the major North Arabian Volcanic Province, which extends from the southern rim of the Damascus Basin in Syria, across Jordan and into Saudi Arabia, along the eastern margin

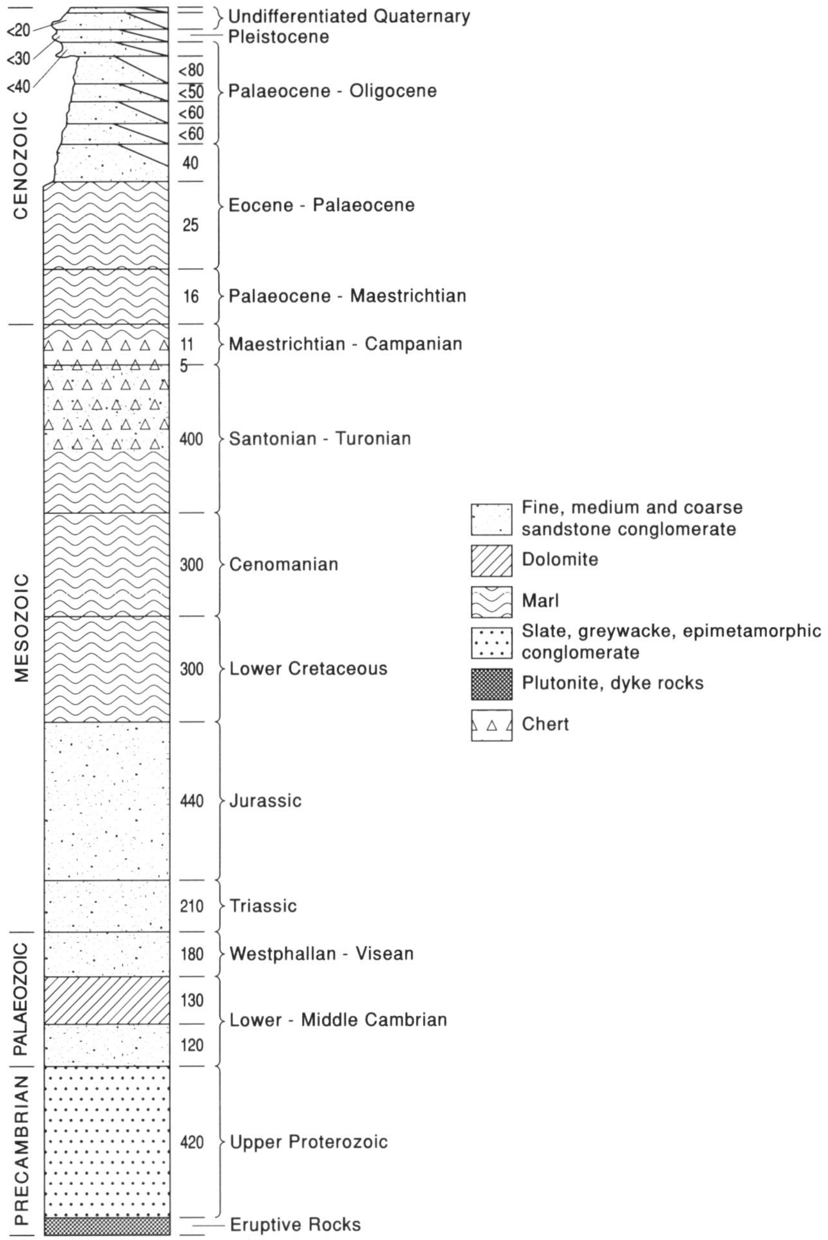

Fig. 3. Stratigraphic column of NE Jordan (after Bender 1968).

of the Azraq and Sirhan basins. Detailed studies have established that the basalt outcrops are alkali-olivine in character. Absolute age determination using K-Ar techniques has identified a number of different basalt lava flows, with ages ranging from 13.7Ma to less than 0.5Ma for the exposed rocks. Some of the unexposed flows date at 23Ma. The younger flows are centred on Jabal al-Arab. Recent studies by the Natural Resources Authority in Jordan (Ibrahim 1992) have classified the exposed basalts as the Harrat ash-Shaam super-group. The Harrat ash-Shaam can be further sub-divided into five major groups known as the Bishriyya, Rimah, Asfar, Safawi and Wisad. The basalts comprise lava flows, dyke systems which usually trend north-west to south-east, pyroclastics and volcanic centres of different types including shield volcanoes and stratovolcanoes. Considerable amounts of material have been erupted from numerous fissures, with basalt and some tuff being extruded from clusters of isolated cones. Bordering the basalt in the south and south-west of the Programme Area is a region of dissected Upper Cretaceous and Eocene sediments. Photogeologic evidence suggests that the area is a broad structural high, with several geomagnetic positive anomalies.

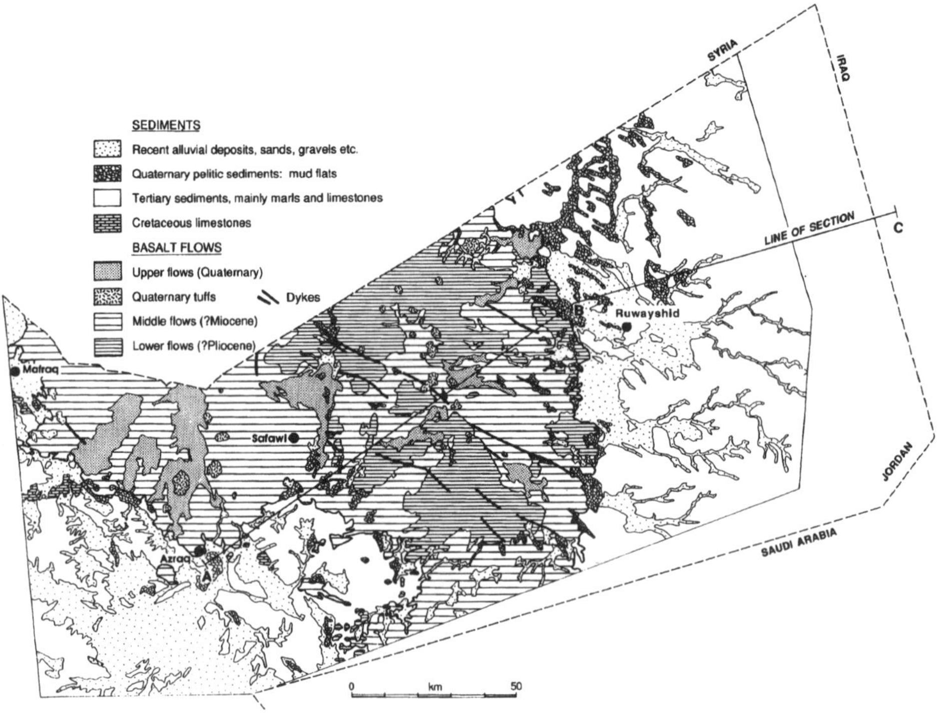

Fig. 4. Major geological divisions of the Badia (after Burdon 1959).

Four of the basalt lava flows are particularly important geomorphologically in the Programme Area: the Abed, Salaman, Madhala and Bishriyya. All are part of the Harrat ash-Shaam super-group. They are different in character and age. The ground cover associated with the four basalts is distinct. The Abed is the oldest basalt at approximately 8.9 million years. It is an olivine picric basalt, part of the Safawi group and characterized at the ground surface by a pavement of angular boulders which have a maximum a-axis of 30cm to 40cm. Orange, loose, fine-grained sediment underlies the basalt boulders and is exposed between clasts. The Salaman is a flood basalt, in the region of 8.45 million years old and again forms part of the Safawi group. It is typified by a fairly continuous and dense cover of small basalt pebbles and fragments at the ground surface, with few large boulders. Little or none of the underlying fine grained sediments are exposed. Boulders covering the ground surface where the Madhala basalt outcrops have an a-axis of 20cm to 25cm. The Madhala forms part of the Asfar group, which is dated to between

1.96 and 3.41 million years and is again olivine picric in character. The Bishriyya is the youngest basalt at 0.1 to 1.45 million years and is characterized by large ridges of intact bedrock with a vesicular structure. Boulders often approach a 50cm a-axis length, with small, angular, basalt stones between the larger clasts. The Bishriyya is the most recent lava flow and the drainage network across its surface is poorly developed. Finally, there are scoracious volcanic deposits which are usually found around recent volcanic cones. Their distribution is linked to contemporary dyke systems and the major structural features of the region.

The dominant structural feature in Jordan is the north-south trending Dead Sea Rift. Left-lateral strike-slip displacement across the transform boundary is estimated at slightly more than 100km. Several east-west faults are traceable from the broader faults of the rift, among them is the Wadi Zarqa Main Fault, traceable for approximately 50km. The east-west faults and several north-west striking fracture zones have acted locally as conduits for Neogene-Pleistocene basalt dykes and flows. The volcanism is especially pronounced north of and along the east side of the Azraq basin. Here, extensive fissure-flows have formed 45,000km^2 of plateau basalt that extend northward into Syria (Bender 1975). The whole of the north-east Badia exhibits major faulting systems, many of which can be identified from Landsat TM images. Block faulting over the central and north-east part of Jordan contributed to the formation of broad swells and basins including the Azraq-Wadi Sirhan Basin. There are numerous ground surface lineaments and the basalt dykes all lie along distinct linear zones. Three dominant fault trends can be recognized: east to west, north-west to south-east, east-north-east to west-south-west. The north-west striking fault systems, caused by deep tensional forces, are very long. Vertical displacements are frequently small. The most important is the Ramtha-Wadi Sirhan graben fault system, which runs through the Azraq area and into Saudi Arabia. The Fuluk fault system is responsible for an increase in thickness of the Mesozoic and Tertiary deposits of the Azraq basin.

3. GEOMORPHOLOGY

The Programme Area has an overall altitudinal variation of approximately 800m. The highest ground, of around 1,150m, occurs in the north-west, on the foot-slopes of Jabal al-Arab. The lowest ground, of around 400m, is found along the border with Saudi Arabia. Throughout much of the region and around Safawi in particular, the mean altitude is approximately 700m. There is a ground surface gradient away from the Jabal al-Arab, which has significant effects on precipitation and surface runoff. The topography is dominated by extensive tracts of gently undulating, low hills, introducing local elevation differences of typically 25m to 30m between drainage divides and drainage lines. Gradients are seldom steep and there are few sudden breaks of slope except at some lava flow boundaries and where major wadis have incised significantly. Almost everywhere the topography is characterized by slopes, with a variety of shapes from concave to convex forms. The topographic variation and differences in slope form within the basalt area depend very much on the age, physical and mineralogical characteristics of specific basalt flows. Broadly speaking, on the older flows the landscape is more rounded, with well developed drainage patterns and a fine colluvial rill network. The more recent flows, on the other hand, show highly irregular topography with many silt filled depressions and a poorly connected drainage network.

Most of the ground surface is boulder-strewn. The size of individual clasts varies. The size distribution is principally controlled by the ground surface geology, with some characteristics of sorting, ground cover and relative exposure being influenced by slope position. In some of the basalt fields large boulders are predominant, while in other places small, angular clasts are more common. Towards the north-west the basalt boulders become increasingly covered by lichen, an indication of the increased moisture availability. The basalt clasts generally form a mantle of stones overlying fine grained, brown to orange coloured sediment. Such of the basalt types, such as the Abed, have boulder cover characteristics where the underlying sediment is conspicuous from a distance because of the spacing of the boulders. Other basalts, such as the Bishriyya, appear to be much darker from a distance. Many of the boulders are sorted into crude polygonal patterns. The orange sedimentary deposit is unlithified, laterally extensive and of variable depth. Exposures in basalt quarries suggest that the sediment is a product of basalt weathering, although there may be an input of fines from aeolian processes. There is a decrease in the volume of sediment down-section but the transition is gradual, rather than forming a distinct boundary. Also, at depths in excess of 3m to 4m from the ground surface, the distribution of the fine material and the surface of the *in-situ* basalt suggests that the former is a weathering product of the latter, rather than on entirely separate deposit.

Another noticeable geomorphological feature which occurs through much of the Programme Area are fine grained, water-lain sediment deposits. They vary in size and character. Three main morphological types can be identified, based on size, shape and drainage characteristics.

1. Extensive deposits of large surface area, which are fed by wadis but do not have major outward drainage. Locally termed *qa'a*, they are often saline and display many of the characteristics of playa deposits.
2. Deposits which are linear, fed by and drained by wadis. They usually represent zones where channelled wadis spread laterally over more extensive ground surface areas. The deposits are seldom saline or are of low salinity and are known locally as *marab*.
3. Small features which occur in depressions. Some receive and deliver water. Most act as sediment traps due to an absence of drainage and surrounding topography. For the purposes of clarity these have been termed mini-*qa'a*.

Important distinctions are drawn by the local population between *marab*, which are well drained, and *qa'a*, which often form closed basins.

The wadi systems which drain the Programme Area are extensive. The general flow direction is from the north to the south and south-west, with much of the water entering the Azraq basin. Some of the wadi systems extend for many kilometres. Wadi Rajil, for example, stretches from north of the boarder with Syria into the Azraq basin and Saudi Arabia. Peak flows during wet seasons can be considerable although gauging records are limited. Indeed, information from the local population suggests that during parts of the year, lack of water is not a problem. However, the surface water resource is poorly managed, much rapidly drains away and is not retained or utilized for the benefit of the area. There have been some attempts to drain surface waters into the underlying aquifers.

The other key geomorphological zone in the Programme Area is characterized by wind-blown sediment deposits, which are encroaching on the southern boundary of the

Badia. The aeolian deposits represent the margin of an extensive sand-sea in Saudi Arabia. Landsat TM image analysis suggests that the cover is thin and that the dunes are not extensive.

Image analysis of the Programme Area has resulted in its sub-division into twelve geomorphological zones. The main characteristics of each of the zones is as follows.

Zone 1: Limestone bedrock, densely covered with angular chert fragments on the ground surface. Numerous wadi systems, some of which are incised but with few *marab* deposits. Much of the drainage appears to be radial, draining to and terminating in a large *qa'a.* The area includes some small basalt inliers. A gypsum crust exists throughout much of the zone, becoming thick in places and building up in fan deposits towards the Azraq basin. Porcellanite outcrops are present and there is a thin aeolian sediment cover towards the south and south-west.

Zone 2: An area of basalt outcrop to the south of zone 1, partially covered in some places with a thin sheet of wind-blown sand.

Zone 3: Vesicular, basalt lava flows, termed Fahda, containing many *qa'a,* some large and small features. Zone 3 forms slightly higher ground than its surroundings and divides wadi drainage to east and west. Topography is subdued and in many places it is difficult to recognize either the presence or flow direction of wadis.

Zone 4: Basalt outcrop of lighter colour and with a small number of *qa'a* deposits than other basalt areas, suggesting more recent lava flows. The zone divides drainage to east and west, indicating that the most recent lava flow forms slightly higher topography than the surrounding area.

Zone 5: The main area of basalt outcrop, with numerous wadi systems, *marab* and *qa'a* deposits. The wadis are highly inter-digitated and frequently incised, particularly as individual channels merge towards the south. Many of the wadis flow to the east and south, with one or two larger wadis flowing directly towards the Azraq basin in the south-west.

Zone 6: Land rising to the Jabal al-Arab, comprising basalt outcrops with distinct volcanic centres in places.

Zone 7: A basalt zone with a high density of wadis and numerous small *qa'a* deposits, becoming more predominant to the south. The wadi systems radiate out from the higher ground of the Jabal al-Arab (zone 6) in the north. Lichen growths on ground surface basalt boulders and the presence of vegetation increase from south to north, as precipitation totals rise towards the highlands across the border in Syria. Some areas in the north have had the basalt boulders removed and are used for growing crops. The number of fields increases from south to north. Significant erosion and sediment movement occurs by wind and water.

Zone 8: Towards the far north-east corner of the study area, characterized by predominantly parallel, north-west to south-east trending wadi systems, *marab* and *qa'a* deposits.

Zone 9: A region with a thin cover of aeolian sediments, forming small, linear dunes in places and with frequent rock outcrops due to the lack of wind-blown sediment. The sediment is orange-brown in colour and increases in thickness across the border in Saudi Arabia. There are some major *qa'a* deposits in the north, fed by wadi systems draining from zone 5. The *qa'a* are absent in the south.

Zone 10: Characterized by aeolian sediment cover with signs of east to west wind streaking. Numerous small basalt outcrops can be seen, suggesting that the aeolian cover is thin.

Zone 11: A low lying area of limestone and chert, forming an alluvial depression which receives sediment from many of the major wadi systems draining from the north. Zone 11 forms a major part of the Azraq basin, with highly gypsiferous soils and a well-developed gypsum crust on the extensive alluvial outwash plain. The north-east sector is characterized by low, rolling plains, with chert being by far the most predominant material at the ground surface.

Zone 12: Brown-orange sediment deposits, possibly aeolian, but with a slightly higher elevation. The thickness of the aeolian cover is reduced over much of the area. A small number of *qa'a* are present in the north-east.

4. CLIMATE

Jordan can be divided into four climatic zones.

1. Jordan Valley and Southern Ghor. A narrow, 660km strip of land, about 400m below sea level, stretching from the Lake of Tiberius in the north to the Gulf of Aqaba in the south.
2. Highlands. A mountainous area adjacent to the Jordan Valley, which vary from 600m to 1,500m above sea-level.
3. Eastern Hills. An area to the east of the Highlands, known for its relatively plain lands, with a gradual slope towards the east.
4. Eastern Badia. An area known for its dry climate.

Climatically, the Programme Area is recognized to be a transition zone between the environment of the Jordan valley and the arid interior desert areas of eastern Jordan, beyond the town of Ruwayshid. A generally arid desert climate exists, dominated by low precipitation (Table 1) and high potential evaporation (Table 2) between 1,500 and 2,000mm per annum. The consequence is scarce water resources.

Table 1. Rainfall distribution in Jordan

Rainfall (mm / yr)	Area (1000m^2)	% total area in Jordan
< 50	55,700,370	63.2
50 - 100	13,851,000	15.0
100 - 200	11,390,500	12.4
200 - 300	3,948,000	4.3
300 - 400	1,788,000	1.9
400 - 500	1,253,000	1.3
> 500	979,000	1.0

Table 2. Available class A-pan evaporation stations from the Azraq Basin

Station name	Station code	Elevation (m)	Length of record
Azraq	F0009	533	1961-1988
Safawi	F0002	712	1963-1988
Bayir	J0001	900	1968-1988
Al Ruwayshid	H001	686	1963-1988

Table 3. Available rainfall stations for the Azraq Basin

New code	Station name	Elevation (m)	Type*	Record from-to
F0001	Umm al-Quttayn	886	D/R	1947-1988
F0002	H5 Evap. Station	715	D/R	1968-1988
F0003	Azraq Police Post	533	D	1934-1970
F0004	Dayr al-Kahf	1025	D	1963-1988
F0006	al-Qaryatain	800	D	1963-1988
F0007	al-Wisad	700	D	1963-1988
F0009	Azraq Evap. Station	533	D	1963-1988
F0010	ash-Shaumari	525	D/R	1962-1988
F0011	al-Umari	525	D	1962-1988
F0012	Oasr Tuba	750	D	1968-1988
F0013	Wadi Salahb	700	T	1967-1988
F0014	Tullul al-Kuraysha	550	T	1968-1988
F0015	Tullul al-Ghar	750	T	1968-1988
F0016	Qa'Samiqa	760	T	1968-1988
F0017	Gadir al-Mallah	630	T	1968-1988
F0018	Jabal Aseeikhim	640	T	1968-1988
F0019	Jabal Mudysisat	900	T	1968-1988
F0020	Qasr al-Kharrana	650	T	1968-1988
H0004	Tullul al-Ashqaf	950	T	1967-1988

There are seasonal temperature variations, with summers tending to be hot and dry and winters cool and wet. Mean annual maximum temperatures reach 35° C to 38° C in August but absolute maximum values regularly exceed 40° C. Temperatures can fall below zero degrees Celsius in winter. Annual mean minimum temperatures drop as low as 2° C to 9° C. When cold, continental air penetrates into the area, the temperature can drop as low as 1.6° C. The highest and lowest recorded temperatures are 46.4° C and 12.0° C respectively.

Rainfall is subject to drastic fluctuations in place and season. The mean annual rainfall in Jordan varies from less than 50mm per annum in the south of the Programme Area, to more than 600mm over the Ajloun Heights (Fig. 5). Rainfall occurs mostly between the months of November and May, with 80 per cent of the annual precipitation arriving between December and March. The overall mean annual rainfall for the Kingdom is approximately 65mm (Tables 1 and 3). Individual storms frequently result in wadi flow since many are of high intensity. The estimated potential evapotranspiration can be in excess of fifty times mean annual rainfall and solar energy supplied to the earth surface is far above that required to vaporize all the water at or close to the ground (Burdon 1982).

There are two dominant wind directions for most of Jordan, including the Programme Area. In the summer, north-west winds predominate, due to frequent seasonal troughs to the north of the country, which represent an extension of the Indian Monsoon. Wind speeds average 8 knots to 15 knots, with velocity increases in mid-afternoon sometimes reaching 20 knots to 25 knots. In the winter, the dominant wind changes to a west to south-west direction due to depressions which travel along the eastern Mediterranean. There are occasional gales.

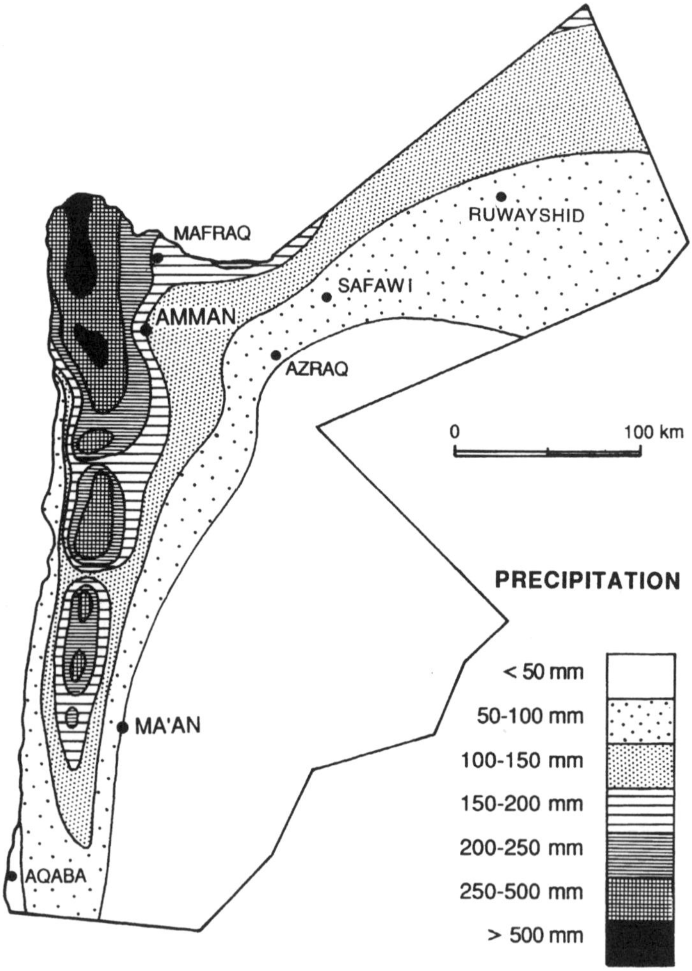

Fig. 5. Precipitation distribution in Jordan.

5. HYDROLOGY AND SURFACE WATER

Water is one of the most important natural resources in arid parts of the world and is critical to sustainable development in the Programme Area. In 1991, the total national water consumption in Jordan approached 730 million m^3. It is anticipated that demand will rise to 2,105 million m^3 by the year 2020. Figure 6 shows the future projection of Jordan's water demand. Approximately 8,500 million m^3 of precipitation falls in Jordan each year, of which somewhere between 85 per cent to 92 per cent is lost to evaporation, 5 per cent to 11 per cent disappears through infiltration and 2 per cent to 4 per cent generates surface runoff.

The drainage systems in Jordan are sub-divided into fifteen major regions (Table 4, Fig. 7). Other than basin delineation, little is known about the characteristics of each region, particularly on parameters which influence runoff and flood events including catchment size and shape, wadi channel patterns, the width, depth and slope of channels and the nature of sediments within the wadis. The hydrological situation in the

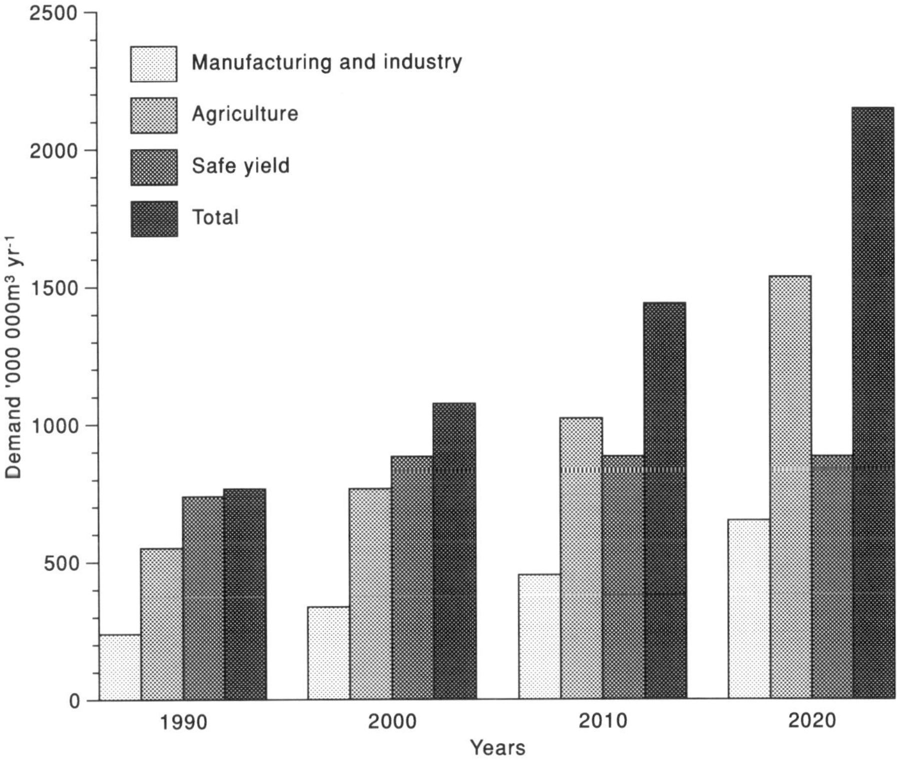

Fig. 6. The projected water demand for Jordan.

Table 4. Drainage systems in Jordan and surface water flow (million m^3/year)

	Basin	Base flow	Flood flow	Spring flow	Total flow
1	Yarmouk at Adasiya	130.00	155.00	21.40	285.00
2	Ghor - Jordan Valley	0.00	2.40	19.30	21.70
3	N. Jordan River (side wadis)	36.07	13.91	48.63	49.98
4	S. Jordan River (side wadis)	24.76	5.58	28.90	30.34
5	Zarqa River	33.51	25.67	38.91	59.18
6	Dead Sea	53.95	7.20	57.60	61.15
7	Mujib	38.10	45.54	16.00	83.64
8	Hasa	27.40	9.04	3.90	36.44
9	N. Wadi Araba	8.99	2.57	15.63	18.20
10	S. Wadi Araba	0.00	3.16	2.44	5.60
11	Southern Desert	0.00	2.15	0.05	2.20
12	Azraq	0.00	26.80	0.60	27.40
13	Sirhan	0.00	10.00	0.00	10.00
14	Hammad	0.00	13.00	0.00	13.00
15	Jafer	0.00	10.00	1.92	11.92
	TOTAL	352.78	332.02	255.28	684.80

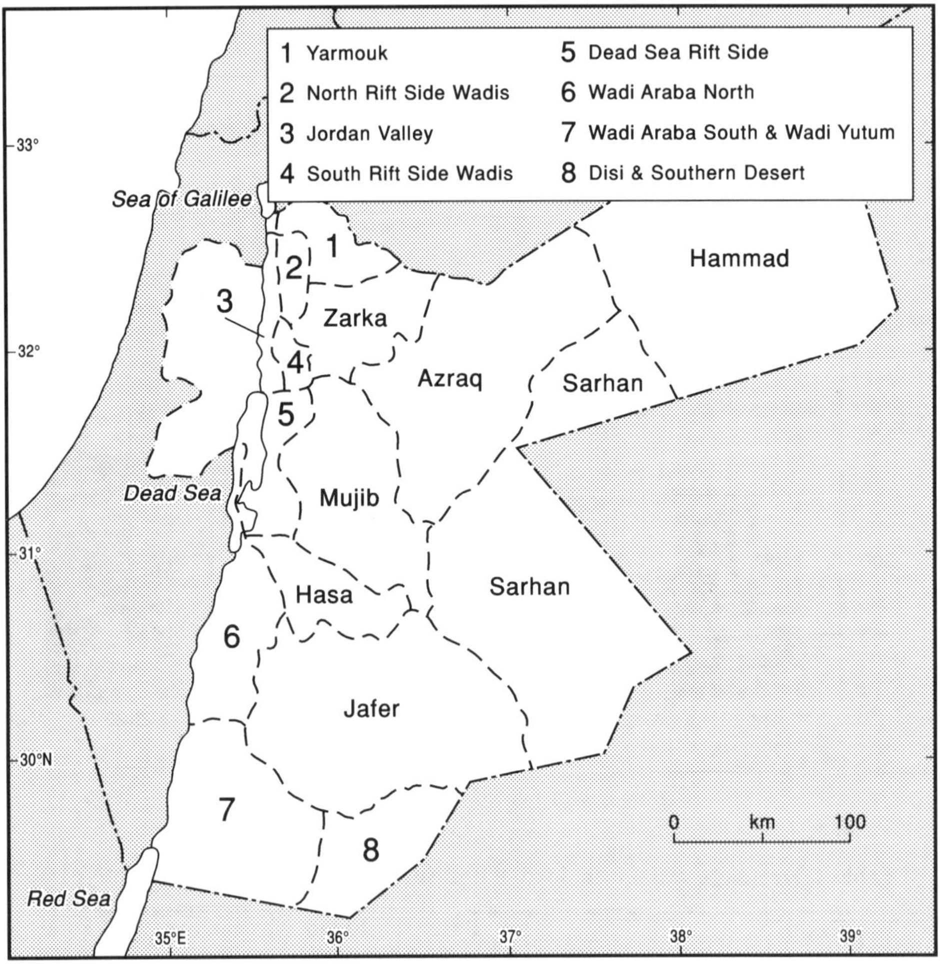

Fig. 7. The main surface water drainage basins in Jordan.

Programme Area is poorly understood and there is a dearth of information on all aspects of the water balance. Runoff producing events are seasonally concentrated and occur every year. Daily rainfall events can be several times the size of the monthly average. However, it is far from clear how frequently rains of sufficient magnitude generate runoff. The amount of precipitation required to generate runoff will also depend on other prevailing conditions such as antecedent soil moisture and the nature of surface materials and vegetation. The water resources problems in terms of their socio-economic significance are outlined in Table 5.

Table 5. Water resources problems in terms of social and economic significance

Physical Manifestation	Consequences	Economic and Social Significance
1 Semi-arid climate, low precipitation, high evaporation.	Fluctuating water supply, limited resource base, periodic drought.	Planning problems, development target uncertainty.
2 High population growth rates.	Water demand and competition increasing, groundwater depletion, pollution.	Reduction in living standards, health problems.
3 Conflicting demands.	Inequitable allocations, regional price structuring.	Emergence of water lobbyists.
4 Riparian conflicts.	Critical supply augmentation not undertaken.	Economy destabilization.
5 No effective conservation program.	Environmental impacts, inefficiency losses, water logging problems.	Productivity decline.
6 Financial constraints.	Demand exceeds supply.	Productivity decline increasing health problems.
7 Lack integral water policy.	Numerous as listed above.	Low social and economic development potential.

6. GROUNDWATER

Blake (1928, 1930, 1937) undertook the first serious hydrogeological studies within Jordan, with later work being conducted by Ionides and Blake (1939), Goldschmidt (1947) and Bender (1968). The distribution of groundwater in Jordan is now fairly well known. There are twelve different basins (Fig. 8). Groundwater flow directions are shown in Figure 9. A number of major aquifer systems can be identified in Jordan. Those which are important in the Programme Area are as follows:

1. Upper aquifer complex:

   (i) Neogene-Quaternary alluvial sediments;
   (ii) Neogene-Quaternary basalts;
   (iii) Lower Tertiary marly and chalky limestones with cherts (the Rijam aquifer or B4)

2. Middle aquifer complex:

   (i) Upper Cretaceous limestone, sand limestone (the Amman-Wadi Sir aquifer or B2, A7);
   (ii) Middle Cretaceous crystalline to chalky limestone (the Hummar aquifer or A4).

3. Lower aquifer complex:

   (i) Lower Cretaceous sandstone (the Kurnub aquifer);
   (ii) Lower Mesozoic-Palaeozoic (the Disi aquifer).

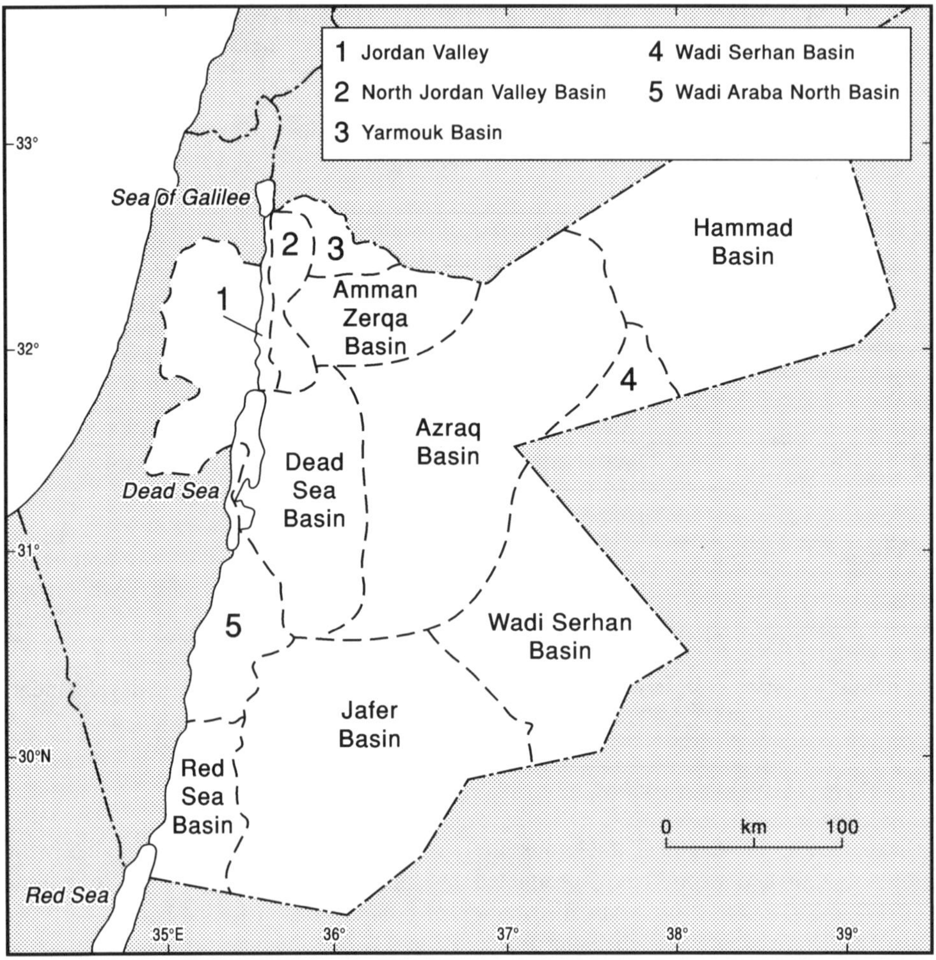

Fig. 8. Groundwater basins in Jordan.

The upper and middle systems are separated by an aquifer of Maestrictian age. The middle and lower aquifers are separated by marls and marly limestone of low permeability. A Silurian shale may separate the Lower Cretaceous Kurnub aquifer, from the deeper, Palaeozoic, Disi sandstone aquifer.

The upper aquifer complex covers most of the Programme Area. Groundwater occurrence is complex due to the variety of lithologies present. Recharge is from the north, via flows between and below the basalt sheets arising from the Jabal al-Arab. Some recharge via runoff occurs from Wadi Ruwayshid in the north-east. Discharge is by human extraction, evapo-transpiration and lateral outflow. Groundwater storage is likely to be moderate, since the saturated formations are not thick. Depth to water table varies with topography, being > 350m deep in the north along the Syrian border and less than 100m deep towards Azraq.

The Neogene and Quaternary component of the upper aquifer sequence includes sandstones and sandy marls comprising fluviatile sandstones, gravels, conglomerates, wadi fans and talus deposits. All are highly permeable, usually unconfined and highly variable in their physical characteristics. The Lower Tertiary chert limestone rocks form

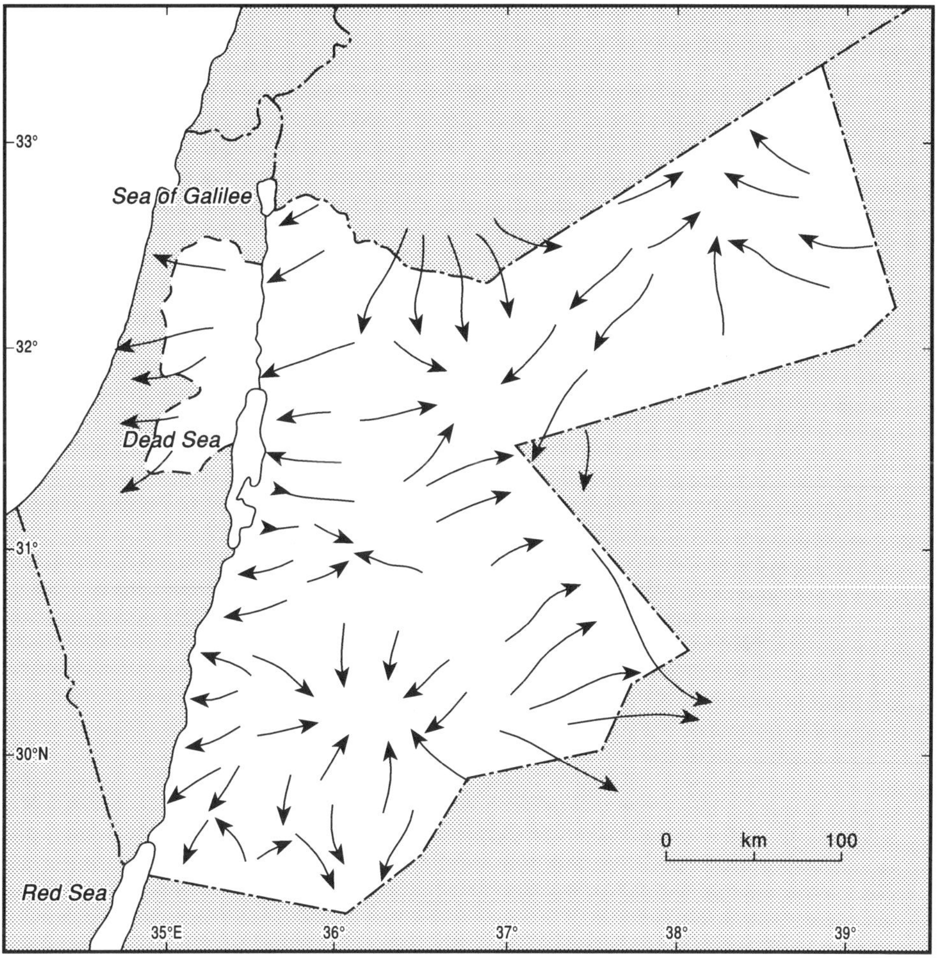

Fig. 9. Main groundwater flow directions in Jordan.

the Rijam aquifer. In the Azraq basin, the Rijam aquifer is hydraulically connected to overlying Neogene sandstones and an extensive upper basalt aquifer (Fig. 10).

Groundwater salinities for the upper aquifer range from a few 100ppm to > 4000ppm. Where rainfall and flow rates are high, the chemical composition of the aquifer matrix has little effect on water quality. However, where rainfall and flow rates are low, the chemical composition of the rock increasingly influences water chemistry as distance increases from the recharge zone. In general, the basalt waters are of good quality and can be readily distinguished from water derived from the marly sequences. There have been recent signs of water deterioration in some areas due to over-extraction.

The deeper, middle and lower aquifers are less well understood. They are mainly confined, fossil resources. Storage is high but water quality is variable, in some cases being saline, geothermal and unusable. Horizontal hydraulic gradients are shallow and flow frequently occurs in structural fissures, sometimes introducing a westerly trend to flow movements. Structure is a major control on regional flow patterns, either directly by fractures facilitating infiltration, or indirectly by the formation of areas of relative

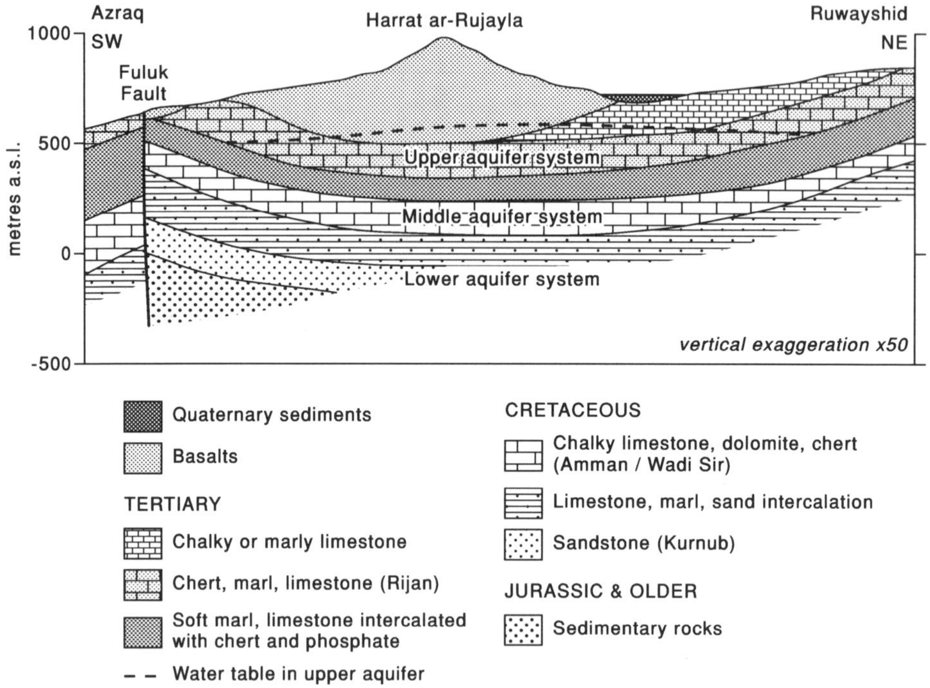

Fig. 10. Cross section showing the main aquifers of the Badia.

uplift and subsidence which control the aerial distribution and thickness of permeable strata and their regional dip (Khouri 1982). The dominant structural control is the Azraq-Wadi Sirhan Basin, bounded in the west by the Jordan Uplift, to the north by the Halab Uplift in Syria and to the east by the Rutbah-Ha'il Arch in Iraq.

7. SOILS

The formation, types and properties of soils in Jordan are closely controlled by parent material, topography and climate. The soils of the Programme Area comprise three distinct types. The major groups broadly follow geological variations, in particular the spatial distribution of basalt lava flows and the outcrop of other, more recent sediments.

There are soils associated with the East Jordan Limestone Plateau. Soil parent materials in the Plateau area include limestone, chalk, calcareous sands, gypsum, marls and Tertiary cherts. The soils are highly variable, as a result of repeated cycles of erosion and deposition and include shallow to deep entisols and aridisols. Down-slope, bedrock surfaces merge into sequences of gravel fan and terrace deposits, with entisols frequently occurring in wadi beds and depressional mudflats. Secondly, there are soils associated with the North Jordan Basalt Plateau. The main soil divisions are related to topographic and weathering differences. The soils include well-developed xerochrepts on older basalt flows, with gypsitic and calcitic horizons and weakly developed xerothents on more recent basalt flows and in wadis. A common feature of the soils is the presence of a thick, secondary caliche deposit on the basalt bedrock. Finally, there are the soils of the north-east Jordan Basalt Plateau, which are classified on the age and physical characteristics of lava flows and eruptive centres. A wide range of soils occur, including gypsiorthids, camborthids and calciorthids on older surfaces, with torriorthents and lithic sub-groups on less weathered terrains. Despite the identification and classification of the different soil types in the Programme Area, the distribution and total spatial extent of each soil type remains to be established in many areas.

8. SEDIMENT EROSION AND MOBILITY

One of the most noticeable characteristics of the physical environment in the Programme Area is the erosion of soil and other fine-grained, loosely compacted sediments. The erosion and redistribution of sediment has a number of components: environmental controls on sediment production and detachment; aeolian mobilization processes; runoff mobilization processes and factors promoting contemporary sedimentation. The same strategic question can be asked of both aeolian and runoff process environments, although the methods of investigation and sampling frameworks will vary. They include erosion initiation thresholds, rates and volumes, spatial and temporal distributions, key variable controls on erosion systems such as vegetation and topography, transport characteristics and finally patterns of deposition. Erosion by water is a consequence of high magnitude, low frequency storms, the absence of a protective vegetation cover across much of the ground surface and the low stability of soil aggregates. A surface crust frequently develops which seals the ground, reduces infiltration, promotes runoff and aggravates erosion. Questions relating to sediment mobility provide an important link between soil and water studies.

Erosion by wind is most frequently aggravated when ground surface disturbance, for example by livestock or vehicles, provides locations where aeolian erosion processes can be initiated. There is extensive evidence of wind erosion, particularly during long, dry summers, when soil surfaces dry out, crack, lose cohesion and begin to fragment. Extensive tracts of uninterrupted land with little or no topographic variation promote rapid wind speeds. It is all too frequently the case that the top, most productive soil layers, are lost at times when little or no vegetation is present and wind speeds are at their most severe. The same is the case where basalt boulders have been removed from the ground surface, exposing the underlying sediments to deflation.

9. DEVELOPMENT STRATEGIES AND IMPLICATIONS

9.1. Physical Resources

The geology and physical resources research is providing important data on extractable resources, including their type, location, spatial extent, quality, extraction viability and potential use. Mineral resources are important in two contexts.

1. Materials for use in construction projects. Included here are rocks for building on a brick-by-brick basis and aggregates, such as sand and gravel, suitable for foundation materials, concrete structures and in extending the regional infra-structure, such as roads.

2. Materials with industrial uses. Preliminary field investigations undertaken by the Geology Directorate of the Natural Resources Authority have identified the presence of tuff, zeolite, ornamental basalt, diatomite and other clay minerals.

Geothermal water has been located towards the east of the Programme Area, with temperatures approaching 55° C. The Royal Scientific Society (RSS) is presently involved in projects which are examining the potential for exploiting wind and solar energy resources.

9.2. Surface Water Retention, Harvesting and Control

While there is the potential to harvest water from the desert areas of Jordan, no data have been collected to corroborate this, nor has any work been done in the Programme Area to examine the likely consequences of water control and management. There are a number of issues which provide the focus for much of the surface water development strategy work. These are the control of high magnitude runoff events during intense storms, the collection of water, provision of water for irrigation, provision of water for livestock and groundwater recharge.

Key questions include the following. What techniques have been used in the past for surface water harvesting, management and distribution? What techniques are currently being used for surface water harvesting, management and distribution? How effectively is the surface water supply being managed? What can be done to maximize the volume of surface water available for human use? Which areas have the best potential as water collection zones? What spatial differentiation exists between potential new collection zones and locations where water is required for use?

Water management depends mainly on the collection, storage and redistribution of runoff, which is seasonally concentrated and frequently occurs as high magnitude/low frequency events due to short duration, intense rainstorms. The resulting wadi flood peaks can cause severe damage and structures designed to direct or store runoff have to account for the physical dynamics of overland flow. Remnants of ancient water harvesting systems are visible in the Badia and their evaluation will be an important starting point for work on the location and design of new structures. The region has many ancient ponds but most have become silted up. A rapid maintenance programme would bring many ponds back into immediate and effective use, for little cost and a minimum effort. Similar problems exist with ancient dams in the Badia. The structure at Dayr al-Kahf, close to the Syrian border, is an example. There is an estimated 17m of sediment behind the dam wall. Excavating the debris would return the dam to a valuable seasonal water store for the local population. Three types of water harvesting structure are common in remote, semi-arid environments:

1. Desert dams. A small dam built on an intermittent stream or wadi, which retains water during the wet season for the purpose of either direct consumption and/or irrigation and/or groundwater recharge.
2. Ponds. Excavations for the collection of water. Ponds ideally have a capacity of around 100,000 m^3. The collected water is most frequently used for irrigation and watering livestock.
3. Dykes. Structures for diverting and directing water across the ground surface to zones where resources are scarce or need to be supplemented.

Development of water resources in the Badia may include the construction of water retention and water diversion structures as an integral component of the Programme. The opportunity exists to examine the potential which a Geographic Information System (GIS) might have in the selection of sites for the construction of water-retaining structures. Important aspects of the work will include the location of suitable dam, pond and dyke sites, given the base-line hydrological data, and determination of the geotechnical properties of local materials and their suitability as construction materials. The basic hydrological and runoff data will provide the starting point for water management studies. Meteorological and wadi flow data will be particularly important. Once individual sites suitable for implementing water management techniques have been identified, the relevant geological and geotechnical information must be collected. Investigations to be completed at later stages may include the on-site drilling of boreholes and digging of trial pits to obtain samples for geotechnical analysis.

With some projects, for example the construction of dams, soil and rock samples will have to be collected for further laboratory studies and a topographic survey will be necessary to precisely locate structures and calculate the ground surface area to come under the control of any new development. Data will be needed to undertake good estimate reservoir water balance simulation studies. Simulations must account for varying parameters such as dam height, spillway dimensions and yield. A major dam has recently been constructed across a down-stream section of Wadi Rajil, close to the north-east margin of the Azraq Basin. The drainage capture area for the dam is 3,243km^2 Topography in the area is flat and the water body will have a large surface area. Construction is using materials available from the local area. The reservoir capacity is 2.6 million m^3. Some useful data may well be available from this site and should be consulted as a starting point to any new reservoir study.

The excavation of ponds will be one of the most efficient methods of water harvesting in the Badia due to their simplicity of construction, economy and ease of maintenance. Work will be required to identify the best possible location for pond construction and quantify site characteristics such as hydrology, topography, evaporation, infiltration and sedimentation rates. Locations where water can be redistributed across the ground surface to permit its more efficient use need to be located and the best constructions identified. The goal should be to locate, design and construct small, simple structures, for the collection, re-direction and spreading of water for immediate use in agriculture. Water spreading structures are usually simple and easy to construct, following close to contour lines, thereby using the natural topography to move water. Local materials should be used in construction. No detailed or complicated pre-construction studies will be necessary and the final design should be easy to repair following storm events.

9.3. Groundwater exploitation

At present, approximately 940 million m^3 of groundwater is extracted within Jordan per annum, of which 139 million m^3 is used for municipal supply purposes and 26 million m^3 by industry. The remainder is used for irrigation. The potential for the future development of agriculture in the Programme Area hinges on the availability of water supplies. A key component of development will be the effective exploitation and careful management of the groundwater reserve. The basic data collected on groundwater dynamics will be used in assessing the potential of further extraction for development purposes. It is important to recognize that results indicate that the natural sub-surface water reserve is already being over-exploited, that rates of production need to be reduced and strategies for recharge introduced. Scenarios for groundwater decline need to be established and tested in the context of the likely response of aquifers to development strategies in the future.

Recharge estimates will be critical in the assessment of groundwater resource potential, since a minor change in replenishment will have a major impact on resource sustainability. Development strategies will have to take into account the fact that piezometric levels in the shallow aquifers may decline either because of a reduction in recharge due to climatic change, in response to recent increases in extraction or because extraction already exceeds recharge. It will be necessary to examine the potential for artificially promoting groundwater recharge. Local evidence suggests that quality has been declining in recent years and the water withdrawn from some wells is no longer suitable for human consumption, having significant implications for sustainable development.

A further important component of groundwater studies will be the investigation of aggressive salts and associated weathering, their presence, concentrations, precipitation and effects, such as building material degradation, volume changes upon crystallization and associated effects on structures and foundations and implications for soil quality and ground deterioration. The approximate limit of the capillary fringe established, the relative effects of capillary rise in a high evaporation environment ascertained and salt deposition due to irrigation quantified. Measures of salinity will be necessary and, for areas already prone to salt problems, the potential for leaching out crystallized deposits established.

9.4. Soil Geomorphological Management and Quality Enhancement

The development of soil resources in the Programme Area and the identification of good management practices will require the input of data from both soil survey and sediment erosion and mobility studies. Research needs to be focused on establishing the best practices for managing the soil resources, their improvement, maintenance and the restoration of denuded land. The most satisfactory agricultural practices to be undertaken on areas with an adequate soil cover need to be defined. Both water and wind erosion are presently exacerbated due to the removal of basalt boulders, low percentage of plant cover, overgrazing, a long drought season, erratic and uneven rainfall distributions and poor soil quality. The soil-oriented development needs of the Programme Area ultimately focus on the productivity improvement of land and the identification of new areas suitable for agricultural production. Specific needs include the following.

1. The improvement of soil water relationships.
2. The minimization, control and prevention of erosion.
3. The improvement, conservation and development of soil quality and properties pertaining to fertility and productivity.
4. The selection of plant species suitable for integrated cropping systems which should be developed in the area.
5. The improvement of natural vegetation cover.

Soil surveys will be required in areas close to water resources, or in regions where water harvesting and management techniques can be employed to bring land into crop production. Site selection for planting new crops will be based on soil physical and chemical properties and nutrient content.

9.5. Environmental Management

The ultimate goal of the Jordan Badia Research and Development Programme is to provide a framework for planned, sustainable development. From the point of view of geomorphology and physical resources, a key component will be to define good management strategies for all aspects of the environment. A plan for environmental management will act as the focus for all studies. Collected data will be used to achieve the following:

1. The identification of the best location for development.
2. The definition of management strategies in developing areas.
3. The limitation of undesirable environmental impacts as a result of development.
4. The proposal of a strategy for environmental impact assessment, thereby allowing the modification of management strategies through time.

By synthesizing data in an environmental management context, the best possible strategy for maintaining the environment will be proposed. It is important to recognize that this will have to be achieved within national, regional and local planning structures, and that the administrative mechanisms and financial resources available for plan implementation will be constraining factors.

ACKNOWLEDGEMENTS

An earlier version of this overview was published in *GeoJournal*: Al-Homoud, A.S., Allison, R.J., Sunna, B.F. and White, K. 1991. Geology, geomorphology, hydrology, groundwater and physical resources of the desertified Badia environment in Jordan. *GeoJournal*, 37.1, 61-7. Permission to reproduce the paper, in modified form, is appreciated by the editors of this volume.

The advice and assistance of the following people is warmly appreciated. *In Jordan:* Dr. A. R. Al-Fataftah, Department of Agriculture, University of Jordan; Nancy Bakir, The Higher Council for Science and Technology; Dr. M. El-Sallag, Water Authority; E. Fadda, Geology Directorate, Natural Resources Authority, Ministry of Energy and Mineral Resources; K. M. Ibrahim, Geology Directorate, Natural Resources Authority, Ministry of Energy and Mineral Resources; Dr. S. Khattari, Department of Soil Science, University of Jordan; O. Malkawi, Royal Jordanian Geographic Centre; S. Tuffaha, Ministry of Water and Irrigation. *In the United Kingdom*: Professor D. Brunsden, Department of Geography, King's College London; Dr. W. G. Burgess and Mrs. J. Dottridge, Department of Geological Sciences, University College London; Professor R. U. Cooke, University of York; Dr. R. W. Dutton, Centre for Overseas Research and Development, University of Durham; Professor A. S. Goudie, School of Geography, University of Oxford; Professor D. K. C. Jones, Department of Geography, London School of Economics; P. Miller, formerly Centre for Overseas Research and Development, University of Durham; Professor A. C. Millington, Department of Geography, University of Leicester and N. Winser, Royal Geographical Society (with the Institute of British Geographers).

REFERENCES

Bender, F. *et al.* (1968) Geological Map of Jordan 1:25,000 (5 sheets). Geological Survey of the Federal Republic of Germany, Hanover.

Bender, F. (1968) *Geologie von Jordanien.* Beitr. Regionale Geologie erde, Vol. 7. Gebruder Borntraeger, Berlin.

Bender, F. (1974) *Geology of Jordan.* 196 pp. Gebruder Borntrager, Berlin.

Bender, F. (1975) *Geology of the Arabian Peninsula: Jordan.* United States Geological Survey Professional Paper 560-1.

Blake, G.S. (1928) *Geology and Water Resources of Palestine.* Jerusalem.

Blake, G.S. (1930) *The Mineral Resources of Palestine and Transjordan.* Publication No. 2, the Geological Advisor in Palestine.

Blake, G.S. (1937) Old shore lines of Palestine. *Geological Magazine,* 74, 68-78.

Burdon, D.J. (1959) *Handbook of the Geology of Jordan.* 82 pp. Government of the Hashemite Kingdom of Jordan, Amman.

Burdon, D.J. (1982) Hydrological considerations in Middle East. *Quarterly Journal of Engineering Geology,* 15, 71-82.

Goldschmidt, M.J.: (1947) *Geology and Water Resources of Palestine.* Department of Land Settlement and Water Commissioner, Government of Palestine.

Ibrahim, K.M. (1992) *The Geological Framework for the Harrat Ash-Shaam Basaltic Super-group and its Volcanotectonic Evolution.* Natural Resources Authority, Amman.

Ionides, M. G. and Blake, G.S. (1939) Report on the Water Resources of Transjordan *and* their development: incorporating a report on geology, soils and minerals and hydrogeological correlations. 372 pp. Crown Agents for the Colonies, London 1939.

Khouri. J. (1982) Hydrology of the Syrian steppe and adjoining arid areas. *Quarterly Journal of Engineering Geology*, 15, 135-54.

Field Research

CHAPTER 3

A Synthesis of Climatic Data with Specific Interest in the Precipitation Record

ALASTAIR KIRK
(University of Durham)

1. INTRODUCTION

Whilst arid lands can be resource rich, this is not the case in eastern Jordan and the fact that the area suffers from a large water deficit compounds the human and ecological hardship. The removal of what little vegetation there was, associated with population pressure and more specifically over-grazing, is held to be the major cause of land degradation (Farhan and Mikbel 1986; Agnew and Anderson 1992).

Jordan occupies the position of a buffer state, separating on the national scale, the Mediterranean state of Israel to the west and the Arabian desert to the east and on the continental scale, Jordan along with Israel lies on the narrow strip of land which connects the two largest continents on this earth; Asia and Africa (Phillips 1954). Despite its small size, however, the country has a large diversity, both culturally and physically which comes from its being a transition zone. The desert areas or Badia includes the whole of the eastern part of Jordan, making up over 80 per cent of Jordan's total land area (Al-Homoud *et al.* 1995; Ch. 2). It is classified as a semiarid to arid steppe environment and falls almost exclusively in the arid climate zone with rainfall less than 100mm p.a. in most areas (see Fig. 1). The natural response of most human groups to harsh and fragile environmental conditions is to diversify, to rely on a variety of different resources so that if one fails there will always be another (Betts 1992). It is in such a precarious environment that development is being encouraged and provides the context and indeed the necessity for research into the physical environment. More specifically, rainfall is the one environmental variable which can allow a population to survive in such a region. The lives of the communities of the Badia depend absolutely upon it. Therefore, the collection of good climatic data is vitally important as a foundation not only for research looking into surface and groundwater hydrology, but also for all studies of soil-water interactions and their subsequent effect on agricultural productivity.

Long (1957) describes the climate of the Syrian desert, of which the Jordanian Badia is a part, as 'Mediterranean Saharan' because although it is arid, it is less arid than the Arabian or North African deserts and has a smaller seasonal and diurnal temperature range, with all the sparse rainfall concentrated in the winter months. Air temperature is highly variable and although the average is only 17.5° C, the minimum and maximum temperatures lie in the range of -5° C and 46° C.

The general pattern of climate which controls much of the Levant and the Gulf is dominated by certain large-scale atmospheric circulation conditions which would have

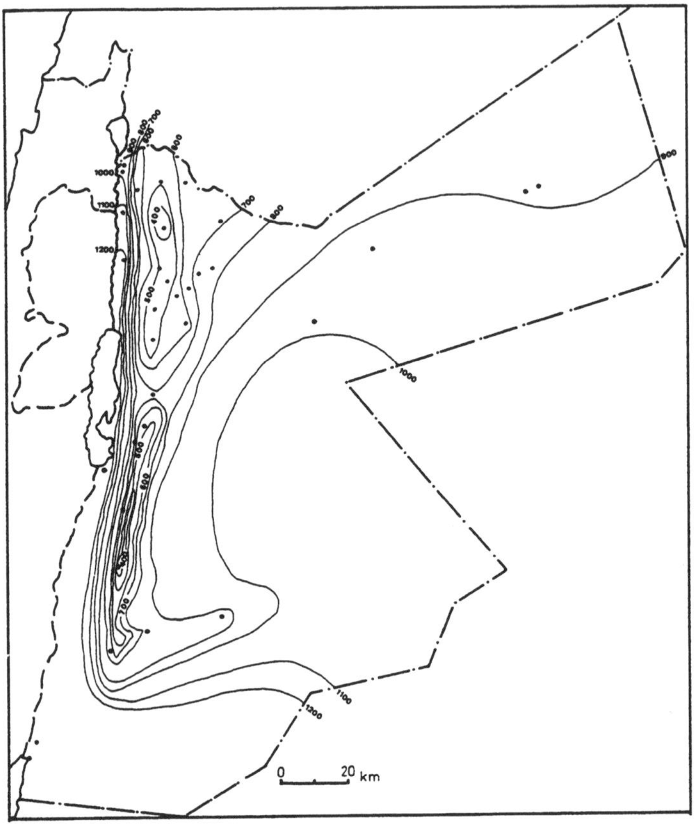

Source: Shehadeh (1985, 34)

Fig. 1. Mean moisture deficiency (mm).

definitely been present in the past, although they could have been geographically displaced and changed in intensity over time (Wigley and Farmer 1982). Winter is characterized by cyclonic disturbances and low mean pressure in the Mediterranean, with higher pressure further east associated with the Siberian high. Unsettled weather in the Mediterranean basin is especially prevalent when the westerlies are in their 'low-index' or 'blocked stages' and the Polar Front Jetstream (PFJ) exhibits a strong oscillatory pattern allowing meridional transport of cold air (Perry 1981). The lower position of the PFJ over central Europe brings with it substantial amounts of rain. During spring the Subtropical Jetstream (STJ) begins to move northwards from its location over the Tropic of Cancer and as the PFJ is pushed northward, precipitation is reduced.

By May, the influence of the polar front and its associated westerlies is negligible while they are replaced with the sub-tropical ridges of high pressure. It is during the summer months that the contracted circumpolar vortex ensures that the Mediterranean is a region of subsidence associated with a subtropical upper tropospheric high (Perry 1981). The effect of the Indian Monsoonal system is considerable: as it advances northwards during June, the easterly jet strengthens due to pressure differentials over continental Asia. The resultant complex set of jets extend over southern Arabia and East Africa, the intensity of which is directly proportional to the rainfall received by the Upper Nile catchment. The jet system is also associated with upper air convergence causing widespread subsidence which is characteristic of the Middle Eastern desert areas.

In the winter and spring, the Cyprus lows are an important key to precipitation over the Levant. Eastern penetration depends on the zonality of the upper flow and by the strength of the Siberian High. There is some debate as to whether the position of the European Trough in the Mediterranean gives more of less rainfall in the Levant. It is generally believed that if cyclonic conditions stagnate over Cyprus, dry winter conditions will tend to prevail over the Middle East, but if the surface cyclonic conditions move eastwards, westerly winds begin to predominate and precipitation occurs especially corresponding to orographic features. Lamb (1968) believes that wetter conditions between the first and third millennium before Christ were primarily due to the easterly position of the trough. Krown (1966), however, relates wetter conditions in the Levant to a more westerly position of the trough in the Mediterranean.

It is during the transitional seasons, and especially in early autumn that year to year variations in the general circulation are greatest causing widely differing weather conditions in the Mediterranean. The combination of the irregular southerly movement of the STJ from its summer position over Turkey down to northern Sudan and the subsequent increase in surface pressure over the Levant causes a somewhat pulsatory seasonal transition (Perry 1981).

2. CLIMATE CHANGE OVER THE HOLOCENE

When analysing contemporary rainfall observations it is important to realize that present-day cycles in climate and precipitation are just a small part of the overall pattern which extends back over thousands of years. If such a context is ignored, it is likely that the observer will consider present fluctuations as more important than they are. Instead small differences over tens of years need to be seen in the context of larger differences over hundreds or thousands of years.

It is evident from archaeological studies of the Badia that the climate has not always been as dry as it is today. The area, named by archaeologists as the southern Haurân (Butler 1919), has been irregularly populated since the prehistoric period (Gerrard *et al.* 1975). Although there is much circumstantial evidence for small fluctuations in the climate over the last two millenia (Butzer 1955; 1958) there is very little quaternary geological research which verifies such evidence. Roberts (1982) could only identify 31 sites in the whole of the Middle East where quaternary climate change research had taken place and much of that has been confined to the Nile and Israel (Butzer 1975). The only research which has taken place in the vicinity of the Badia is that at al-Jafr in the southern Badia (Huckriede and Wieseman 1968) and in the Damascus Basin (Kaiser *et al.* 1973). It is clear that there have been periods of relatively good levels of precipitation, or pluvials, throughout the Pleistocene and Holocene, but what is very unclear is the frequency and magnitude of these periods. The example from Huckriede and Wieseman (1968) gives a date of 27 700 ± 870 for the end of pluvial limestone sedimentation in the El-Jafr depression. Whilst most scientists would agree that the end of the last pluvial was towards the end of the Pleistocene, there is much controversy over smaller fluctuations since that period. Issar and Bruins (1983) propose a series of multi-annual cycles of relative drought and periods of higher humidity, related to deuterium levels in the geochemistry of the Kurnub Sandstone. Bruins and Yaalon (1979) use particle size data from palaeo-eolian deposits in the Negev to suggest changes in the magnitude of summer Hamseen winds which again can be related to climate and more particularly precipitation. However, it remains a significant problem to try and quantify smaller sub-pluvial events where small fluctuations in climate have led to slightly drier or wetter conditions.

Unlike the threshold changes from pluvial to interpluvial which induce significant signals in the stratigraphy, the smaller variations can only be detected in the historical record by using alternative sources of information. The only complete attempt to link such smaller fluctuations over the Holocene period has been that of Butzer (1958).

He takes evidence from a number of sources; historical, archaeological and geomorphological to produce a chronology of what he calls postpluvials and subpluvials which he illustrates by producing an approximation of rainfall fluctuations (Fig. 2).

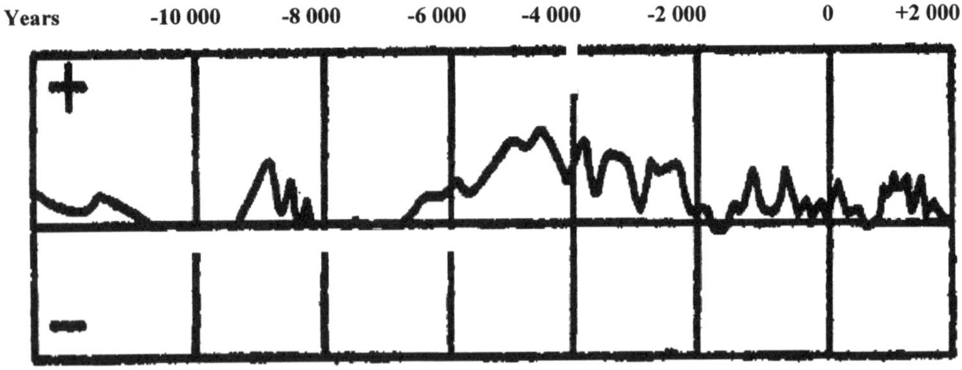

Source: Butzer (1958, 142)

Fig. 2. Relative trend of Near Eastern precipitation during the Late Glacial and Postglacial periods.

3. CLIMATIC CONSIDERATIONS

To gain a thorough understanding of climatic conditions in an area such as the Badia, where data points are few, it is necessary to take data from a number of different temporal scales and various resolutions. In general the majority of data comes from stations such as those in Table 1 which are monitored by the Water Authority of Jordan or the Jordanian Department of Meteorology; some of these monitor daily totals of rainfall such as the station at Dayr al-Kahf, while others record hourly data for rainfall and other climatic characteristics. The most important short-term high resolution data has been obtained from an Automatic Weather Station (AWS), which was loaned to the author from the Natural Environment Research Council (NERC) equipment pool. The AWS was placed close to the author's main field site at Ashrafiyya at the police compound in Manara as it was hoped that some of the soil-water experimentation could be tied into the evaporation fluxes which could be obtained from the station. The station was operational from the middle of August 1994 until the summer of 1996. There are additional data from an AWS owned by Leicester University, which is situated in Be-qa-wier, but this has only been operational from December 1995 and has not been included in the analysis. The weather station (Fig. 3) has six sensors; solar radiation, net radiation, wet- and dry-bulb temperature, wind direction and wind run, in addition to the rain gauge. Each sensor takes a measurement every ten seconds and produces an average for every hour of the day. The rain gauge works as a tipping bucket mechanism whereby the rainfall is measured at a resolution of every half a millimetre and summed per hour. All the data are collected on a data logger which is powered by a small solar panel fixed on to the main mast of the weather station. The data logger must be downloaded every month onto a portable unit which connects to a port on the back of a PC.

Table 1. Rainfall stations within the Azraq Basin which are within 50km proximity to the AWS at Manara

Station name	Station code	Palestinian Grid reference		Altitude (m. asl)	Date established	Type of gauge
		North	East			
Dayr al-Kahf	F4	185°	325°	1025	1963	D
Umm al-Quttayn	F1	192°5	303°5	986	1947	D & R
al-Aritayn	F6	170°	330°	800	1963	D
H5 - As-Safawi	F2	180°	348°5	715	1968	D & R
Tullul al-Khuraysha	F14	156°	303°	550	1968	T
Azraq Evap. Station	F9	141°5	320°	533	1962	D & R

Adapted from Water Authority, Jordan (1989)

[1] D: Manual Daily ; D & R: Automated Daily and Recorder ; T: Totalizer

3.1. Temperature

The climate of the western area of the Badia is distinguished by a marked, but not extreme variation in temperature (see Fig. 4), both in terms of its diurnal nature and annual change. Summer temperatures rarely rise above 40° C while winter temperatures only seldomly dip below 0° C. This would reinforce the fact that the area lies close to the transition from Mediterranean conditions to a more continental climate further east, which would be characterized by more extremes at both ends of the scale. The mean daily temperature for the two years (1994-6) lies between a maximum of 26.2° C in August 1994 and a minimum of 5.2° C in December of the same year. The following year was less extreme with a maximum of 25.5° C in August and a minimum of 7.3° C in January 1996. Concomitantly, there is a reduction of standard deviation in the average temperature from summer to winter. The highest standard deviations occur in late spring and early summer, for example May 1995 was the highest with 7.5° C, and is probably associated with the higher propensity for the hot Hamseen winds to blow from Arabia in those months. Generally, however, the cool months of December through to February have low standard deviations of between 3 and 4° C.

The diurnal variation in air temperature is also cyclical with the highest variation during late spring and early summer as can be seen in Fig. 5. The lowest temperatures occur between 4 and 6 am while the highest occur between 2 and 4 in the afternoon. The lowest diurnal variation occurs in December and January where the mean range will usually stay between about 0° C and 20° C. The range increases in April, because the lowest night temperatures are still below 5° C but the maximum daytime temperatures are over 32° C.

Fig. 3 Automatic Weather Station at Manara.

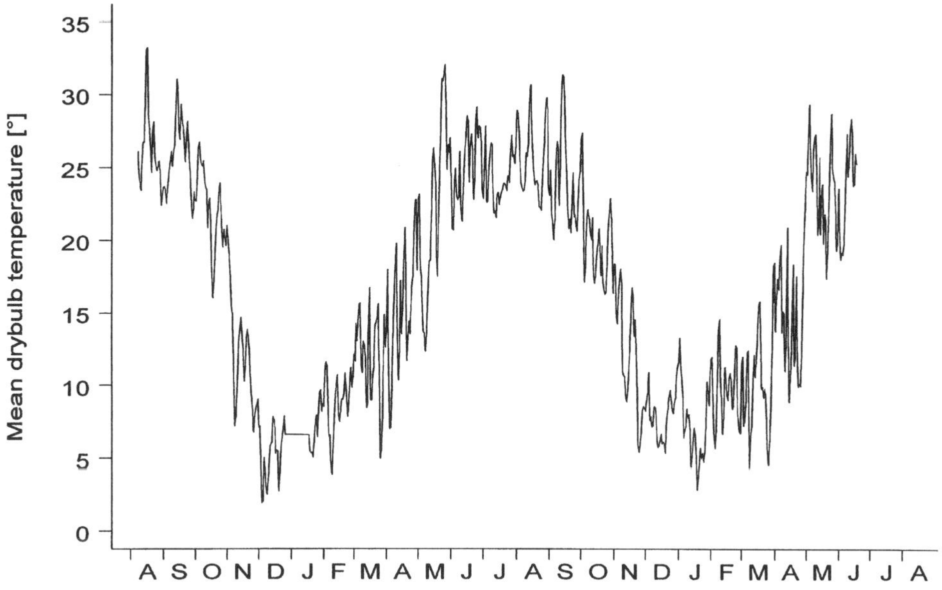

Fig. 4. Median and mean daily temperatures, August 1994 to June 1996.

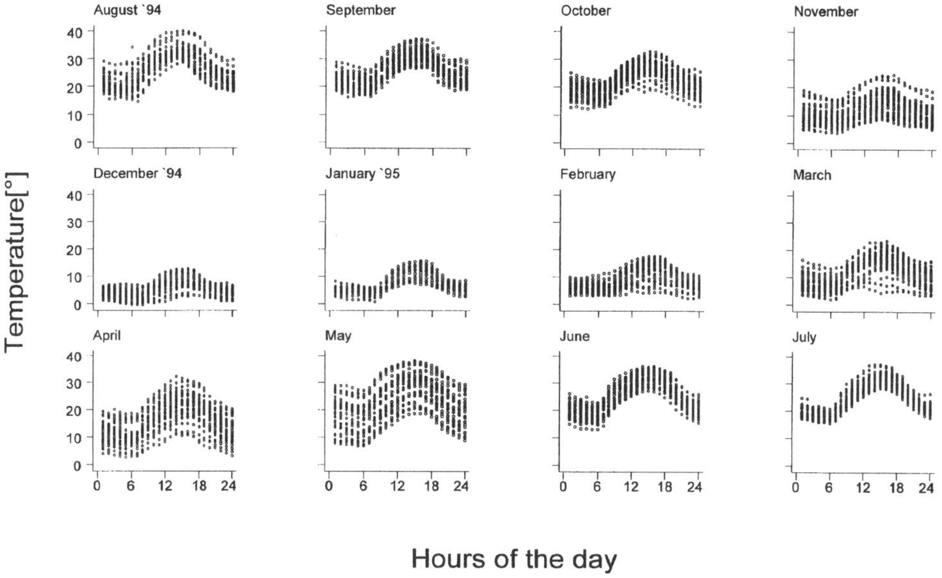

Fig. 5. The change in diurnal temperature ranges, 1994 to 1996.

3.2. Wind Run and Direction

Most days throughout the year it is possible to experience wind coming from all directions of the compass. However, although there is a constant wind in the Badia, there is a fairly consistent pattern in its variation which depends upon a summer-winter split. In the summer (see Fig. 6), the archetype consists of a prevailing north-easterly wind which blows during the evening and night, but in association with the increase in air temperature between 5 am and noon there is a significant shift to the north-west which seems to follow two patterns; usually it shifts slightly via the north, but regularly there is a gradual change via the south. The shifts of around 270° from north-east to north-west via south are indicative of periods of climatic instability and therefore it is often on these occasions that there is a risk of rainfall.

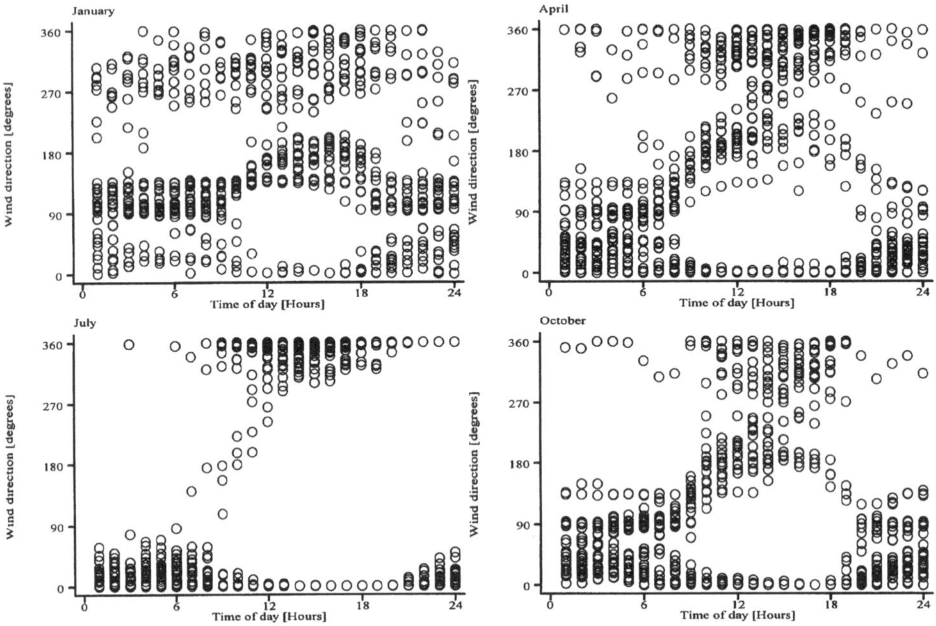

Fig. 6. The relationship between diurnal wind direction and time of year.

The pattern is considerably different during the winter as can be seen in Figure 6 for the months October to April. The wind during the night predominantly comes from the east to south-east, although there are still occasions when there is more of a northerly to north-westerly wind. The only quarter which does not supply the wind during the night is the south-west. As with the summer pattern there is a movement towards the north-west as the daytime temperature increases, although the majority of the shift occurs via the south. The midday wind direction will commonly come from the south, the south-west of the north-west. In the evening there is a shift back via the south. Obviously this summary is over-simplistic because it generalizes about the summer and winter positions. In reality only June to August fit completely into the summer pattern and November through to February fit the winter pattern, while the intervening months are inevitably evolving from one pattern to the other.

The three circular plots (Fig. 7) show the data in a form which is easier to observe although it must be remembered that they represent a mean for the two years of data rather than for specific months. They indicate how the wind direction changes during the day and compare it with the wind direction which accompanies rain events. During the night and morning, when the wind speed is relatively low, the mean wind direction is coming from the north-east (45.8°). However, in the afternoon, as the wind speed increases, there is a significant shift to a northerly wind (2.2°). Such a change probably represents katabatic airflows descending from the Jebel al-Arab due to differentials in pressure. More interesting is the fact that the average wind direction during rainfall events is from the north-west (318.2°) which corresponds to the fact that unsettled weather is dominated by winds coming from the Mediterranean. In this case, it is the larger wind systems and fronts which control the climate and wind characteristics whereas normally they epitomize local variations in pressure gradients caused by the orographic effect of the Jabal al-Arab mountains themselves.

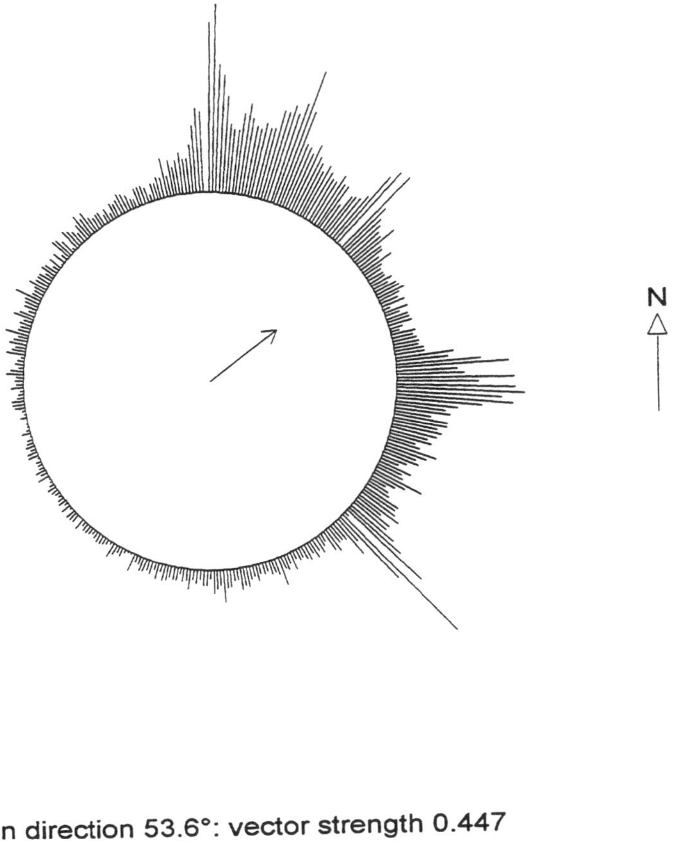

Fig. 7a. Wind direction between midnight and midday.

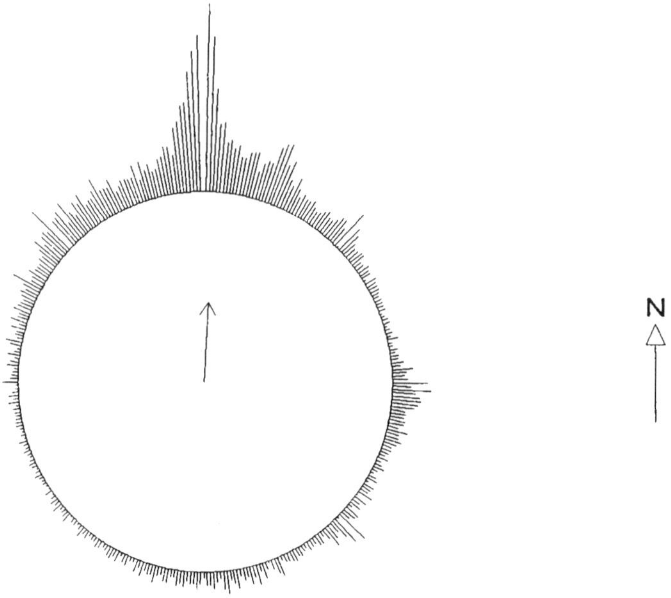

mean direction 3.6°: vector strength 0.425

Fig. 7b. Wind direction between midday and midnight.

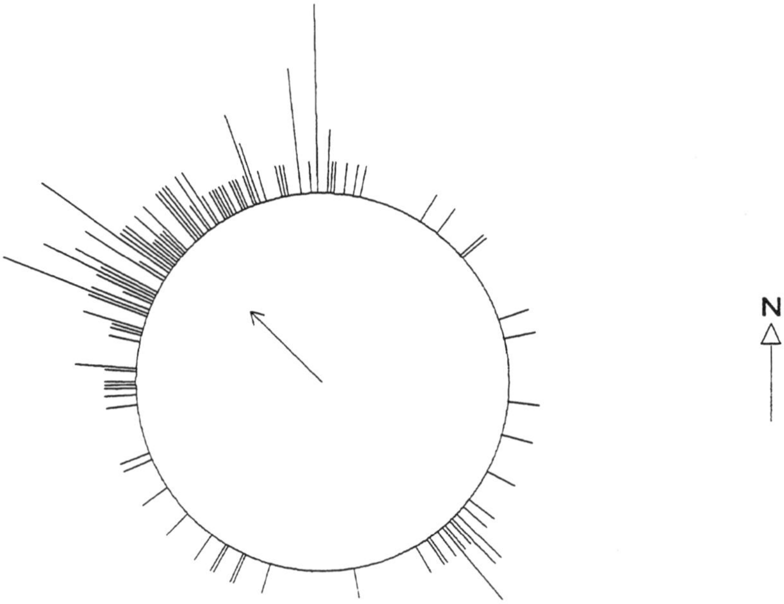

mean direction 314.1°: vector strength 0.529

Fig. 7c. Average wind direction during all rainstorms.

The relationship of windspeed is more difficult to determine due to the lack of reliable data after October 1994, but the main trends can be seen in Figure 8. It is much more difficult to generalize about patterns within the temporal series because of the enormity of variation. Even averages over one hour are liable to hide significant variations in the time-series, especially the importance of gust events where large magnitude wind events could occur and might be important as geomorphological agents. Despite the inherent problems it is possible to make some useful observations. The highest wind speeds occur in the hours following sunset, reaching a maximum between 2000 and 2200 hours. Figure 9 indicates that there is a strong relationship between wind strength and direction with the highest windspeed occurring from the north and north-east and although there is much less wind from the southern and western quadrants, it is also much less strong.

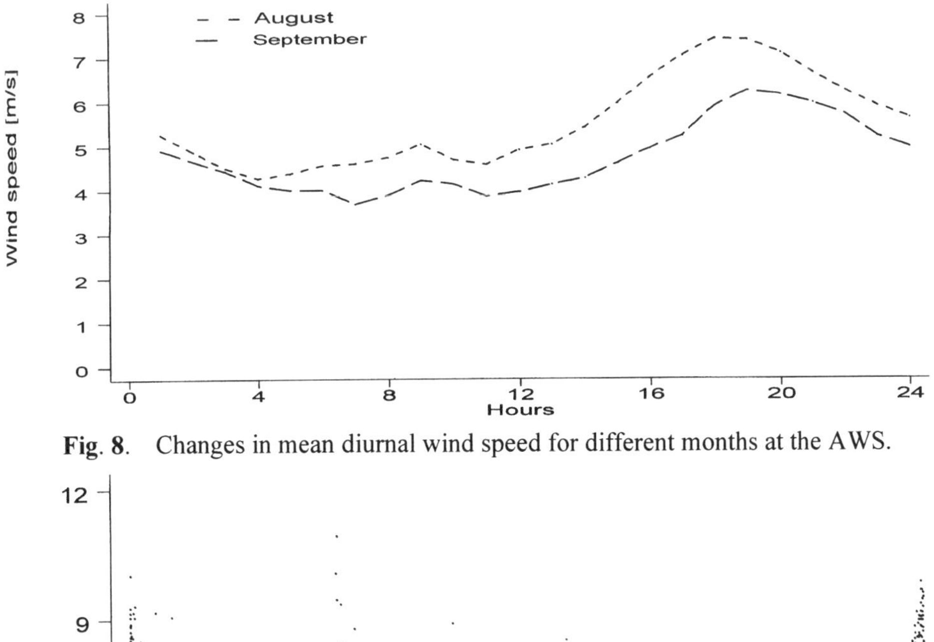

Fig. 8. Changes in mean diurnal wind speed for different months at the AWS.

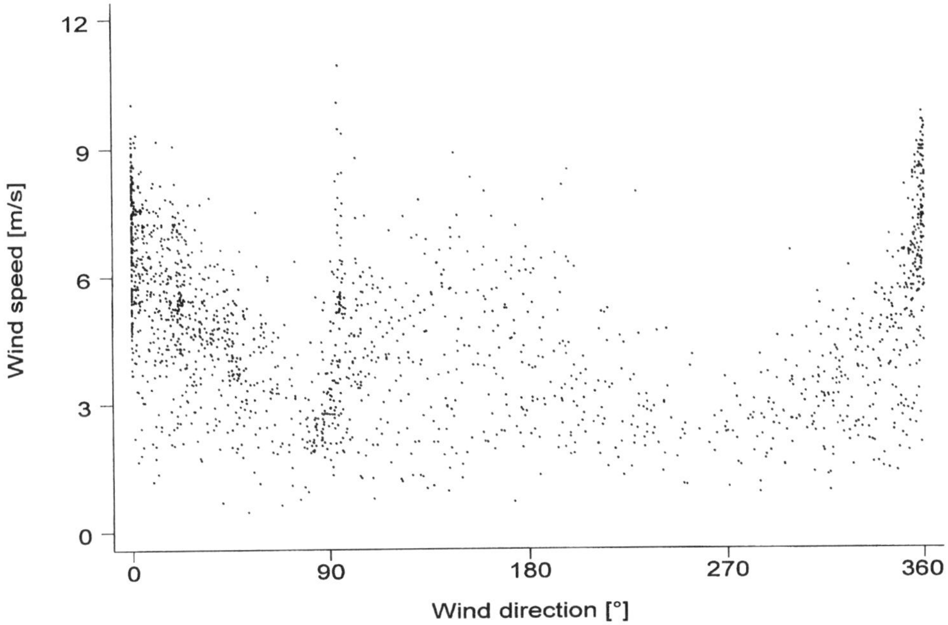

Fig. 9. The relationship between wind direction and speed.

3.3. Evaporation

In an environment such as the Badia, which is characterized for much of the time by high temperatures and moderate winds, the evaporation fluxes become very important in understanding the whole relationship between soil, plants and water. From a very practical point of view, there is a considerable amount of water wastage at farms where water is pumped from the basalt aquifers into large open storage tanks before being used to irrigate the fields. On a more subtle level, over-irrigation can lead to rapid precipitation of salts on the soil surface due to the high evaporation rates. In any arid environment the soil moisture flux remains negative for much of the time, only turning positive for a period of hours or at most days after rainfall or the application of irrigation.

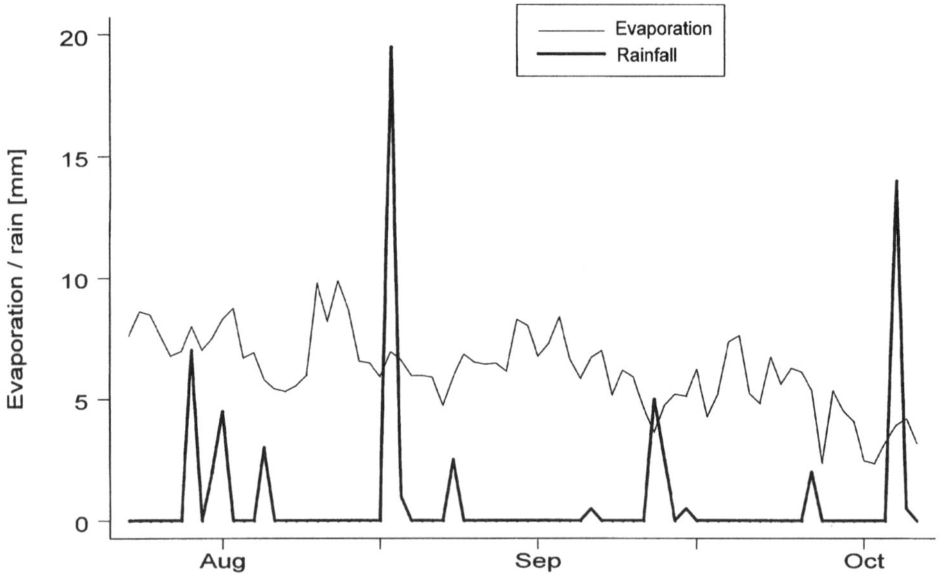

Fig. 10. Total daily evaporation vs rainfall [7th August - 20th October 1994].

In order to calculate the hourly evaporation coefficient from the data obtained from the AWS, a Penman-type equation was used (Penman 1948) to combine the net radiation, wind speed and humidity factors. The Penman equation relies upon two terms, C_1 and C_2, which represent the net radiation and the aerodynamic terms respectively and are constants which add up to 1. From the daily totals of evaporation and rainfall in Figure 8 for the first ten weeks, it is clear that during the summer months there is a very strong negative water balance with daily evaporation rates commonly above 7.5mm/day. Only on days of extreme rainfall, which do occasionally occur at the close of the summer season, is there a small positive water balance which only lasts for a few hours. However, by the middle of October the evaporation rates are usually below 5mm/day which gives a much greater potential for there to be a positive water balance albeit only for a few days.

3.4. Rainfall

This is unquestionably the key factor controlling all aspects of the soil-water balance and undoubtedly has the biggest influence on soil processes such as erosion and surface seal formation and on the viability of rain-fed agriculture. The frequency of rainstorms, the intensity and duration of those rainstorms and the pattern of rain events are all factors which must be taken into account. Rainfall is the one climatic variable about which there is a reasonable amount of data at different spatial and temporal scales (see Table 1).

The stations which are monitored by an automated system (D & R) contain the most detailed information including hourly readings of rainfall which correspond to the accuracy of the AWS information. The manual daily sites can only give information about the total rainfall which has fallen during a 24 hour period, but are still useful in ascertaining the days of the year in which rain events occurred. The least useful type of station is the totalizer, which in this case is situated at Tullul al-Khuraysha, which only gives an annual estimate of rainfall and so can only be used for looking at long-term trends in rainfall.

Taking the monthly data averaged over the years 1980 to 1990 for the five rainfall stations in the north-western part of the Azraq Basin, it can be seen that there are some interesting trends which are summarized in Figure 11. Umm al-Quttayn has the highest annual rainfall of approximately 122mm p.a. with a rapid increase in rain from October to a maximum in December before slowly decreasing until March and then dropping off quickly in April and a negligible amount in May. Dayr al-Kahf and al-Aritayn show a similar pattern with total averages of 113 and 89mm p.a. respectively and their highest rainfalls in December. These three stations give credence to the idea that frontal systems which continue over the hills to the east of the Rift Valley will tend to give precipitation only as they are forced to rise over the higher ground of the Jabal al-Arab. The amount of rainfall at these stations during the winter is highly dependent upon two variables; the altitude of the station and its distance eastwards. Umm al-Quttayn has the highest rainfall because it is the most westerly of all the stations on the Jabal al-Arab even though Dayr al-Kahf has a slightly higher elevation. Umm al-Quttayn receives the rainfall first whereas Dayr al-Kahf, which lies on the south-eastern side of the hills is slightly in the rain shadow and therefore receives less.

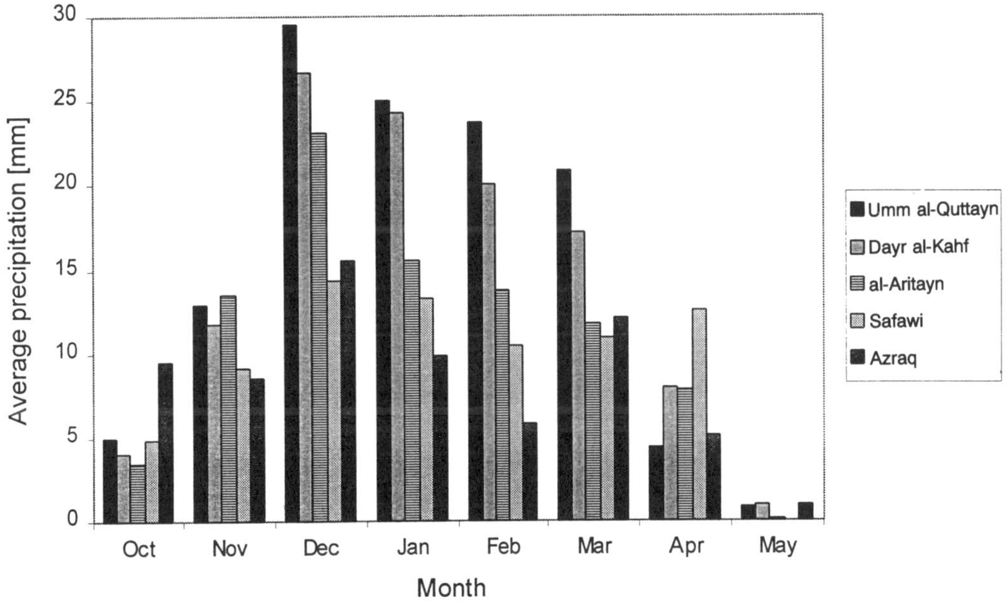

Fig. 11. Average monthly rainfall for five stations in the Azraq Basin (1980-90).

While the three stations on the footslopes of the Jabal al-Arab mountains experience rainfall due to an orographic effect, the other three stations, which lie at much lower elevations, seem to be dominated by other climatic patterns which allow rain to fall in a more random distribution. These stations are dominated by an arid desert rainfall which often bears no relation to the precipitation falling on the hills to the north and west. Azraq, for example, has an October rainfall average which is almost double that of the other stations and is only one of three stations to experience rainfall in May. The Safawi and Azraq graphs, showing no distinctly intra-seasonal trends and no particular maximas, is very dissimilar to the more traditional shape of graph exhibited by the first three sites.

Looking at the longer term variation in annual rainfall, it can be seen that there is much variation between years and in most cases a visible decrease in the amount of precipitation from the onset of data recording in the early 1960s to the late 1980s and early 1990s. First, it can be observed on both Figures 12a and b that there is a large coefficient of variation between years for all stations; the standard deviation for each station is of the order of 45 per cent of the average total. There is a reduction of standard deviation for Umm al-Quttayn and Dayr al-Kahf from the earliest years to the 1980s. The standard deviation for the whole period for both stations is 76.7 and 63.4mm respectively whereas if the standard deviation for 1980-90 only is 34.2 and 51.3mm. Safawi and al-Aritayn do not have a significant decrease, but Azraq actually increases from 31.4 to 37.1mm.

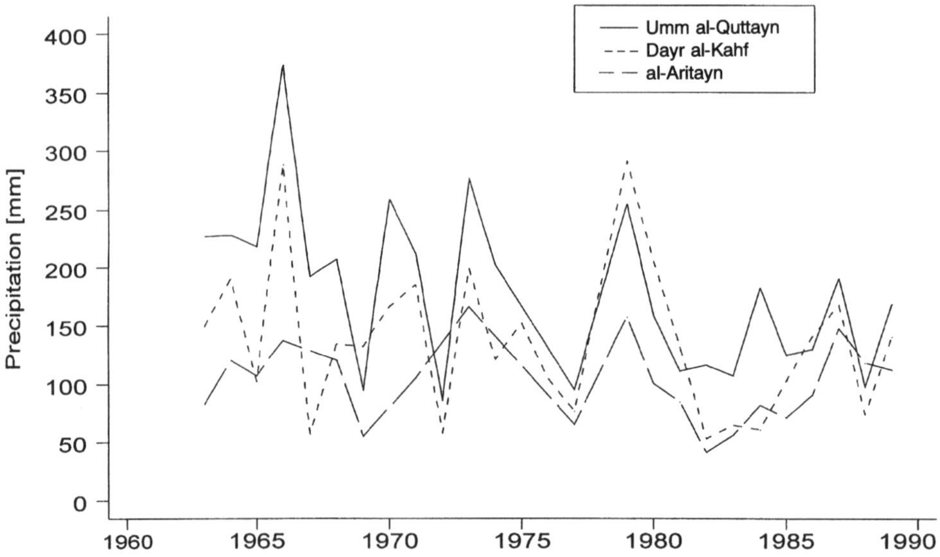

Fig. 12a. Annual changes in rainfall for the Jabal al-Arab sites (1963-89).

The long-term changes can again be separated by site as before. The Jabal al-Arab stations in Figure 12a experience a considerable reduction in annual precipitation when the first ten years is compared with the last ten years. The averages for Umm al-Quttayn, Dayr al-Kahf and al-Aritayn over the whole period are 169.4, 136.4 and 100.9mm p.a. respectively, but if the 1980-90 average is taken, those figures decrease to 137.4, 110.1 and 87.6mm p.a. respectively. In the case of Umm al-Quttayn, the period 1963-74 saw only two years with annual rainfall significantly less than 200mm p.a., while for the

1980s period there was not a single year when the rainfall reached 200mm p.a. Additional data for the winters 1992/3 and 1993/4 of 12.8 and 79.8mm p.a. respectively would suggest a continuation of this trend. Dayr al-Kahf suffers a similar fate and there are indications that al-Aritayn has also had a similar reduction although the data are more limited. The annual totals in 1994/95 and 1995/96 for the Manara automatic weather station, which lies between the three Jabal al-Arab stations, were 203.5mm p.a. and 122mm p.a. This does not take into account rainfall which could have fallen between the end of December and the third week of January when the instrument failed to record the climatic parameters. The winter of 1994 was in fact the wettest year since 1979, which would account for the high levels of precipitation in that year, but the winter of 1995 was back to around the long-term mean.

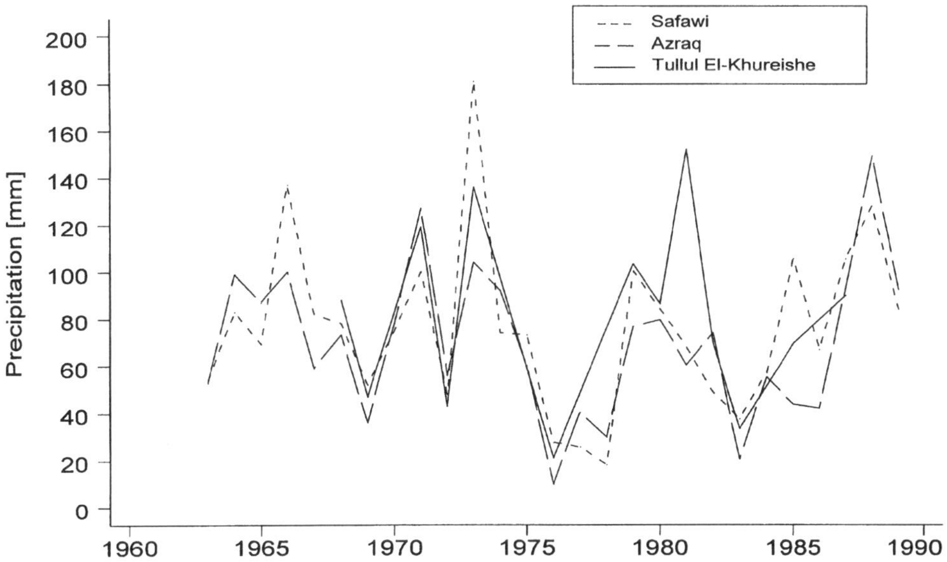

Fig. 12b. Annual changes in rainfall for the arid desert sites (1963-89).

The desert sites represent a rather different picture (see Figure 12b). The averages for Azraq and Safawi for both the total series (1963-90) and the shorter period (1980-90) are similar: the average for Safawi being 76.1mm p.a. and 76.7mm p.a. for both periods while Azraq decreased slightly but not significantly from 70.0 to 67.3mm p.a.

The statistics show that the rainfall pattern becomes more settled the further away the site is from the Jabal al-Arab. The desert sites of Azraq and Safawi display a more consistent, if limited, pattern of rainfall throughout the winter season and over the last thirty years there has been more continuity in the amounts of rainfall. Paradoxically, the sites on the Jabal are less consistent and indeed seem to be within a cyclical system whereby there are years of plenty followed by years of relative drought. The stations would seem, therefore, to be more profoundly affected by the larger-scale meteorological conditions which control the climate further to the west.

So far, the chapter has dealt with the general climatic conditions which lead to a position of being able to model evaporation throughout the year, but to complete the analysis it is important to look at the frequency and intensity of the rainfall in order to identify moisture budgets at varying temporal scales. These can then be used for further research into infiltration, erosion and evapotranspiration fluxes at the soil interface. The final section of the chapter seeks to discuss more detailed aspects of the rainfall patterns over the winter seasons of 1994 and 1995 using the data from Manara.

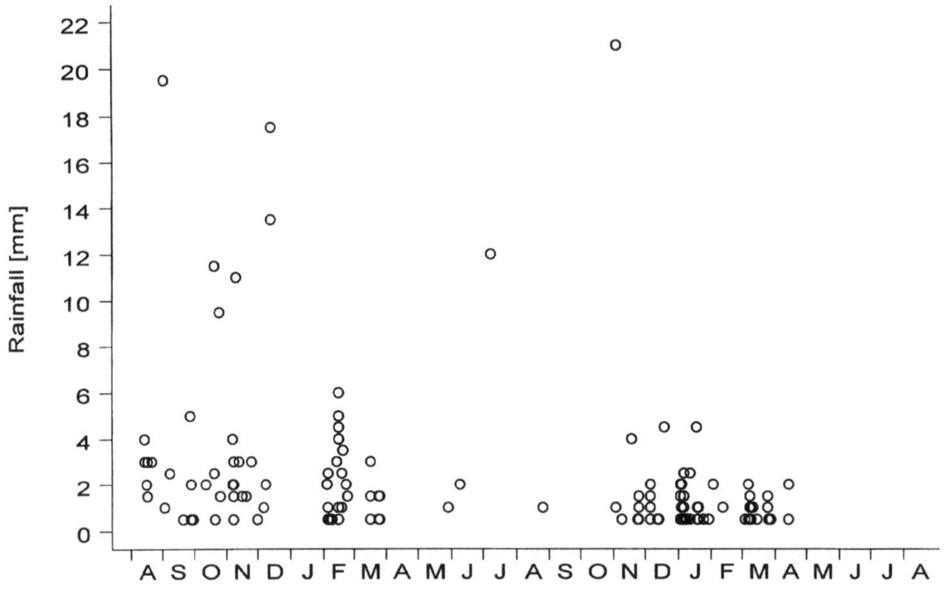

Fig. 13a. Individual rainfall events - per hour [August 1994 to June 1996].

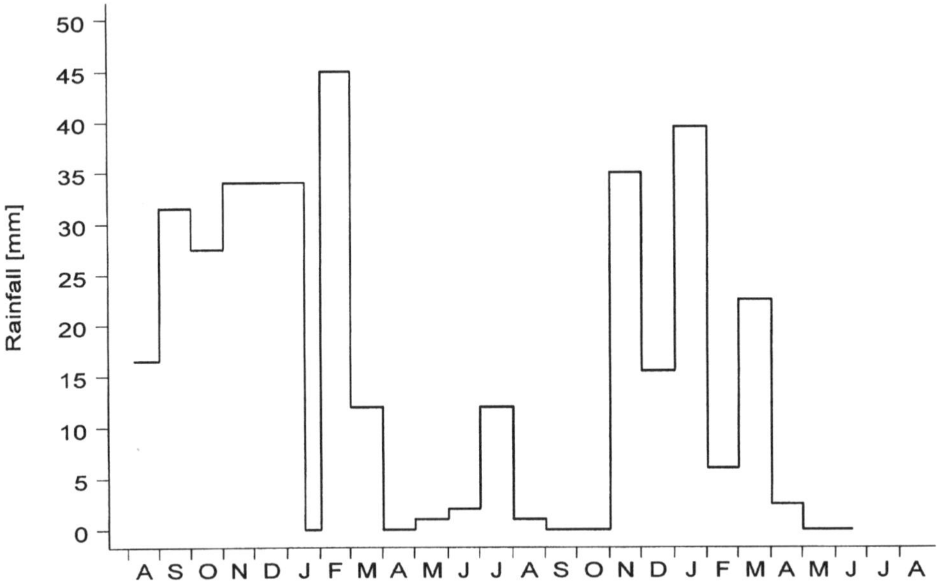

Fig. 13b. Cumulative monthly rainfall [August 1994 to June 1996].

Figures 13a and b show the individual rainfall events which have occurred since the setting-up of the AWS in August 1994 and a calculation of the cumulative monthly rainfall respectively. Note that the gap of rainfall events which accounts for a reading of 0mm in January 1995 was due to the failure of operation of the data-logger during the first three weeks of that month which therefore suggests that the total of 0mm is likely to be wrong. It can be seen in Figure 13a that the majority of rainfall events give less than 5mm of precipitation, but in a high rainfall year such as 94/95 there are several significant storms in which 10mm or more fall in one event. It is interesting to note that the high magnitude events occur towards the beginning of the season, while the rest of the season is dominated by lower magnitude and higher frequency patterns of rainfall. Comparing the two annual totals, it seems as though the 95/96 total is made up of a majority of events of 0.5-2mm events where low intensity drizzle would continue over several hours. The winter before, however, has much less low intensity rainfall with a significant amount occurring over 2mm.

Obviously it would be unwise to make too many generalizations about rainfall patterns from this very limited data set which covers only two years at one location in the Badia, however, it does emphasize the relative unpredictability and random nature of the rainfall even in close proximity to the Jebel al-Arab. The data also shows that there are some high magnitude events, often with intensities with return intervals of a number of years, which can be the most important in modifying the landscape. In arid and semi-arid environments it is extremely difficult to obtain data which focuses upon landscape change without taking measurements over a long period of time. To give an example from the Muwaqqar experimentation station in Jordan, where the annual rainfall average is 150mm per year, five winter seasons were needed to obtain enough rainfall data to carry out adequate experimentation on the effect of different soil amendments because some years had as little as three rainstorms in the whole winter season as can be seen in Table 2 (Abu-Sharar, 1996). This sort of research cannot be done within the restraint of a three year doctoral study and there is no guarantee that the storm required will occur during the research and therefore it is often more effective to use rainfall simulation to model the activity during certain rainstorms.

Table. 2. Selected parameters of the rainfall events in the Muwaqqar Experiment Station for the seasons 87/88 - 91/92

Season	Date		Total rainfall (mm)	Number of rain storms	Rain storms (mm):	
	First	Last			Maximum	Minimum
87/88	17/10	19/3	114	18	20.4	0.3
88/89	19/11	14/3	123	11	49.8	0.5
89/90	2/1	2/4	74	8	21.2	0.9
90/91	22/10	16/2	29	3	11.6	6.4
91/92	13/10	31/12	76	7	22.8	2.3

Source: Abu-Sharar (1996)

Intensity is the only primary variable, assuming wind is a secondary one, that is commonly recorded along with amount of rainfall at meteorological stations. However, even intensity is at best usually measured once an hour as with the AWS at Manara. Very few studies have been carried out to look at the rainfall intensities during storm events due to a lack of data and yet high-intensity rainfall is responsible for the majority of erosion (Hudson 1961; Kowal and Kassam 1976). The lack of resolution is often to due to the management of data because finer resolutions of every minute would produce an inordinate amount of data. However, to assume that the average intensity over an hour is a true picture of the overall intensity would be wrong, especially in a non-temperate context. Storms in semi-arid, arid and even tropical areas are often short but with a large magnitude, however, if average intensities are only measured hourly, 3mm falling over five minutes will be recorded in the same way as 3mm falling over 50 minutes. In addition, storms themselves are temporally variable with lower intensities at the beginning of an event, before increasing to a climax and then decreasing once again. In considering the rainfall which has most effect in terms of erosivity, it should be the period of maximum intensity which is concentrated upon.

3.5. Rainfall Parameters

No rainfall data are available in Jordan for the intensity or drop size distribution of individual storms. The highest resolution data for rainfall intensity are hourly figures for the meteorological stations which are jointly administered by the Water Authority of Jordan (WAJ) and the Meteorological Department. In addition to the six stations which are located within 50km of the AWS (see Table 1), there are hourly data from the author's AWS.

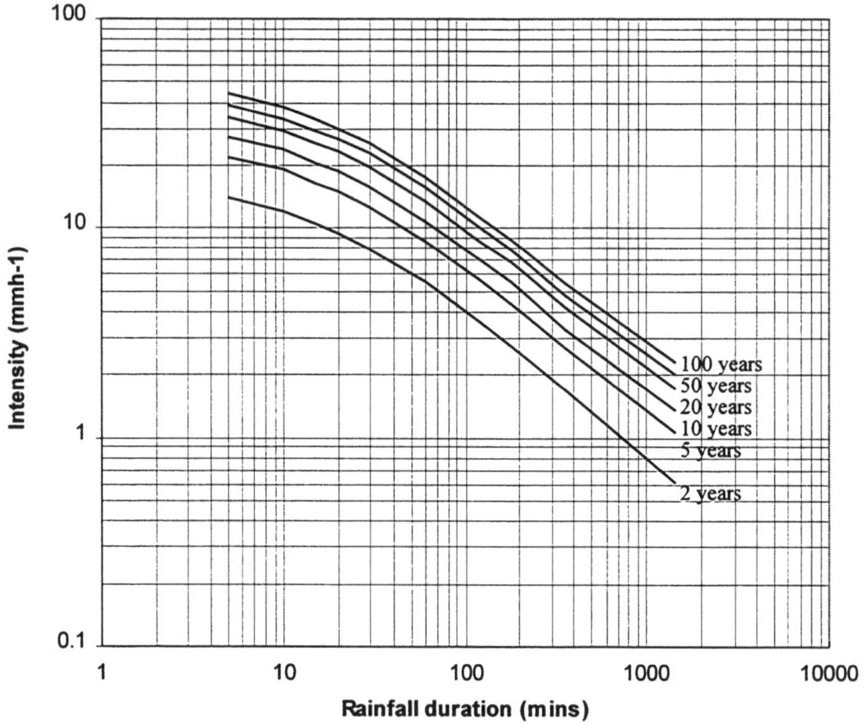

source: Water Authority (1989)

Fig. 14. Rainfall intensity diagram for Azraq rainfall station calculated over the years 1965-85.

The most accurate data from Umm al-Quttayn from the winters of 1992/3 and 1993/4 would indicate that the maximum rainfall intensities lie around 50mm per hour with a storm of 42mm per hour (14mm in 20 minutes) on 18th December 1993 and 60mm per hour (20mm in 20 minutes) on the 17th January 1994. Such data seems to compare unfavourably with rainfall intensity data for the Azraq station, a few km to the south of the field site (Water Authority of Jordan 1989). Data from the AWS for the winter seasons 94/95 and 95/96, which only gives rainfall per hour (Figure 14), similarly suggest that the highest storm intensities lie between 40 and 60mm per hour, but that most of the rainfall events consist of between 0.5 and 5mm which gives intensities of between 1 and 15mm per hour. It would seem that Jordanian intensity graphs, such as the one shown in Figure 14 for Azraq, are calculated upon rainfall amounts per hour, which give much lower intensities because the storms rarely last more than 20-30 minutes.

From observation of rainstorms in the Badia, it can be seen firstly that there are a few short highly intense rainstorms and a larger number of longer very low intensity events. Second, by judging the marks of the first raindrops on the soil surface, it can be suggested that the drop size is usually large. Lastly, it appears that, due to the very poor aggregate stability, even the smallest and least intense amount of rain will cause mechanical breakdown of the surface soil aggregates and crust formation which has severe implications for soil erosion and degradation.

REFERENCES

Abu-Sharar, T.M. (1996) Modification of hydraulic properties of a semiarid soil in relation to seasonal applications of sewage sludge and electrolyte-producing compounds. *Soil Technology* 9, 1-13.

Agnew, C. and Anderson, E. (1992) *Water Resources in the Arid Realm*. Routledge, London.

Al-Homoud, A.S. Allison, R.J. Sunna, B.F. and White, K. (1995) Geology, geomorphology, hydrology, groundwater and physical resources of the desertified Badia environment in Jordan. *GeoJournal* 37, 51-67.

Betts, A.V.G. (1992) Eastern Jordan: economic choices and site location in the neolithic periods. In: *Studies in the History and Archaeology of Jordan. Vol. IV*. (Ed: Hadidi, A) Department of Antiquities, Amman, 111-14.

Bruins, H.J. and Yaalon, D.H. (1979) Stratigraphy of the Neviot section in the desert loess of the Negev. *Act. Geol. Acad. Sci. Hung.* 22, 161-9.

Butler, H.C. (1919) *Publication of the Princeton University archaeological expedition to Syria 1904-5 and 1909*. Division II ed. Vol. Section A, Architecture, South Syria. Leiden.

Butzer, K.W. (1955) *Some aspects of postglacial climatic variation in the Near East considered in relation to movements of population*. M.Sc. Thesis, McGill University, Montreal.

Butzer, K.W. (1958) *Quaternary stratigraphy and climate in the Near East*. Ferd. Dümmlers Verlag, Bonn.

Butzer, K.W. (1975) Patterns of environmental change in the Near East during late pleistocene and early holocene times. In: *Problems in Prehistory: North Africa and the Levant*. (Eds: Wendorf, F and Marks, A.E.) Dallas, South Methodist University Press, 389-410.

Farhan, Y. and Mikbel, S. (1986) Applied geomorphological mapping and surveys in Jordan. *Transactions of the Japanese Geomorphological Union* 7, 91-102.

Gerrard, A.N. Stanley Price, N.P. and Copeland, L. (1975) A survey of prehistoric sites in the Azraq Basin, Eastern Jordan. *Paléorient* 3, 109-26.

Huckriede, R. and Wieseman, G. (1968) Der jungpleistozäne Pluvial-See von El-Jafr und weitere Daten zum Quartar Jordaniens. *Geologica et Paleontologica* 2, 73-95.

Hudson, N.W. (1961) An introduction to the mechanisms of soil erosion under conditions of sub-tropical rainfall. *Proceedings and Transactions of the Rhodesian Scientific Association* 49, 15-25.

Issar, A.S. and Bruins, H.J. (1983) Special climatology conditions in the deserts of Sinai and the Negev during the latest pleistocene. *Palaeogeography, Palaeoclimatology, Palaeoecology* 43, 63-72.

Kaiser, K. Kempf, E.K. Leroi-Gourhan, A. and Schütt, H. (1973) Quartärstratigraphische Untersuchung aus dem Damaskus-Becken und seiner Umgebung. *Zeitschrift für Geomorphologie* 17, 263-353.

Kowal, J.M. and Kassam, A.H. (1976) Energy load and instantaneous intensity of rainstorms in Samaru, Northern Nigeria. *Tropical Agriculture* 53, 185-97.

Krown, L. (1966) Approach to forecasting seasonal rainfall in Israel. *Journal of Applied Meteorology* 5, 590-4.

Lamb, H.H. (1968) The climatic background to the birth of civilization. *Advancement of Science* 25, 103-20.

Penman, H. (1948) Natural evaporation from open water, bare soils and grass. *Proceedings of the Royal Society A* 193, 120-45

Perry, A. (1981) Mediterranean Climate - A synoptic reappraisal. *Progress in Physical Geography* 5, 107-13.

Phillips, P.G. (1954) *The Hashemite Kingdom of Jordan: Prolegomena to a Technical Assistance program.* Ph.D. Dissertation, The University of Chicago, Illinois.

Roberts, N. (1982) Lake levels as an indicator of Near Eastern palaeo-climates: a preliminary appraisal. In: *Palaeoclimates, Palaeoenvironments and Human Communities in the Eastern Mediterranean Region in Later Prehistory.* (Eds: Bintliff, J.L. and Van Zeist, W.) B.A.R. International Series 133, Oxford, 235-267.

Shehadeh, N. (1985) The climate of Jordan in the past and present. In: *Studies in the History and Archaeology of Jordan. II* ed. (Ed: Hadidi, A.) Department of Antiquities, Jordan, Amman, 35-8.

Water Authority of Jordan (1989) *The North Jordan water resources investigation project: Azraq Basin.* Amman, October 1989.

Wigley, T.M.L. and Farmer, G. (1982) Climate of the eastern Mediterranean and Near East. In: *Palaeoclimates, Palaeoenvironments and Human Communities in the Eastern Mediterranean Region in Later Prehistory.* (Eds: Bintliff, J.L. and Van Zeist, W.) B.A.R. International Series 133, Oxford, 3-39.

CHAPTER 4

Water Resources Quality, Sustainability and Development

JANE DOTTRIDGE
(University College London)

1. INTRODUCTION

Scarcity of water has traditionally restricted development in the Jordanian Badia and will remain a limiting factor in the future. Sustainable development in this vast, arid area depends on the provision of adequate, long-term supplies of good quality water for human consumption, livestock and agriculture. Consequently, water has been one of the themes of the Geomorphology and Physical Resources component of the Badia Research and Development Programme (BRDP) since the Programme started in 1992. The work during Phase 1 concentrated on the evaluation of groundwater resources, which are very substantial in the Programme area and provide the most widespread and reliable source of water. However, for sustainable supplies, it is essential to avoid over-abstraction, because 'the exploitation of most groundwater aquifers beyond their safe yield is causing depletion and salination of the water and this practice should be stopped before permanent and irreversible damage occurs' (HCST 1993).

Although the average rainfall over most of the Programme area is only 50 to 150mm per year, the north-eastern Badia benefits from large volumes of surface water, flowing in wadis draining the Jabal al-Arab in Syria. At present, most of this water is lost by evaporation from the *qa'as*, thus the wadi flows potentially provide a valuable source of additional water. However, wadi flows are localized, infrequent and of short duration, thus utilizing the surface water is not a simple task. Further research and data collection are essential to determine the size and reliability of this resource and to optimize methods of water harvesting and storage. In 1995, the BRDP initiated work on the surface water resources of the project area.

This paper summarizes the current knowledge of the water resources of the pilot study area of the BRDP, highlighting the contributions made by the Programme's researchers, and considers the options for future development of water resources in the area.

In hydrological terms, the pilot study area of the BRDP lies mostly within the Azraq basin, but also extends into the Sirhan and Hammad basins as shown on Figure 1. Although these basins are surface drainage basins defined topographically, the same sub-divisions are conventionally used for discussion of groundwater resources, but groundwater flow may cross some of the boundaries. Geologically, the Programme area coincides approximately with the extensive outcrop of basalts of Neogene to Quaternary age, but in the south, extends beyond the basalt outcrop to the border with Saudi Arabia.

Despite the relatively uniform surface geology, the deeper geology and hydrogeology are more complex and varied.

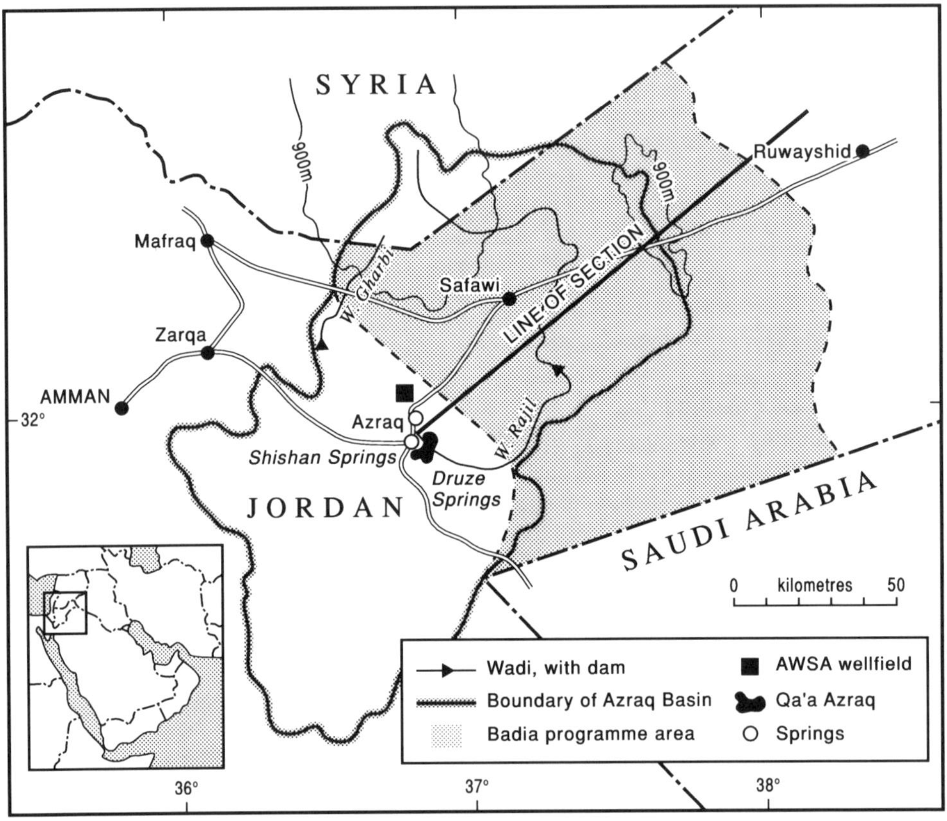

Fig. 1. Location of study area and drainage basins.

2. PREVIOUS AND RECENT INVESTIGATIONS

Over the past thirty years, numerous studies of water resources have been conducted in the north-eastern Badia, concentrating particularly on the upper aquifer in the Azraq basin (e.g. UNDP 1966; Mudallal 1968; Parker 1971; Howard Humphreys 1977 and 1982; Ayed 1986; Core Lab 1987; WAJ 1989; Al-Kharabsheh 1991; Ta'any 1996). Recently, the focus of water resource investigations has widened to include the middle aquifer of the Azraq basin (Salzgitter 1994) and the Sirhan and Hammad basins (GITEC *et al.* 1996). Several other organizations and individuals are actively working on projects and collection of data relevant to the water resources of the Badia Programme area, including: the Water Authority of Jordan (WAJ), the Natural Resources Authority (NRA), the Meteorological Office, the Azraq Oasis Conservation project, the Hammad Basin project and the University of Jordan. The data supplied by these organizations provided an essential contribution to the project and is gratefully acknowledged.

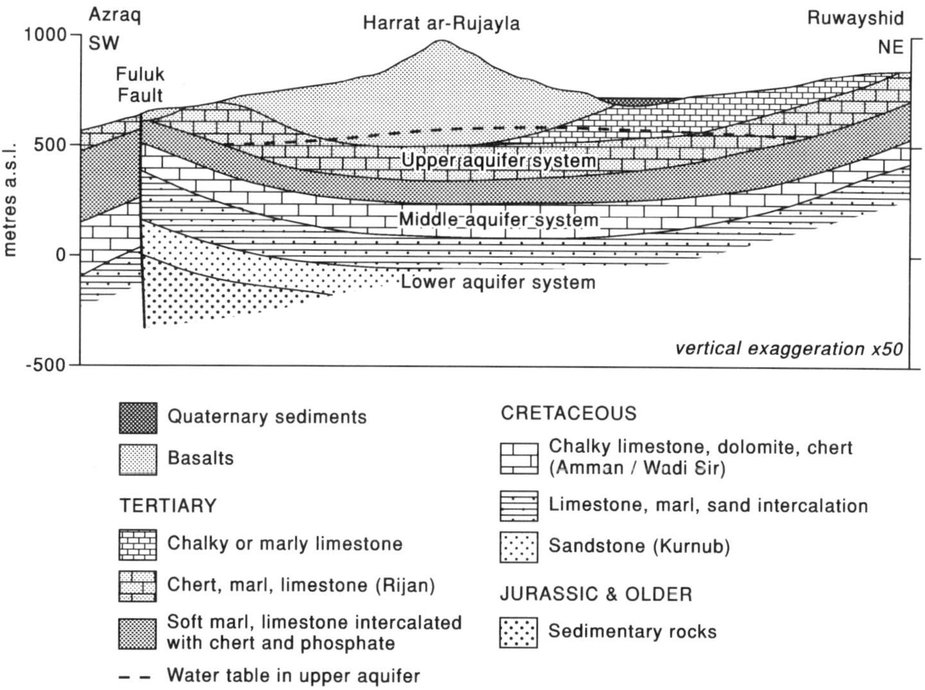

Fig. 2. Simplified cross section.

The previous reports and existing data formed the background information for the BRDP's work on water resources, which combined collation, review and evaluation of available data with fieldwork to collect additional data on water chemistry, isotopic composition and dating, water levels, recharge, abstraction rates and water use. The results of this work has been summarized in two reports (Dottridge 1994 and 1995) and described in detail in a series of M.Sc. theses (Wilkinson 1992; Drury 1993; Gibbs 1993; Noble 1994; Waddingham 1994; Holden 1995; Haaland 1996). In addition, Dr. Mohammed Sellag of WAJ has compiled a map of boreholes in the project area and studied the ancient water harvesting structures.

3. GROUNDWATER RESOURCES

3.1. Hydrogeology

Groundwater in north-eastern Jordan occurs in three aquifer complexes, Upper, Middle and Lower. A simplified cross-section is shown in Figure 2. The exact sequence varies with location, but the typical aquifer units for the three basins are shown in Table 1. In all three basins, the Upper and Middle aquifer systems are separated by an aquitard, the Muwaqqar (B3) marls, through which significant leakage is thought to occur. The direction of flow varies, depending on the relative piezometric heads. The Middle aquifer system dips under the basins from outcrops in the western highlands of Jordan and in Iraq, occurring at depths of 400 to 700m in most of the study area. However, the geology is complicated by large throws on the major faults, especially the Siwaqqa, Fuluk and Azraq-Sirhan fault zones (Andrews 1992; NRA 1993). The Khreim shales form an aquitard between the Middle and Lower aquifers. In the project area, the Lower aquifer is little known and occurs at depths of 1.3 to 3.4km.

In the Upper aquifer system of the Azraq basin, groundwater flow is radial towards the Azraq topographic depression, principally in the fissured basalt and the Rijam limestone, with piezometric levels between 560m above sea level (asl) near the Syrian border to 500m asl in the centre of the basin. Some groundwater flow also occurs in the upper Shallala formation but the marls of the lower Shallala formation partially restrict the hydraulic continuity between the basalt and Rijam limestones. The saturated thickness of the aquifers decreases towards the centre of the basin from over 300m in both north and south to about 50m around Azraq.

Transmissivity is highly variable, showing the heterogeneity characteristic of fractured aquifers, with values ranging from 270 to 65,000m^2/d and averaging 11,000m^2/d at the Azraq wellfield (Howard Humphreys 1982). Parker (1971) recorded a range of 2 to 113,000m^2/d for the basalt, much higher than the values of 6 to 230m^2/d in the Rijam limestones. Estimates of specific yield for the unconfined aquifers also vary widely, from low values of 1 to 2 per cent, which are typical of fractured aquifers (Gibbs 1993) to 12 to 22 per cent from test pumping (Howard Humphreys 1982).

In the North Sirhan and Hammad basins, the basalt is only locally saturated and does not form a significant aquifer. The upper aquifer in North Sirhan comprises the Qirma and upper Shallala formations. No test results are available, but borehole yields are good, up to 1,500m^3/d. In the Hammad basin, only the western part of the Shallala outcrop is saturated. The proportion of marl and degree of fracturing control the hydraulic properties and borehole yields, which vary from 8 to 1,700m^3/d. Estimates of permeability range from 0.1 to 5m/d, with a typical specific yield of 0.5 per cent. The flow patterns approximately follow the surface drainage, with flow towards the east in the Hammad basin, towards Azraq where the aquifer is developed in the Azraq basin and west or south towards Saudi Arabia in North Sirhan

The marls of the lower Shallala formation act as an aquitard confining the underlying Rijam aquifer, which is divided into three blocks by major faults. As in the Azraq basin, the permeability depends on fracturing, varying from 10^{-4} to 50m/d (GITEC 1996). Recent data (GITEC 1996) suggest that the flow pattern is different to the overlying Shallala, with flow towards Azraq from North Sirhan and the southern Azraq basin, and easterly flow towards Ruwayshid from the north-eastern extremity of the Azraq basin and the Hammad basin.

The Middle Aquifer Exploration Project (Salzgitter 1994) explored the properties of this aquifer in the Azraq basin, and their results were reviewed by Noble (1994). The investigation drilled and tested 12 exploratory boreholes, which showed that both transmissivity and water quality in the aquifer are highly variable. The transmissivity is mostly low (35 to 450m^2/d), but higher in the Sirhan fault zone to the north-west of Azraq. Good quality water occurs in the northern part of the basin, but elsewhere quality is poor, with total dissolved solids reaching 11,950mg/l. Unfortunately in most of the Badia Programme area, the aquifer is deep, with low transmissivity and poor water quality, thus the middle aquifer does not represent a significant source of usable water. Outside the investigation area, information is extremely limited, although WAJ records show that two boreholes in the Hammad basin reach this aquifer at depth.

Table 1. Aquifer units in the north-eastern Badia

Aquifer system	Azraq basin	North Sirhan basin	Hammad basin	Approximate age
Upper	Basalt (BS)	Basalt (BS)	Basalt (BS)	Neogene - Quaternary
		Qirma sandstone/ limestone		Miocene
	Wadi Shallala chalky limestone (B5)	Wadi Shallala chalky limestone (B5)	Wadi Shallala chalky limestone (B5)	Eocene
	Rijam cherty limestone (B4)	Rijam cherty limestone (B4)	Rijam cherty limestone (B4)	Eocene
Middle	Amman-Wadi Sir limestones (B2/A7)	Amman limestone (B2)	Amman limestone and Ajlun dolomites (B2)	Cretaceous
		Hamza dolomite		Cretaceous
		Rajil sandstone		Cretaceous
		Wadi Sir/Lower Ajlun limestone and dolomite (A1/A2)		Cretaceous
	Kurnub sandstone (K+)	Kurnub sandstone (K+)	Kurnub sandstone (K+)	Cretaceous
		Ramtha sandstone and dolomite	Ramtha sandstone and dolomite	Triassic
Lower	Disi sandstone	Disi sandstone	Disi sandstone	Palaeozoic

3.2. Hydrochemistry and Water Quality

In the upper aquifer of the Azraq basin, the distribution of Total Dissolved Solids (TDS), shown in Figure 3, confirms the radial flow pattern (Drury 1993). Contouring reveals two zones of low concentration coincident with the highly transmissive al-Aritayn and Sirhan fault zones (Holden 1995). An area of high TDS in the east is attributed to greater mixing with water from the Middle aquifer, due to enhanced vertical permeability along the Fuluk fault. The shallow groundwater under Qa'a Azraq is a hyper-saline brine, which appears to be hydraulically contained by the bounding faults.

The regional hydrochemistry is largely explained by the processes of weathering and hydrolysis of minerals in the basalt aquifer, dissolution of surface evaporites during recharge and mixing of low TDS basalt waters with more saline waters in the Rijam limestone. The source of salinity is solution of surface evaporite crusts, during the recharging of the limestone aquifer. Most chemical constituents also have mineralogical sources within the aquifer, but nitrate may have an organic origin. The waters are ubiquitously $NaHCO_3$ or NaCl type.

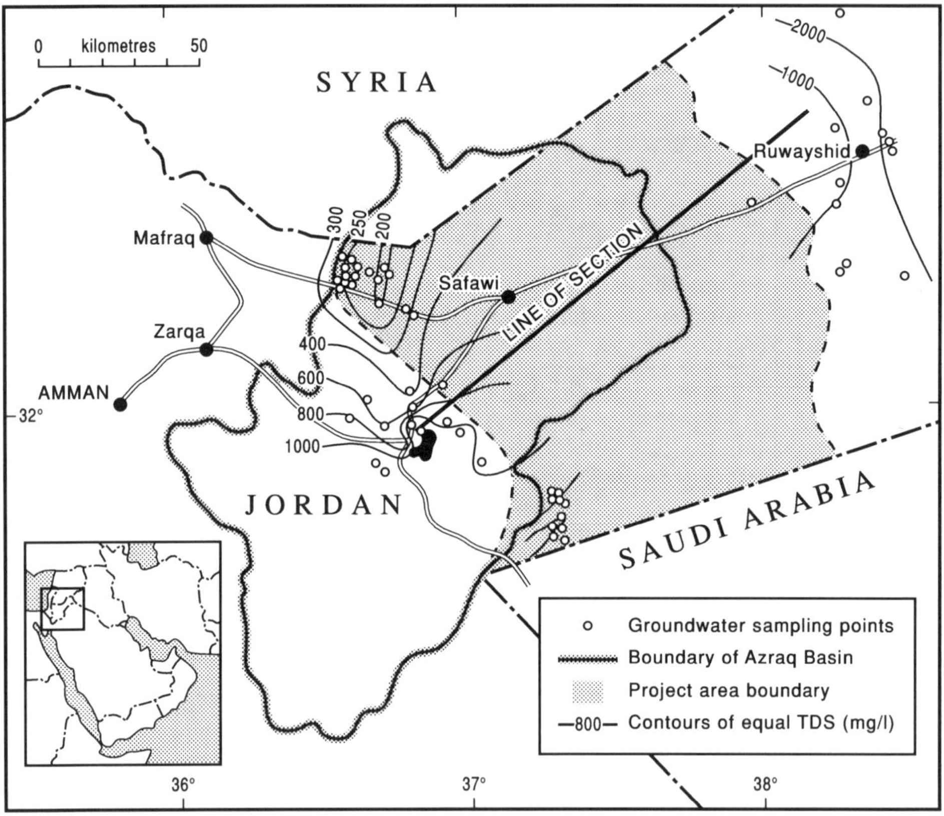

Fig. 3. Salinity map of upper aquifer.

Historical data indicate no changes in the composition of the end members of the basalt-limestone mixing process, and show no evidence of any movement of the hypersaline brine body. Changes in the chemistry of the Druze springs and in some boreholes can be explained by a hydraulic control on the mixing process. A steady increase in abstractions between 1955 and 1980 resulted in a gradual increase in salinity as more water was provided by the Rijam limestone. However, when the abstraction rose suddenly in 1980, a greater proportion of water was derived from the highly permeable basalts, causing a decrease in TDS.

An assessment of the source of the groundwater and recharge mechanism was made by comparing the composition of the stable isotopes in the groundwater with Jordanian rainfall. The recent analyses are consistent with previous results (Al-Momani 1991), which indicate that the temperature of recharging water is 6-8° C, at an altitude in excess of 850m. The observed trend down the flow path is increasing altitude of recharge and decreasing temperature. This was explained by considering the basalt aquifer as a wedge, thinning southwards, where the water recharged in the far north flows deeper and is discharged to the south, while the recharge which occurs in the middle zone has a shorter, shallower flow path.

Carbon-14 uncorrected ages show ageing of water down the hydraulic gradient. In the northern part of the Badia Programme area, around Umm al-Quttayn, the average age is 7,000 years, with only one result sample showing a young age of 1,240 years. Correcting the ages using water bicarbonate $\delta^{13}C$ and $\delta^{13}C$ of soil and vegetation samples suggests that this is modern water, and the average age reduces to 5,000 years. Tritium data indicate significant recent recharge only in Wadi Rajil.

The water quality in the northern part of the Badia Programme area is of exceptionally good quality for drinking and irrigation, with very low TDS and no contamination. Most of the basalt-limestone water conforms to WHO guidelines for inorganic constituents in drinking water. The waters fall mostly in irrigation classes C2-S1 or C3-S1, and are appropriate for consumption by livestock. Due to the thick unsaturated zone, the risks of contamination are very low, except through well heads, which are often poorly completed and may allow entry of diesel from pumps or bacteriological pollution.

Recent analyses of water samples from the North Sirhan and Hammad basins (Haaland 1996) shows that the water quality in these areas is inferior to the Azraq basin. TDS vary between 600 and 2,800mg/l in the Hammad basin, increasing eastwards following a trend identified by previous studies (GITEC 1996). In North Sirhan, salinities are higher, from 1,400 to 3,800mg/l, but no spatial trends are apparent and rapid variations are observed between closely spaced boreholes. In both areas, there is rarely more than 50 per cent of any constituent, but in general, sodium and calcium are the dominant cations, with chloride and sulphate as the major anions. Evaporites in the Qirma formation and in surface crusts may be a source of these constituents. Most of the boreholes contain a low concentration of dissolved oxygen (<1 per cent) and smell of sulphide, indicating a reducing environment, even in the shallow Qirma aquifer.

The water from the boreholes in the North Sirhan and Hammad basins is mostly unsuitable for drinking, with concentrations of at least four constituents exceeding the recommended maxima in all samples. Most boreholes contain excessive quantities of several substances which pose a health hazard: cadmium, fluoride, lead, manganese and nitrate. The water is however suitable for consumption by livestock and for irrigation.

3.3. Recharge

The average annual rainfall on the Azraq basin decreases southwards from 350mm in the north to less than 50mm in the south, with significant inter-annual variation. In this arid climate potential evapotranspiration is high, from 1,800 to 1,980mm/year. Recharge through direct infiltration of rainwater occurs mainly in the wetter, northern part of the basin, above 1,000m. This is consistent with the stable isotope data which indicates an altitude of recharge in excess of 850m. In wet winters, some recharge may also occur in the larger, more arid area below 1,000m.

Limited data from well hydrographs and tritium analyses show that indirect recharge occurs through infiltration of runoff. No gauging of wadi flow in the Azraq basin has been conducted to date, but the runoff was estimated by WAJ for each storm from 1965-88 (Otova and Ayed 1988) and up to 1993 by Noble (1994).

The estimates of total modern recharge vary considerably, but mostly average between 22 and 36 million m^3/year, approximately 3 to 5 per cent of the rainfall. The most reliable estimates based on daily meteorological data are those by WAJ, which average 34 million m^3/year. Due to the variability of the rainfall, the range of values is high, from 0 to 67 million m^3/year, with a standard deviation of 21 million m^3/year. A reliable minimum value is 16 $\pm$ 3 million m^3/year, which is consistent with both the value calculated for recharge above 1,000m (Noble 1994) and the historical spring flows (Gibbs 1993).

Evidence from carbon dating, stable isotopes and observation boreholes suggest that modern recharge is unlikely to occur in the North Sirhan and Hammad basins. These areas are much drier than the Azraq basin, with mean annual rainfall of 40 to 50mm in North Sirhan and 60 to 90mm in the Hammad. Although runoff occurs occasionally, especially from the basalt hills of Tullul El Ashaqif, wadi beds are mostly lined with fine-grained material.

In all areas, leakage from water pipes and return flow from irrigation provides an additional source of recharge, which will be most important in the areas where the water table is shallow. The return flows are only significant if furrow irrigation is used; infiltration from drip irrigation is insignificant.

3.4. Discharge, Abstraction and Groundwater Use

The main natural outlets from the Azraq basin are the springs at Azraq, the Druze springs fed by water flowing from the north and the springs at Azraq Shishan fed from the south. The spring locations coincide with the intersection of fault zones and lithological boundaries. In the mid 1950s, the discharges were measured as 3.5 million m^3/year from the Druze springs and 22.8 million m^3/year from the Shishan springs. The spring flows have declined over the last forty years, until the Druze springs ceased to flow in 1987 and the Shishan springs dried up in 1992. This has led to deterioration of the Azraq Oasis and environmental derogation (GEF 1993; Jones 1990). The declining trend in spring flow and piezometric levels, illustrated in Figures 4 and 5, has been accompanied by increasing pumped abstraction since the 1940s. With rapid growth since 1980, the total abstraction from the Azraq basin in 1994 was estimated to be between 58.9 and 65.1 million m^3 (Waddingham 1994; Dottridge and Gibbs 1996).

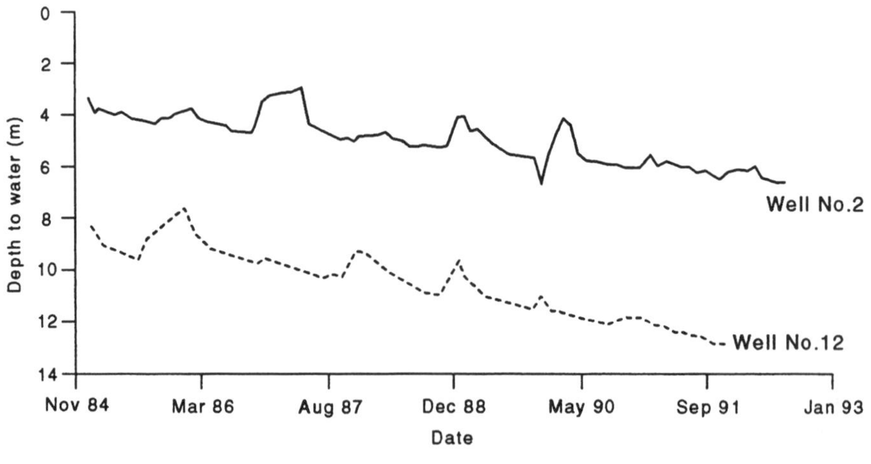

Fig. 4. Hydrographs of observation wells.

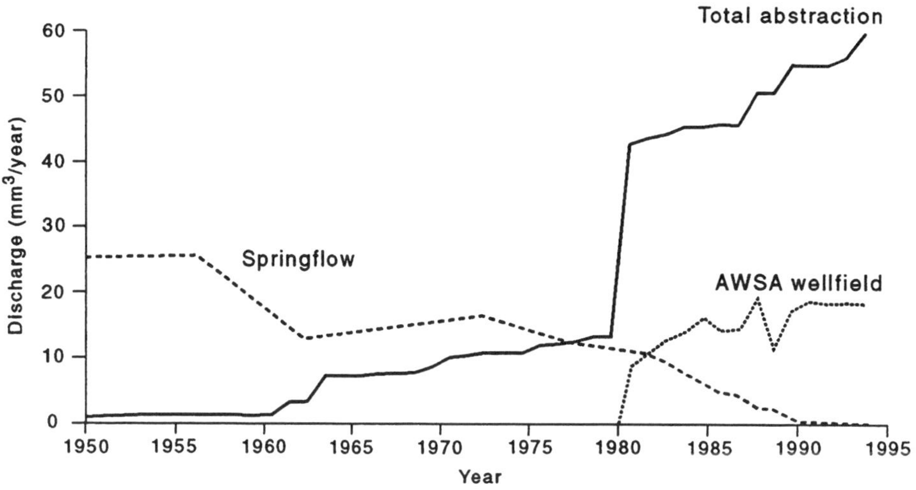

Fig. 5. Variation in abstraction and springflow with time.

Groundwater is used for:

- large-scale public water supply to the Amman area, and on a smaller scale to Mafraq;
- local small supplies for domestic use and livestock;
- industrial users, with low water demands of only 56,000m^3 /year at present;
- intensive irrigation concentrated around Azraq;
- scattered smaller scale irrigation in the north of the project area.

The single largest abstractor (18 million m^3/year since 1988, 19.2 million m^3/year in 1994) is the Amman Water and Sewerage Authority (AWSA) wellfield, north of Azraq. In 1994, only 0.87 million m^3 of this total was used for local supply, with the remainder exported from the area. A further 5.15 million m^3 from wells along the Mafraq-Safawi road was supplied to Mafraq. In the northern part of the basin, local supplies in 1994 accounted for 2.6 million m^3, 18 per cent of total abstraction, but drilling of new boreholes for local public supply continued in 1994 and 1995. Losses in the distribution system are very high, estimated to be 61 per cent (Waddingham 1994).

Irrigated agriculture, with pumping from over 600 boreholes, is concentrated in the agricultural areas to the east of Azraq and the north of the Badia Programme area. The abstraction for irrigation in the northern agricultural area was calculated as 6.6 million m^3 in 1994. Although drip irrigation is used, poor design and maintenance results in moderate distribution efficiencies, represented by a measured value of 68 per cent. This suggests that the crops' peak requirements are only just satisfied by the present systems. In the Azraq area, abstraction for irrigation was estimated to be between 25.35 million m^3 (Dottridge 1994) and 32.55 million m^3 (Waddingham 1994) with much lower efficiency, below 50 per cent, due to the coarser soil and continuing use of channel and basin techniques.

In North Sirhan, groundwater abstraction has increased since 1993 from 0.04 million m^3/year (GITEC 1996) to 1.41 million m^3/year (Haaland 1996). The sudden growth is due to the development of irrigated farming in Wadi Hazim, where a total area of 1,150 dunums is used to grow two crops per year of water melons and vegetables. In the

Hammad basin, the main use of groundwater is for water supply, and the pumping rates have remained relatively constant, approximately 0.8 million m^3/year, although the abstraction has become distributed more widely through the drilling of new wells for rural supply.

3.5. Resource Assessment and Water Balance

In the Azraq basin, the upper aquifer contains substantial volumes of excellent quality water, which is easily exploited. In the Badia Programme area, water levels are deep, thus borehole construction costs are high, but yields are invariably good. However, on a basin scale, current abstraction exceeds recharge, and groundwater is being 'mined' from storage. The total volume of groundwater removed from storage is between 40 and 205 million m^3/year, depending on which estimates of recharge and abstraction are adopted. Consequently, the existing abstraction from the Azraq basin is already unsustainable, with a significant detrimental impact on the unique environment of Azraq oasis. Thus, although technically feasible, any increase in development is both undesirable and illegal, due to a ban on the issue of new abstraction licences.

The groundwater resources of the North Sirhan and Hammad basins are extremely limited. Estimates for the two basins vary between 5 and 13 million m^3/year (CES and Arabtech 1993). From more detailed studies, GITEC (1996) propose 1.5 million m^3/year for North Sirhan and 5 million m^3/year for the Hammad basin. If there is no modern recharge, then long-term development may be unsustainable. An additional problem is the poor water quality, which limits the use of this groundwater for domestic supplies, although it is acceptable for agricultural use. The best quality water is in the basalt highlands of Tullul El Ashaqif between Safawi and Ruwayshid, and although the volumes are too small to be used for large-scale development, this groundwater could provide a good source for small, rural supplies.

4. SURFACE WATER RESOURCES

Although there are regular, short duration floods during winter in the wadis flowing south from Jabal al-Arab, flows have not been measured. Total flows have been estimated from rainfall data to average 27 million m^3/year. Although some of the surface water flows through the *marabs* and supports natural vegetation and pasture, most of the water runs onto the mudflats of the *qa'as* and is lost through evaporation. It is reported that flood frequency and magnitude have declined markedly in recent years, due to the construction of dams on the upper reaches of the wadis in Syria. A Jordanian dam was constructed on the lower reach of Wadi Rajil in the 1980s, but the reservoir has only filled once, in the winter of 1994/5.

In 1995, the Azraq Oasis conservation project constructed gauging structures on six wadis to the north of Azraq, which are to be equipped with automatic recorders. The BDRP has recently purchased flow gauges to be installed on Wadi Rajil and other wadis in the north-eastern part of the Azraq basin.

5. CONCLUSIONS AND RECOMMENDATIONS

To conform with the Badia Programme's objectives of *sustainable development,* any additional exploitation of water resources to supply future development must also be sustainable, and fit within Jordan's overall strategy and legal framework for management of water resources. This provides a dilemma for the BRDP, as groundwater in the upper aquifers of the Azraq, North Sirhan and Hammad basins provides the main source of water, which is reliable and widely available. This dilemma arises because the upper aquifer of the Azraq basin is already over-abstracted beyond its sustainable limit, and abstraction must be reduced if spring flows are to be restored. However, in contrast to the central part of the basin, the Badia Programme's project area still has ample supplies of groundwater, recharge exceeds local abstraction and the water quality is excellent. Alternative sources of groundwater are not promising, as it is doubtful whether there is any modern recharge in the Sirhan and Hammad basins and to the deeper aquifers, therefore these resources may be non-renewable and thus not sustainable for long-term development. However, in all three basins, provision of additional water supplies from groundwater, especially from widely scattered boreholes with modest yields, is technically feasible, with only localized decline in piezometric levels.

An additional restriction on future groundwater development is a ban on new abstraction licences in the Azraq basin, as part of the implementation of RAMSAR agreement to safeguard Azraq Oasis. At times (e.g. 1992-5), this prohibition extended to the whole of Jordan, thus new boreholes would contravene Jordanian regulations. The Jordanian government has tough policies on licensing and controlling groundwater use, but these are not always enforced when other considerations receive a higher priority. Enforcement of controls and firm resource management are essential to define the strategy and priorities for water use, co-ordinate different groups with conflicting aims and to implement the management plans to safeguard the resource in the long term. As these goals are entirely consistent with the aims and objectives of the BRDP, the Programme's strategy must be to support existing policy, facilitate implementation in the Programme are and not to undermine it by proposing unsustainable exploitation of groundwater.

The BRDP will therefore concentrate on maximization of water availability by evaluating alternative sources of water, including use of surface flows, artificial recharge of the upper aquifer, reuse of waste water and leakage control. The most substantial and under-used resource is surface flow, which is capable of providing significant additional supplies through water harvesting. There will be no significant impact downstream because, at present most of the surface water flows onto the fine silty surface of the *qa'as* and then evaporates. Only a small proportion of the water is used, mostly by natural vegetation on the *marabs* (playas with outflow), which provide pasture, and in a few localities water is diverted and stored for local use. The largest of these systems is at a farm on Wadi Gharbi, to the north-west of Azraq, where the wadi flows are captured by a small dam and pumped into a deep storage reservoir. The extensive fields of olive trees and presence of water in the reservoir in October 1995 demonstrate that surface water harvesting can provide a year round supply of water, providing the winter rainfall is sufficient, but a borehole is used as a reserve supply. The feasibility and optimum methods of utilizing the surface water resources form the main objectives of the water theme for the second phase of the BRDP (Dottridge 1995). However, groundwater will continue to form an essential source of water, due to its reliability and widespread availability.

ACKNOWLEDGEMENTS

This work was made possible by the co-operation and generous contributions of data, expertise and practical help from the staff and researchers of the BRDP in Amman, Safawi, and the UK, WAJ, Ministry of Water and Irrigation, Natural Resources Authority (NRA), Ministry of Agriculture and many others in the water sector in Jordan. I am extremely grateful for their help and support. Thanks to the M.Sc. students for their hard work and dedication during their projects. Financial support from the RGS, HCST, Royal Society, Natural Environment Research Council and Royal Jordanian is gratefully acknowledged.

REFERENCES

Al-Kharabsheh, A.A.H. (1991) *Hydrogeological and hydrochemical study of the upper aquifer system in Azraq Basin (Jordan).* Geology M.Sc. thesis, Yarmouk University, Jordan.

Al-Momani, M. (1991) *Relationship between A.S.W.A. wellfield and North East desert ground water (upper aquifer) and sources of salinity in deep B2/A7 aquifer, Azraq Basin, Jordan.* Presented in the workshop of 'Isotope Hydrology in the Middle East Countries' held in Ankara, October 1991.

Andrews, I.J. (1992) Cretaceous and Paleogene lithostratigraphy in the subsurface of Jordan. *Subsurface Geology Bulletin*, Natural Resources Authority, Jordan.

Ayed, R. (1986) *Initial study for water resources in the Azraq ground water basin.* WAJ, Jordan.

CES and Arabtech (1993) *Review of Water Resources Development and Use in Jordan*, WAJ, Jordan.

Core Lab (1987) *Azraq Basin Study. Volume IIID, Reservoir Geology Hydrodynamics.* Prepared for the NRA of Jordan.

Dottridge, J. (1994) *Groundwater resources and quality*, Jordan BDRP report.

Dottridge, J. (1995) *Water resources: progress and priorities*, Jordan BDRP report.

Dottridge, J. and Gibbs, B.R. (1996) *Water for sustainable development in the north-eastern Badia, Jordan.* in: Carter and Howsam (eds.) Water Policy: Allocation and Management in Practice, pp.38-45.

Drury, D.M. (1993) The hydrochemistry of the Azraq Basin, N.E. Badia. unpublished M.Sc. thesis, University College London.

GEF (1993) *Arid region nature and natural resources conservation and rehabilitation - Azraq component* - Project Document. Global Environmental Facility, UNDP and Government of Jordan. Unpublished report.

Gibbs, B. (1993) The hydrogeology of the Azraq Basin, N.E. Badia, Jordan, unpublished M.Sc. thesis, University College London.

GITEC Consult Gmbh, Hydrogeological Services International and GEMT Co (1996) *Groundwater investigation in the Hammad and Sirhan basins*, WAJ, Amman.

Haaland, A. (1996) Investigation of new groundwater development in the north-eastern Badia, Jordan, unpublished M.Sc. thesis, University College London.

HCST (1993) *The Badia, rural development: investing in our common future.*. Higher Council for Science and Technology.

Holden, W. (1995) Modelling and isotope study of the Azraq basin, north-east Jordan, unpublished M.Sc. thesis, University College London.

Howard Humphreys (1977) *North Jordan water use strategy - Azraq - proposals for water resources investigations.*

Howard Humphreys (1982) *Amman Water and Sewerage Authority Azraq wellfield evaluation.*

Jones, T.A. (1990) *Azraq Oasis, Jordan.* RAMSAR Convention Monitoring Procedure Report No. 16.

Mudallal, U.H. (1968) *Water balance study in Azraq area* Ground water division, Dept. of Research and Investigation, Natural Resources Authority, Jordan.

Noble, P. (1994) Quantification of recharge to the Azraq basin, north-eastern Badia, Jordan, unpublished M.Sc. thesis, University College London.

NRA (1993) Al Azraq 3353 1. 1:50,000 Geological map. Geology Directorate, Natural Resources Authority, Jordan.

Otova,T. and Ayed, R. (1988) *Surface water inventory and artificial recharge* Report No. 1, Azraq Basin, Wadi al-Butum. North Jordan Water Resources Investigations Project, WAJ, Amman.

Parker, D.H. (1971) The hydrogeology of the Mesozoic-Cainozoic aquifers of the Western highlands and plateau of East Jordan. Ph.D. thesis, University of Nottingham.

Sa'ad, A. (1993) Rainfall intensity duration curves for Azraq. Unpublished figure. Water Authority of Jordan, Amman.

Salzgitter (1994) *Middle Aquifer exploration project*, Azraq basin. Final report.

Ta'any, R.A. (1996) Hydrological and hydrogeological study of the Azraq basin, Jordan. Unpublished Ph.D. thesis, University of Baghdad.

UNDP (1966) *General report on the ground water investigation of the Azraq Basin.*

Waddingham, J. (1994) Water demand in the north-east Badia, Jordan, unpublished M.Sc. thesis, University of Newcastle-upon-Tyne.

WAJ (1989) *Azraq Basin water resources study draft final report.* Water Authority of Jordan, Amman.

Wilkinson, A.F. (1992) The Hydrogeology of Jordan (with special reference to the North-Eastern Badia), unpublished M.Sc. literature review, University College London.

CHAPTER 5

Baseline Hydrochemical Study of the Azraq Basin

DAVID DRURY
(Golder Associates, Nottingham)

1. INTRODUCTION

Although much work has been undertaken on the chemistry of the groundwater in the Azraq Basin, no comprehensive regional survey has been completed since the work by Parker (1971). In the last 20 years there has been a significant increase in the number of boreholes abstracting water. The increase in abstraction and number of potential sampling points warranted a re-evaluation of the chemistry of the groundwater. This paper describes a regional survey of the major ion chemistry (Drury 1993) of groundwater in the Upper aquifer of the Azraq Basin.

2. BACKGROUND

The flow of groundwater in the basalt and Rijam limestone in the north-east Badia region of Jordan is concentrated in an area that is approximately coincident with the Azraq Basin as defined by the flow of surface water. Like surface water, groundwater flows radially towards the centre of the Basin at Qa'a Azraq.

The groundwater flow in the Upper aquifer occurs principally in two lithologies - a sequence of basalt lava flows which decrease in thickness from the north (in the Syrian Druze mountains) to Azraq, in the centre of the Basin. The basalt is stratigraphically underlain by the Rijam limestone. Boreholes have proved that the two lithologies are separated in places by marl of the Wadi Shallala Formation.

3. SAMPLING AND ANALYSIS

Groundwater sampling was undertaken in May and June of 1993. The majority of the boreholes that were sampled in this study are located in the northern part of the Basin (Fig. 1).

Acidified and unacidified filtered water samples were collected from each borehole and stored at approximately 5° C prior to analysis. Water temperature, alkalinity, pH, dissolved oxygen and electrical conductivity were determined at the point of sampling.

The unacidified samples were analysed for anion concentration using a Dionex series 2000 ion chromatograph. ICP emission spectrometry was used to determine the cation concentration in the acidified samples. Silica concentration was determined by a spectrophotometric method.

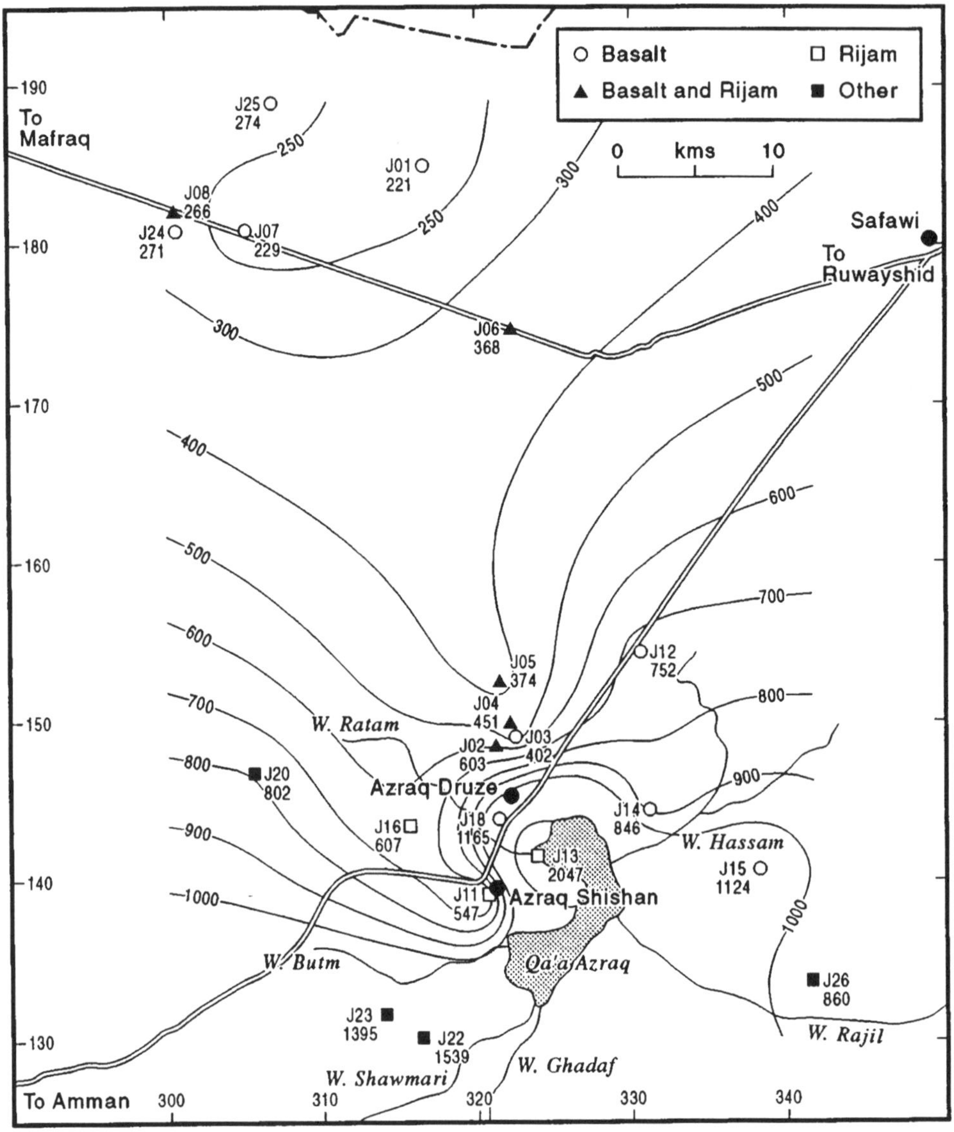

Fig. 1. Map of sampling locations with contours of TDS.

The results of the analyses are given in Drury (1993) including brief details of the sample location and aquifer of origin. Samples J06 and J27 were collected from the same borehole on different occasions and the results of the analysis may be therefore be used as a combined measure of the accuracy and precision of sampling and chemical analysis.

4. RESULTS OF CHEMICAL ANALYSIS

The ionic balance error of all of the analyses is less than 10 per cent. Considering this and the similarity of the results from analysis of J06 and J27, there can be confidence in the accuracy and precision of the sampling and chemical analyses.

4.1. Major Ion Chemistry

The waters are all either sodium chloride type (meaning that sodium is the dominant cation and chloride is the dominant anion) or calcium carbonate type. Plots of the molar concentrations of calcium against sulphate and sodium against chloride for the borehole derived samples are shown in Figure 2. Both plots suggest that the ratio of sodium to chloride and calcium to sulphate in the groundwater is 1:1.

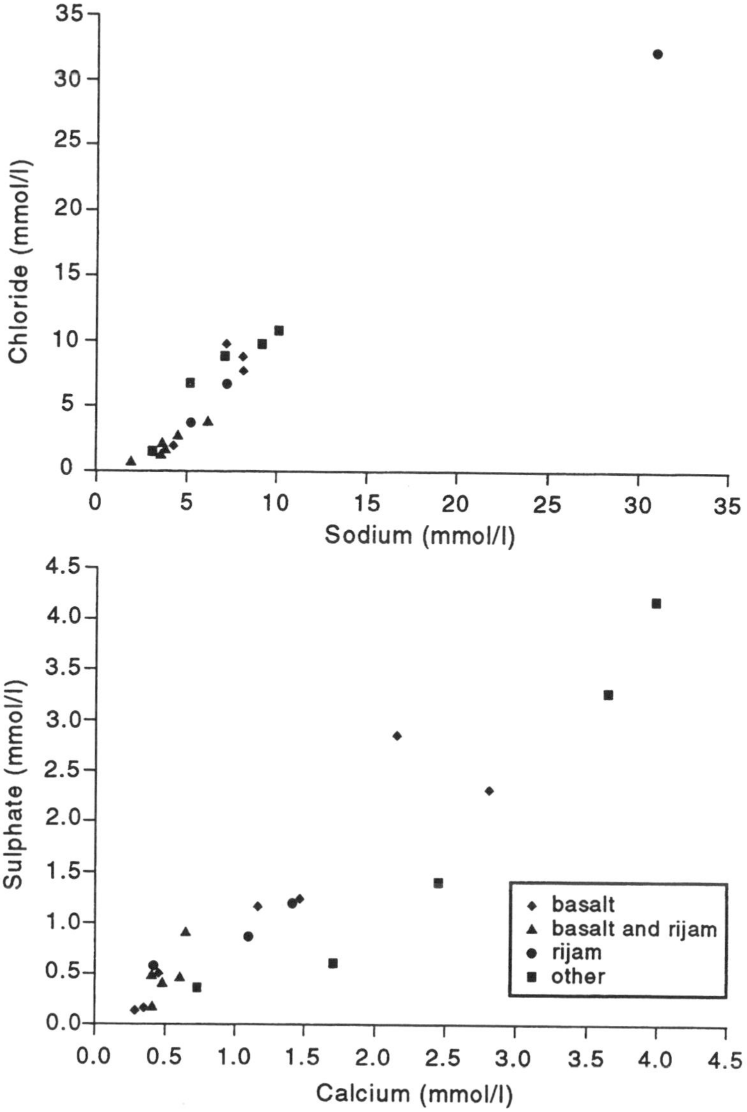

Fig. 2. Plots of major ion chemistry.

The equal molecular masses of sodium and chloride in the groundwater suggests that dissolution of halite (NaCl) is the source of these ions. Similarly, the ratio of calcium to sulphate is indicative of dissolved gypsum or anhydrite ($CaSO_4.2H_2O$ or $CaSO_4$, respectively) controlling the concentration of these ions.

The minerals may have been dissolved in the basalt, or the limestone and/or at the surface during infiltration of rainfall. Halite, gypsum and anhydrite are not reported in the mineralogical descriptions of either the basalt or limestone and therefore these ions must have entered groundwater when rainfall infiltrated at the surface. It is likely that the presence of these minerals at the surface can be attributed to evaporitic crusts that form in arid climates due to the continual evaporation of rainwater.

4.2. Total Dissolved Solids

The concentration of total dissolved solids (TDS) in water taken from boreholes completed in the Upper aquifer has been calculated and the values plotted and contoured in Figure 1. Lithologically, the basalt has the lowest TDS water (200 to 1,200mg/l) and that in the Rijam limestone is slightly higher (550 to 2,000mg/l).

There is a general increase in TDS towards Qa'a Azraq from all directions (the pattern of increasing TDS from the north is particularly clear). Of note are the two low TDS tongues, one extending from the Mafraq - Safawi road down to the ASWA well field in a NNW-SSE orientation and the second, trending NW-SE to the west of Azraq Druze and Azraq Shishan. The two tongues are geographically coincident with the al-Aritayn fault zone and Shishan fault zone, respectively.

5. DISCUSSION

Increasing TDS in the direction of flow is usually associated with dissolution of minerals within the aquifer. However, the major ion content of the groundwater suggests that the water chemistry is dominated by the process of dissolution of evaporitic crusts at the ground surface.

In the Azraq Basin, it is expected that surface evaporitic crusts are better developed in the south where the average annual rainfall is lower and the average annual temperature is higher.

After rainfall events in the north of the Basin, water enters the unsaturated zone and percolate to the water table. The climate in the north is unlikely to favour the development of evaporitic crusts at the ground surface and the water will therefore remain fresh. In the centre of the Basin, low intensity rainfall is unlikely to percolate the ground surface but instead will evaporate, leaving salts (principally sodium chloride and calcium sulphate) on the ground surface. However, infiltration may follow intense storms, but the water that percolates to the water table could become saline after dissolution of the evaporitic crust. In summary, the high volumes of rainfall that infiltrate the aquifer in the north become more saline further south as lower volumes of more saline water infiltrate from the surface. Ultimately, the water table becomes close enough to the surface under Qa'a Azraq for direct evaporation to increase the salinity to near saturation with respect to many salts (for instance the concentration of sodium chloride in sample J19 is 294g/l, very close to the solubility of sodium chloride at 25° C which is approximately 308g/l).

But what of the low TDS tongues? Following the model for the development of the major ion water chemistry presented above, the low TDS tongues must be attributed to areas of increased recharge of fresh water, or reduced development of surface evaporitic crusts. Wadis are often the surface manifestation of geological faults in the Azraq Basin and they collect rainwater that has run off the surrounding land. It is postulated that infiltration of rainwater is enhanced by the wadis associated with the al-Aritayn and Shishan fault zones. This may be due to the geometry of the wadis and the associated depth of rainwater that collects in them, the soil type, or fracturing associated with the faulting. Further, because wadis flow with water more often than the surrounding land, the development of surface evaporitic crusts is restricted. The presence of the low TDS tongues may therefore be explained by increased recharge of water from wadis which has low salinity due to the limited development of surface evaporitic crusts.

6. CONCLUSIONS

This study of the baseline quality of water in the Upper aquifer of the Azraq Basin has provided a reliable chemical dataset that has better areal coverage than previously published work.

The evolution of the major ion chemistry of the water is explained by relatively high volumes of recharge of rainwater in the north that become more saline further south due to infiltration of smaller volumes of water that are charged with salts (sodium chloride and calcium sulphate) derived by dissolution of evaporitic crusts.

A contour plan of total dissolved solids in the groundwater shows that there are two areas of low total dissolved solids. It is suggested that these may be areas of enhanced recharge of relatively fresh water associated with wadis.

ACKNOWLEDGEMENTS

I would like to thank the management and staff at Safawi and HCST for all the support given me during my time in the field, and my supervisor, Jane Dottridge, at UCL for her advice and encouragement.

REFERENCES

Drury, D.M. (1993) The hydrochemistry of the Azraq Basin, N.E. Badia. Unpublished M.Sc. thesis, University College London.

Parker, D.H. (1971) The hydrogeology of the Mesozoic-Cainozoic aquifers of the Western highlands and plateau of East Jordan. Unpublished Ph.D. thesis, University of Nottingham.

CHAPTER 6

Modelling and Isotope Study of the Azraq Basin

WAYNE HOLDEN
(Dames and Moore, Manchester)

1. INTRODUCTION AND OBJECTIVES

Research was undertaken over the summer of 1995 on the hydrogeology of the Azraq Basin, north-east Jordan, as part of the groundwater phase of the Jordan Badia Research and Development Programme (Holden 1995). It represents the fourth project completed by University College London Hydrogeology Group for the Badia Programme.

At present groundwater abstraction exceeds recharge in the Basin's main aquifer system and piezometry is continuing to fall with no indication of stabilizing. Previous work has highlighted the fact that aquifer parameters and quantity of recharge are poorly constrained preventing the implementation of a effective management strategy for the resource. This project sets out to address these two issues further through the use of groundwater isotope analysis and numerical modelling of the aquifer system.

The project included six weeks fieldwork in Jordan in order to collect groundwater samples for hydrochemical and isotope analysis. An area around Umm al-Quttayn in the north of the Basin was chosen for the detailed sampling programme as it has undergone little prior investigation and is currently experiencing active groundwater development. The report also outlines the first detailed interpretation of the groundwater-mineralogical interactions in the aquifer using the hydrochemical data.

2. SCOPE OF WORKS

In order to satisfy the objectives, the following tasks were undertaken:

- compilation of a well inventory in the area around Umm al-Quttayn;
- sampling of groundwater for hydrochemical analysis, and interpretation of the results with respect to hydrochemical processes in the aquifer;
- sampling of the groundwater for oxygen and hydrogen stable isotope analysis, and interpretation to determine the conditions and nature of recharge;
- sampling of the groundwater for Carbon-14 analysis, and interpretation to determine the age of groundwater abstracted from different sections of the aquifer and to allow for groundwater travel times to be calculated. This will have important implications on quantifying modern recharge to the aquifer;
- construction of a numerical model of the aquifer system to improve the definition of permeability distribution, storativity and modern recharge.

3. FIELDWORK

Fieldwork was concentrated in the area to the north of the main Safawi-Mafraq road in the vicinity of the village of Umm al-Quttayn and at the AWSA well field near the town of Azraq. At each borehole location a 'site record sheet' was completed with the assistance of the well owner to allow for compilation of the first detailed well inventory of the north of the Basin. Data collected included temporal variation in abstraction quantities, static water levels, pumped water levels, well depth and construction date. Fourteen of the 32 boreholes in the area were also sampled for major ion chemistry and stable isotope analysis. A further nine boreholes were sampled for carbon for Carbon-14 analysis.

Previous attempts at extracting carbon from the Upper Aquifer system's groundwater using barium chloride have proved unsuccessful with precipitation of barium sulphate in preference to the required barium carbonate. To increase the quantity of carbon a new sampling technique was developed by passing CO_2, liberated from the groundwater by hydrochloric acid, through a solution of Barium Chloride. The new method produced sufficient amounts of carbon to allow an accurate determination of groundwater ages.

4. HYDROCHEMICAL INTERPRETATION

The majority of boreholes in the Umm al-Quttayn area are completed in the unconfined Upper Aquifer, the principal groundwater resource of the Azraq Basin. The northern part of the basin is comprised of a series of Neogene-Quaternary continental basalts that are subdivided into five volcanostratigraphic groups of the alkali basalt series. Modally Na-rich plagioclase is the dominant phase (50 per cent), followed by olivine (30 per cent), pyroxene (10 per cent) with K-feldspar and Fe-Ti oxides constituting the remainder. It is possible to interpret variations in hydrochemistry by chemical processes involving the basalt's mineralogy.

4.1. Sodium

All the waters sampled in the Umm al-Quttayn area have a very low total dissolved solid content (190-300mg/l) (Fig. 1) and sodium, as the dominant cation, is closely correlated. A plot of molar sodium against molar chloride often reveals the source of the sodium as there are no significant precipitation reactions to remove either ion. The basalt groundwaters indicate a ratio Na:Cl ratio of 1:0.5 indicating that the principal source of sodium is not from dissolution of evaporites as previously considered (expected ratio 1:1) or from mixing with connate sea water (expected ratio 1:1.2). The process controlling the concentration of sodium is most likely to be the alteration of minerals in the plagioclase solid solution between albite and anorthite found in the aquifer. The breakdown of Na-rich plagioclase to montmorillonite is considered to be the most important of these reactions, and is supported by a positive correlation of Na concentration with pH, and a Na:Si ratio of 1:2 in the basalt's groundwaters.

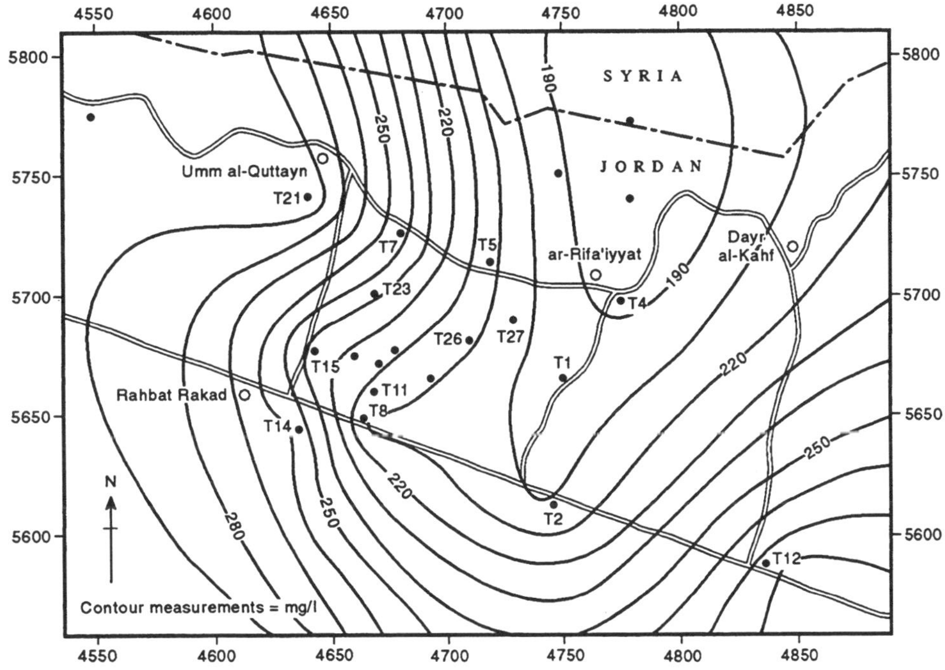

Fig. 1. Total dissolved solids, mg/l.

4.2. Calcium

The principal source of calcium in the groundwaters can be determined by a plot of molar sulphate against molar calcium. The observed Ca:SO_4 ratio of 1:0.55 falls off the expected trend if the principal source of calcium was connate seawater (ratio 1:2.7) or calcium sulphate (ratio 1:1). The most likely additional source is calcium-bearing pyroxenes and Ca-Na feldspars. If it is assumed that both calcium and sodium concentrations are controlled by alteration of plagioclase solid solution, a plot of molar sodium against molar calcium would reveal the Ca:Na ratio of the plagioclase. The plot of basalt waters has an observed ratio 1:5 indicating the formula $Ca_{0.16}Na_{0.84}Al_{1.16}Si_{2.84}O_8$. This is in close agreement with the plagioclase solid solution observed in the basalt's mineralogy.

4.3. Potassium

A minor constituent of the ionic composition of the basalt groundwater, the principal process controlling its concentration is considered to be the alteration of K-feldspar. It is closely correlated to sodium with a Na:K ratio of 1:0.07 which is further evidence that mixing with connate sea water is not a principal control on the basalt's hydrochemistry.

4.4. Magnesium

The principal process controlling magnesium concentrations in basalt groundwaters is typically the alteration of olivine to serpentinite. A plot of molar silica against molar magnesium reveals an expected Mg:Si ratio of 4:1 for such a reaction, but indicates an additional source of silica which is considered to be the plagioclase previously discussed. The concentrations suggests that plagioclase contributes four times as much silica as olivine.

4.5. Chloride and Sulphate

A plot of molar chloride against sulphate reveals that the basalt waters do not show the expected trend if mixing with connate seawater were the principal source controlling their concentrations. The most likely source is from dissolution of surface evaporites.

5. ISOTOPE INTERPRETATION

5.1. Stable Isotopes

The new set of analyses supports previous studies (Al-Momani 1991), indicating recharge of the aquifer at altitudes of 1,000-800 m asl and at temperatures of 7-8° C.

5.2. Carbon-14

The average uncalibrated age of waters sampled in the Umm al-Quttayn area is 7,380 years before present, and 13,145 years before present in the AWSA wellfield. This suggests travel times from recharge to discharge at the centre of the Basin of approximately 10m/year. The results do not indicate conclusively whether modern recharge is a reality in the Basin, and further analysis of the data is required.

6. MODELLING OF THE UPPER AQUIFER

A three-dimensional steady-state and transient numerical model of the Upper Aquifer system was developed in order to shed light on the possible permeability, storativity and recharge distributions in the Azraq Basin. The model was generated using the finite difference modelling package MODFLOW and the pre-processor MODELCAD[386].

6.1. The Conceptual Model

Although the Upper Aquifer comprises several different lithologies it was modelled as a single hydrostratigraphic unit due to constraints on data and on the assumption that flow is predominantly through fractures which are distributed throughout the different units. The model area covered the northern half of the Basin to include the area of supposed recharge in Syria and the main abstraction areas of the AWSA wellfield.

6.2. Assigning Boundary Conditions

The northern and western edges of the model coincided with the edge of the aquifer's groundwater catchment and were therefore represented by no-flow boundaries. The eastern edge of the model was also assigned a no-flow boundary based on observed piezometry. The southern edge was modelled as a general head boundary. The Azraq spring was simulated using the Drain package in MODFLOW.

6.3. Assigning Aquifer Parameters

A review of existing data was undertaken of previous estimates of aquifer parameters and their distribution in the Basin. Information on hydraulic conductivity was garnered from pump tests in the AWSA wellfield, and calculations on specific capacity data using Logan's approximation. Storativity estimates were based on experience of the lithological characteristics of the aquifer, previous modelling work, and the limited number of pump tests. As one of the main aims of the modelling works was to gain a better understanding of the aquifer parameter distributions the values obtained were used as a loose constraint and a guide during calibration of the model.

6.4. Assigning Water Balance Components

Modern recharge of 18 million m^3/yr above elevations greater than 900m was assumed based on previous water budget calculations by UCL Hydrogeological Group (Noble 1994).

Data on abstraction from the AWSA wellfield indicates that following completion in 1980 abstraction had reached 20 million m^3/yr by 1987 and subsequently remained approximately constant. Poor records of abstraction from the rest of Basin necessitated estimation based on the Water Authority of Jordan lists of licensed wells detailing dates of completion and pumping capacity. Time steps were typically of six months duration divided into two main seasons: a wet winter (recharge, low abstraction) and a dry summer (no recharge, high abstraction).

6.5. Calibration

The steady-state model was calibrated against observed piezometry collected from 1949 to 1969 and a springflow discharge of 26 million m^3/yr recorded in 1956. Following calibration the model's simulated piezometry matched closely the observed values and produced an outflow at the springs of 28 million m^3/yr.

The transient model was calibrated against observed water levels in one of the four regularly monitored AWSA wells. It was not possible to calibrate all observation wells due to the large scale of node spacing in relation to the nature of fractured flow in the aquifer. The model was also calibrated against monthly spring gauging which commenced in 1981.

6.6. Calibration Results

The model simulated piezometry and springflow which closely matched observations using a distribution of hydraulic properties that was closely related to the regional geology of the Basin. The major fault zones of Sirhan and Bisawayya were modelled as major conduits to groundwater flow and the Fuluk fault and a secondary fault zone to the north of Safawi as groundwater barriers. The remainder of the aquifer was assigned a hydraulic conductivity of 2.5m/day.

The calibrated transient model used a storativity of 0.01, typical of such fractured aquifers.

7. SUMMARY

The main aims of this project were to reduce uncertainty in the Upper Aquifer parameters and to understand the mechanisms of recharge to the system. In particular it set out to determine whether recharge to in the Basin is a reality. This was achieved through a combination of computer modelling and an isotope study of the aquifer's groundwater.

7.1. Aquifer Parameters

Modelling was the principal tool used to constrain aquifer parameters. Although the model's results are not conclusive, the representation of the aquifer system adequately simulates the system with groundwater flow being controlled by the major faults in the region with negligible spatial variability of storativity. However the values of hydraulic conductivity are not consistent with travel times calculated from the Carbon-14 results. Further work is required to reconcile these discrepancies with more detailed modelling of the area around the wellfield and reinterpretation of the Carbon-14 results.

7.2. Recharge

The calibrated model simulates both steady-state and transient conditions with modern recharge of 16 million m^3/yr above elevations of 900m. Stable isotopes are consistent with this approach. However it is as yet not possible to use the Carbon-14 data to assist in this respect.

7.3. Basalt Hydrochemistry

The project also presents the first detailed study of the hydrochemistry of basalt waters in the north of the Basin, through sampling of newly drilled wells. The low salinity 'recharge mound' protruding across the Syrian border has been better defined, and the processes controlling the groundwater chemistry have been determined. This work can be used in conjunction with further sampling studies to investigate the nature of mixing between the basalt and underlying limestone and marl units. Due to the absence of good quality pump test data it is considered that hydrochemistry will prove a powerful tool in the understanding of regional flow in the future.

ACKNOWLEDGEMENTS

I gladly acknowledge the contribution made to the fieldwork by the staff of the Badia Programme at Safawi. My thanks also to HCST, RGS-IBG and CORD for creating this opportunity for field research. Finally, I would also like to thank Jane Dottridge, my supervisor, as well as staff and colleagues in the Hydrogeology Group at University College London. In Jordan information was provided by the Water Authority, NRA and the Meteorological Department.

REFERENCES

Al-Momani, M. (1991) Relationship between ASWA wellfield and north-east desert ground water (upper aquifer) and sources of salinity in deep B2/A7 aquifer, Azraq

Basin, Jordan. Presented in the workshop of 'Isotope Hydrology in the Middle East Countries' held in Ankara, October 1991.

Holden, W. (1995) Modelling and isotope study of the Azraq Basin, north-east Jordan. Unpublished M.Sc. thesis, University College London.

Noble, P. (1994) Quantification of recharge to the Azraq Basin, north-eastern Badia, Jordan. Unpublished M.Sc. thesis, University College London.

CHAPTER 7

Hydrogeology and Hydrochemistry of the Sirhan and Hammad Basins

ANDERS HAALAND
(Southern Science, Crawley)

1. INTRODUCTION

Groundwater in the Hammad and Sirhan has traditionally been used by the Bedouin, and until recently only abstracted from shallow hand dug wells. Various reports on Jordan's water resources have covered the two basins (Chemonics 1992, CES 1993, Agrar 1977). In essence they all agree with the latter: '...the areas XIII (Sirhan) and XV (Hammad) are without any importance regarding water resources and water use' (Agrar 1977). In the last few years a number of new boreholes have been drilled in the area. Most of these are operated by the Water Authority (WAJ) or the Ministry of Agriculture to supply the Bedouin. In addition, a number of new unlicensed farms have recently started to operate in North Sirhan. This apparently unrestricted development caused some concern among staff and researchers at the BRDP base in Safawi, and it was decided to commission a report (Haaland 1996) to investigate the new developments. The main points of that report are presented in this paper.

2. PHYSIOGRAPHY AND HYDROLOGY

The Badia is in the transition zone between the Mediterranean environment of the Jordan valley and the fully arid environment which characterize the interior desert areas of the Syrian steppe. The region is mainly influenced by the Eastern Mediterranean frontal depressions (Ch. 3). Both the Sirhan and Hammad basins can be classified as arid, with hot dry summers and relatively cold winters. Rainfall show large temporal and spatial variations, and is characterized by short, intense rainstorms. Mean annual precipitation is 40-50mm in Sirhan and 60-90mm in Hammad. Annual potential evaporation varies from 1500mm to 2000mm, thus exceeding precipitation by up to 40 times. The temperature varies between 17° C and 50° C in July, and 0° C and 16° C in January.

The Sirhan basin extends into Saudi Arabia, and only the northern and western margins, comprising 15,300km^2 of an approximate total of 50,000km^2 lie within Jordan. (Fig. 1). It is a closed basin, with drainage converging towards the low lying salt lakes and mud flats of north-western Saudi Arabia. In North Sirhan elevations vary from 800m to 600m asl, with an additional fall of 100m to the mud flats, approximately 30km from the border. The main catchment is Wadi Qutafi (2350km^2), which is relatively well defined. It has a slope of 2-5m/km and a main channel width varying from 50-100m.

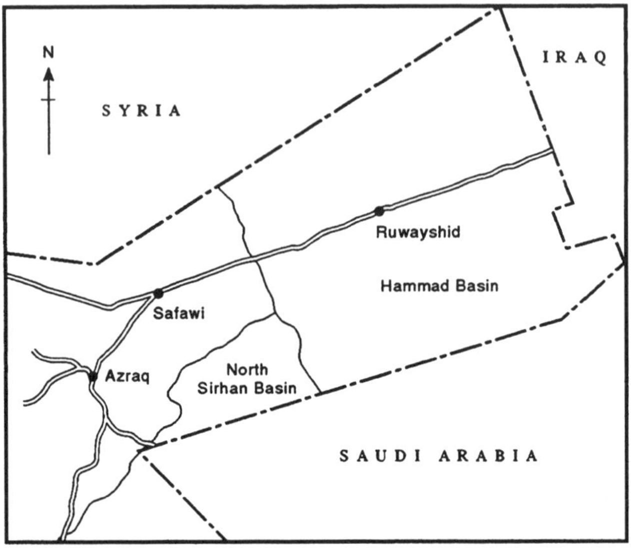

Fig. 1. Location map, Hammad and north Sirhan basins.

The Hammad is much bigger and includes the southern part of Syria, western Iraq and north-west Saudi Arabia, as well as the north-western pan-handle. In Jordan it covers an area of 18,250km² east of the Azraq/Hammad catchment divide. From the basalt hills of Tulul al Ashaqif at 800-1000m asl, water drains in an easterly direction towards the mud flats around Ruwayshid. A number of smaller wadis have been identified in the basalt and less defined but bigger catchments are present in east Hammad.

3. HYDROGEOLOGY

Although the hydrogeology of the Azraq Basin has been studied in great detail, the Sirhan and Hammad basins have often been ignored. When they have been considered, it has only been as components of a greater hydrogeological framework (Burdon 1982; Khouri 1982). The first major study to concentrate on these areas was published as recently as 1996 (GITEC 1996).

The aquifers of Jordan have been divided into three units, separated by significant aquicludes or aquitards. They are referred to simply as the Upper, Middle and Lower aquifer systems. As all formations are not present all over Jordan, important geographical differences exist. Exploratory drilling by NRA in the last few years have greatly expanded the knowledge of the sub surface geology, and led to changes in the classification system.

Figure 2 includes the most recent modifications, and these have been incorporated into this paper. The wells in the study area pump only from the Upper aquifer, which includes the Rijam, Shallala, Qirma and basalt formations, ranging in age from upper Cretaceous to Quaternary. These are described below.

Rijam - The Rijam Formation outcrops in eastern Hammad and is present at depths between 100m and 200m in North Sirhan. Thickness varies generally between 100m and 200m. The formation consists of chalky limestone, chalk, chert and limestone with chert. Locally phosphate beds and limestone with bitumen occur. The Siwaqa fault and the

Azraq-Sirhan fault system divide the Rijam aquifer into three hydrogeological blocks: West Sirhan, the Azraq-Hamza graben and Hammad. In the Hamza, saturated thickness is up to 200m, while in the Hammad it varies from being dry in the east to 150m in the west. Its hydraulic properties are dependent on the presence of fissures and fractures. As a result it is both isotropic and heterogeneous, and thus very variable in terms of productivity and hydraulic properties. Permeability values between 10^{-4} to 50m/day, and storativity values typically around 10^{-5} have been quoted (GITEC 1996).

Aquifer system	**Hydrogeological unit**		
	Name		**Designation**
UPPER	Qirma sst/1st		Aquifer
	Basalt	BS	Aquifer
	Shallala 1st	B5	Aquifer
			Aquiclude
	Rijam ist	B4	Aquifer
	Muwaqqar Marl	B3	Aquiclude/Aquitard
MIDDLE	Amman 1st	B2	Aquifer
	Hazim Marly 1st		Aquitard
	Hamza Dolomite		Aquifer
	Rajil Claystone		Aquiclude
	Rajil sst		Aquifer
	Wadi Sir / Lr Ajlun	A1/A2	Aquifer
	Kurnub sst		Aquifer
	Ramtha		Aquiclude
	Khreim shales		Aquifer
LOWER	Disi sst		

Shading = aquiclude or aquitard
from GITEC, 1996

Fig. 2. Sub-surface geology.

Shallala - Thicknesses of up to 550m had been reported in Saudi Arabia, but in North Sirhan it varies between 250 and 300m. The formation outcrops in parts of eastern Hammad. It consists mainly of chalk and chalky marl, but with thin beds of marly limestone, which can be bituminous. Locally, chalk beds contain glauconite, phosphate and chert. Its depositional environment is the same as the Rijam. It is divided into two units: an upper, thin (<100m) limestone aquifer and a thicker marl aquiclude which locally may act as an aquitard (GITEC 1996). Both in North Sirhan and the Hammad outcropping Shallala is discontinuously saturated and perched water tables can be encountered locally. Hydraulic properties are to a large extent dependent on the presence of marl and the amount of fracturing. Few boreholes have been tested, but reported permeabilities range from 0.1 to 5m/day, and specific yields around 0.5 per cent. Borehole yields are also variable, and vary from less than 0.1l/s to 20l/s. However, these findings are obscured by the fact that WAJ does not seem to acknowledge the existence of the Qirma formation; wells with high yields reported in North Sirhan may not, or only partially, penetrate the Shallala.

Qirma - Present in the Sirhan basin, this sequence is thought to have been deposited in a brackish lake environment. It consists of sandstone, limestone and marl with chert and quartz pebbles. Thickness varies from less than 40m to a maximum of 211m in the Hamza oilfield area. In North Sirhan, the Qirma is in hydraulic continuity with the underlying Shallala. However, the border between the Qirma formation and the sandy upper part of the underlying Shallala formation is not well defined. Most boreholes in the area are probably pumping both the Qirma and the Shallala. Yields are high, according to some operators upwards to $60m^3$/hour.

Basalt - The basalts consist of flood lavas, dyke systems, pyroclastic sediments and volcanic centres. Age determinations using K-Ar techniques suggest ages from 13.7Ma to less than 0.5Ma. The basalt overlies parts of the North Sirhan and the western part of the Hammad. Thinning towards the south into Saudi Arabia, the basalt sheets are up to 500m thick in northern Jordan (Ibrahim 1993). The formation is discontinuously saturated, and water table variations are dependent on topography (GITEC 1996). It is in hydraulic continuity with the Shallala in Hammad. Basalts are by nature highly anisotropic and heterogeneous due to fracturing and weathering between the lava flows, and this is reflected by the hydraulic characteristics. Studies in the Azraq Basin suggest great variability in hydraulic conductivity and specific yield (Gibbs 1993). For modelling purposes, permeability values up to 340m/day in fault zones, and around 2m/day for the rest have been used. Storativity of around 0.01 has been quoted as an average (Holden 1995).

In the Hammad the general flow direction is towards the Ruwayshid area, but swings in an south-easterly direction in the south, which may be due to structural features. Existing information in North Sirhan does not provide a clear view of the flow pattern. A southerly gradient towards the piezometric low in Saudi Arabia has been proposed, but due to the complicated structure of the area there are local inconsistencies. The flow gradient may be due to recharge occurring along the elevated margin, but it can also be due to fossil gradients. This may be caused by a higher rainfall in the past, but a number of other mechanism may also be in operation (Burdon 1977).

There are four possible sources of recharge: rainfall (direct infiltration), runoff (indirect infiltration), vertical leakage and lateral inter-aquifer flow. Traditionally, direct recharge has been estimated by the soil moisture deficit method. However, the technique has proved inappropriate in arid areas where potential evaporation can be orders of magnitudes larger than precipitation. Studies in the Azraq Basin suggest direct recharge of between 3 and 5 per cent of total rainfall (Noble 1994). This is probably higher than can be expected in North Sirhan and the Hammad due to lower precipitation and a smaller percentage of basalt on the surface. Recharge through wadi beds where these are permeable is possible, but commonly they consist of fine grained material with a high content of clay and silt. An example of this was observed in Wadi Abu Hoffna. A 30 minute rainstorm in the middle of May 1996 filled the wadi, and at the end of June there were still a number of water-filled ponds. The continuing presence of water suggests that the wadi bed is only to a small extent permeable, if at all. The fractured nature of the basalt at outcrop will lead to some infiltration, but the amount is impossible to quantify. Where the basalt overlies the Rijam, some transfer of water may occur, but where it is underlain by the Shallala, the marly beds almost certainly impede the water. Moreover, the preferred flow direction in the basalt is horizontal between the different lava sheets, and only small quantities of water flow downwards. In the Azraq Basin there is significant vertical leakage both ways through the Muwaqqar marls (Gibbs 1993), but no information is available in North Sirhan and the Hammad. Some throughflow into the basins is possible, but available data are insufficient to determine this with certainty.

4. HYDROCHEMISTRY

Most water in the study area is mildly brackish, but with some fresh and moderately brackish water. In North Sirhan total dissolved solid (TDS) content varies between 1,400mg/l and 3,800mg/l, and in the Hammad between 600mg/l and 2,800mg/l. Boreholes close to each other can have very different salinities. The best example is in North Sirhan, where two farm wells of the same depth situated only 200m apart have TDS values of 3,270mg/l and 2,020mg/l respectively. Electrical conductivity (EC) and total dissolved solids show a good linear relationship, and measurements can be beneficial in determining water quality. The average TDS/EC ratio is 0.59 with a standard deviation of 0.1.

Most of the wells both in North Sirhan and the Hammad have a distinct smell of sulphide, indicative of a reducing environment. It may seem surprising that the groundwater in North Sirhan, being unconfined and very shallow, should be anoxic, but only small amounts of organic carbon in the aquifer matrix are needed to consume all free oxygen.

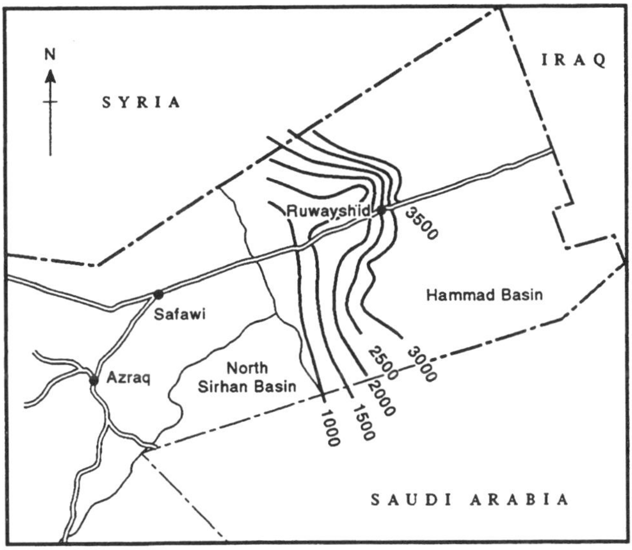

Fig. 3. Hydrochemical contours.

Most dissolved solids in the samples are due to dissolution of evaporites in the aquifer matrix or at the surface. The aquifers are very heterogeneous, with large variations in mineralogy over small distances. This is especially the case in north Sirhan where no flow pattern or direction could be deduced from the water chemistry. In the Hammad hydrochemical contours have been identified which correlate very well with previously published piezometric contours (see Fig. 3). There are substantial local deviations in this basin as well, and the findings can only to some degree be helpful in indicating suitable borehole locations.

Piper diagrams for both basins are shown in Figure 4. Calcium and sodium are the dominant cations, whereas sulphate and chloride dominate the anion distribution. There is in general not more than 50 per cent of any constituent. Sodium is the most important cation, with concentration from 62mg/l to 756mg/l. There is some sodium in the rain, but dissolution of halite is the most important source. Calcium ranges from 64mg/l to 326mg/l. Its distribution is similar to sodium. The principal sources of calcium are the dissolution of evaporites, notably gypsum and anhydrite, and weathering of Ca-bearing aluminosilicates in the aquifer.

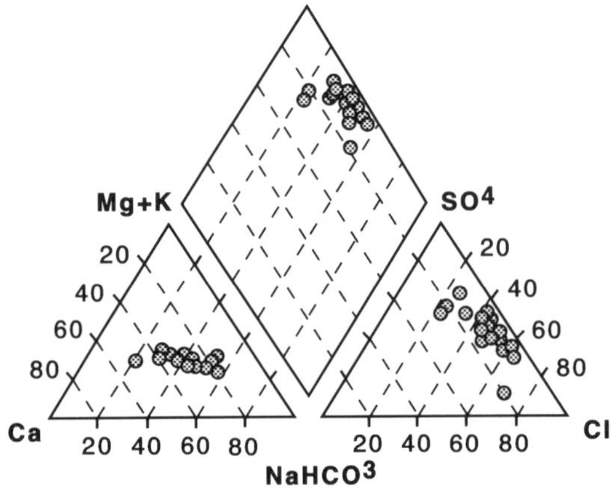

Fig. 4. Piper Diagram: north Sirhan and Hammad.

Chloride is the most important anion with concentrations from 534mg/l to 1,237mg/l in North Sirhan, and between 85mg/l and 728mg/l in the Hammad. It is closely related to sodium concentrations. A plot of sodium vs. chloride is parallel to the halite dissolution line, but with excess chloride due to the surplus from rainwater. Mixing with connate water is unlikely considering the depth of the watertable. Sulphate is the most important constituent by weight, with concentrations up to 1,349mg/l in North Sirhan and 1,056mg/l in the Hammad. Minimum concentrations are 295mg/l and 65mg/l respectively. The principal source of sulphate is dissolution of evaporites, notably gypsum and anhydrite. These are present in the Qirma aquifer, but have not been reported in neither the Shallala nor the Rijam. However, evaporites are present in abundance on the surface, and sulphate in the Hammad may be sourced from infiltrating water. An additional source is oxidation of sulphide minerals in the basalt.

5. WATER RESOURCES

Total abstraction in the Hammad has only increased marginally since the early 1990s. Current abstraction is estimated to 0.84 million m^3. A number of new rural wells have been drilled, but it seems that instead of increased abstraction rates, pumping has merely been spread over a larger area. Abstraction from North Sirhan has increased by more than 30 times since the early 1990s, and is now estimated to 1.41 million m^3. This is almost entirely due to the introduction of irrigation farming in Wadi Hazim. Seven farms of a total of 1,150 dunums grow two crops of water melon and vegetables per year. There are no piezometers in the area, but according to the farmers no major changes in the water table have taken place over the last few years.

An estimate of exploitable resources in the Rijam is 1.5 million m^3/year in North Sirhan, and up to 5 million m^3/year in the Hammad (GITEC 1996), but as modern recharge is small, abstraction relies on storage reserves, and possibly inflow from Saudi Arabia and Syria. Modelling suggests that a wellfield situated in the Ruwayshid area can abstract around 4 million m^3/year, however there are large uncertainties regarding hydraulic properties and especially the expected behaviour of the Muwaqqar marl.

An additional problem of potential development is the unfavourable water quality, which is not generally suitable for human consumption. This is due not only to a high total dissolved solid content, but also to dangerous concentrations of trace elements. The only area where the water quality is acceptable for human consumption is in the basalt highlands between Safawi and Ruwayshid. However, exploitable volumes here are too small to permit a major development.

6. CONCLUSIONS

The annual precipitation falling in the Sirhan and Hammad is between 40mm and 90mm, with potential evaporation at least 40 times higher. Groundwater flow is in general from the elevated margin towards low-lying salt lakes and mud flats. In addition to the WAJ wells supplying the Bedouin, a number of farms have started to operate in North Sirhan, using water from the Qirma and Shallala aquifers. Current abstraction is 1.4 million m^3 per year, almost 30 times more than in 1990. Total abstraction from the Upper aquifer in the Hammad is 0.84 million m^3 per year. This is only marginally higher than in the early 1990s, despite the increase in the number of rural wells in the area.

Most of the groundwater in North Sirhan and the Hammad is mildly brackish. The majority of dissolved solids are sourced from evaporites in the aquifer matrix or at the surface. The main ions are sulphate, chloride, sodium and calcium, but there is in general not more than 50 per cent of any cation or anion. Most water has a distinct smell of sulphide, and no oxygen was detected in any of the sampled wells. The aquifers are very heterogeneous, and especially in North Sirhan huge variations in hydrochemistry can be observed over short distances. In the Hammad a trend of increasing TDS from west to east consistent with aquifer flowlines has been identified.

In terms of water quality, the most promising area for further development is in the basalt highlands on the western margin of the Hammad. However, the quantity of available groundwater is probably too small for large-scale abstraction schemes.

ACKNOWLEDGEMENTS

Thanks are extended to the management of the Badia Programme as well as staff in Durham, Amman and Safawi, especially to Odeh, my guide and translator. I would also like to thank Jane Dottridge, my supervisor, as well as staff and colleagues in the Hydrogeology Group at University College London. In Jordan information was provided by the Water Authority, NRA and the Meteorological Department.

REFERENCES

Agrar (1977) *National Water Master Plan of Jordan.* - Agrar und Hydrotechnik, Amman.

Burdon, D.J. (1977) Flow of fossil groundwater. *Quarterly Journal of Engineering Geology*, 10, 97-124.

Burdon, D.J. (1982) Hydrological conditions in the Middle East. *Quarterly Journal of Engineering Geology*, 15, 71-82.

CES (1993) *Review of Water Resources Development and Use in Jordan.* Water Authority of Jordan, and Arabtech, Amman.

Chemonics (1992) *A Water Management Study for Jordan.* Project in development and the environment. International, USAID/Jordan, Amman.

Gibbs, B. (1993) The hydrogeology of the Azraq basin, NE Badia, Jordan. Unpublished M.Sc. thesis, University College London.

GITEC Consult Gmbh, Hydrogeological Services International Ltd, GEMT Co. (1996) *Ground Water Investigation in the Hammad and Sirhan Basins.* Water Authority Jordan, Amman.

Haaland, A. (1996) Investigation of new groundwater developments in the Badia Region, Jordan. Unpublished M.Sc. thesis, University College London.

Holden, W. (1995) Modelling and isotope study of the Azraq Basin, *NE Jordan.* Unpublished M.Sc. thesis. University College London.

Ibrahim, K. (1993) *The Geological Framework for the Harrat Ash-Shaam Basaltic Supergroup and its Volcanotectonic Evolution.* Natural Resources Authority, Amman.

Khouri, J. (1982) Hydrogeology of the Syrian steppe and adjoining areas. *Quarterly Journal of Engineering Geology*, 15, 134-54.

Noble, P. (1994) Quantification of recharge to the Azraq basin, NE Badia, Jordan. Unpublished M.Sc. thesis, University College London.

CHAPTER 8

Quantification of Recharge to the Azraq Basin

PAUL NOBLE
(Dames and Moore, Manchester)

1. INTRODUCTION

1.1. Project Background

The sustainable development of the desertified Badia Region of Jordan is inextricably linked to the availability of a reliable potable water supply. Groundwater can often provide such a supply if available groundwater resources are appropriately managed.

The first modern groundwater explorations in the Azraq Basin took place about 30 years ago. Since then there have been repeated attempts to describe the prevailing hydrogeological conditions with a view to implementing a suitable water management strategy.

Gibbs (1993) concluded that even in the 'best case' scenario the current estimate of Upper Aquifer discharge considerably exceeds that of recharge. It therefore appears that the aquifer is being exploited beyond its 'safe yield' which can lead to 'permanent and irreversible damage in the form of water depletion and salination' (Higher Council for Science and Technology, 1993).

The annual groundwater recharge is the amount of rainfall that travels through the soil and unsaturated zones to the water table. Total annual recharge in arid zones can be divided into two on the basis of mechanism. 'Direct recharge' is immediate infiltration of rainfall and 'indirect recharge' is infiltration of runoff and ponded water into ephemeral wadi beds and mudflats respectively.

Previous recharge studies can be divided into two types on the basis of the information they provide as follows :

- quantitative recharge estimates, providing an average annual recharge volume, or studies producing quantitative data such as recharge temperature or altitude;
- qualitative studies which may discuss recharge mechanisms or other unquantifiable recharge information.

Since it is not possible directly to measure recharge at the scale of interest (i.e. basin-wide), there is a continual requirement to extrapolate from sparse data points to areal data or to introduce simplifying assumptions. Furthermore, the magnitude of aquifer recharge of arid zone basins is usually lower than temperate zone basins, thus the degree of confidence in a recharge estimate would be greater in temperate zones than arid zones for the same absolute error.

In addition to these problems, it has been shown that present day groundwater discharge patterns and hydraulic gradients in regional arid zone basins can be entirely driven by ancient recharge events or 'pluvials', tens of thousands of years ago (Lloyd and Farag, 1978). In such cases, exponential hydraulic head decay and substantial groundwater discharges driven by 'fossil' hydraulic gradients can occur during an epoch of negligible recharge.

Table 1. Quantitative recharge estimates for the Azraq Basin, 1956-93

Study	Approach	Recharge Estimate (MCM/Yr)	Mechanism proposed or assumptions	Comments
Baker *et al.* 1956	Hortonian	32-240		Too high
UNDP 1966	Penman ET at wetland	20.9	Steady state	
Barber and Carr 1973	Penman ET at wetland	35.0	Steady state and capillary rise too high	Too high
Lloyd and Farag 1978	Water budget	10.0	Infiltration ignored	Need >3yrs data
WAJ 1989	Water budget & SCS method	34±21	Direct recharge	SCS problems
Mudallal 1968	Flow net	30.0	Steady state	Large range
Parker 1971	Flow net	18.1	Steady state	Large range
Barber *et al.* 1973	Flow net	30.0	Steady state	Large range
Howard Humphrys 1982	Flow net of Azraq wellfield	33.1	Steady state	Large range
Arsalan 1976	Water balance, includes flooded *qa'as*	35.0	Steady state	Large range
BGR 1987	Water balance	28.4	Steady state	Large range
Gibbs 1993	Water balance using spring discharge	16±3 (Max.)	Steady state	Not a max. due to leakage
Drury 1993	Stable isotopes altitude & temp	>850m & 6-8 °C	Direct recharge	

Previous estimates of recharge for the basin (Table 1) vary from a mean of 34 million m^3/yr) using a water balance approach to 16-36 million m^3/yr for a 'throughflow' approach and 18-33 million m^3/yr using flow net analysis, while negligible recharge is a possible scenario if a fossil recharge mound exists.

None of the previous studies considered leakage to or from the Upper Aquifer, which is a further potential source of error.

^{14}C data (radiocarbon dating of dissolved carbon in groundwater) indicate that groundwater in the centre of the basin (near the former natural discharge point of Azraq springs) is 12,000-25,000 years old. Similar data for the most likely potential recharge area towards the Syrian Druze mountains in the north of the basin are absent or equivocal, although recharge altitude and temperature were determined as over 850m above sea level and 6-8° C respectively using stable isotope data (Drury 1993).

1.2. Objectives and Methodology

The key objectives of the project were as follows :

- to quantify average annual recharge to the Azraq Basin;
- to characterize potential recharge mechanisms within the Azraq Basin.

Evaluation of independent methods of estimating recharge was undertaken as the best overall approach to constraining the average annual recharge volume for the Azraq Basin and characterizing potential operative mechanisms.

2. HYDROLOGICAL ASPECTS

2.1. Infiltration Rates

Rainfall in the Azraq Basin generally occurs during low frequency, short duration, high intensity storms of limited areal extent which result in localized rapid runoff in the region's ephemeral wadis and flooding of flat lying, internally drained mudflats or *qa'as*.

Determination of infiltration rates under meaningful conditions was not possible due to practical difficulties, summer fieldwork and the huge range of potential infiltration rates. However, a limited fieldwork programme allowed the following general conclusions to be made:

- infiltration into many of the region's *qa'as* is absent; they behave as evaporation pans;
- 'indirect recharge' through transmission losses during surface water runoff in wadis is possible given periodic aquifer outcrop in wadi beds and potential fault-wadi coincidence;
- commonly observed 'evaporation tide' marks suggest very low infiltration rates;
- double ring infiltrometer tests suggest highly variable infiltration rates throughout the Badia. In comparison with rainfall intensities, infiltration (direct recharge) appears plausible at certain locations tested;
- intense storms followed by ubiquitous surface water runoff suggests regional infiltration rates are generally low.

2.2. Storm by Storm Analysis using a Water Budget Model

The Water Authority of Jordan applied a water budget approach to individual storms to produce a recharge estimate for the Azraq Basin of 34±21 million m^3/yr but details of the calculations in the model were not available.

The model discussed below aims to calculate average annual direct recharge using daily rainfall data, rainfall simulation tests and the United States Soil Conservation Service (SCS 1964) (USSCS) method for calculating runoff from rainfall data by the 'initial abstraction' concept (Fig. 1).

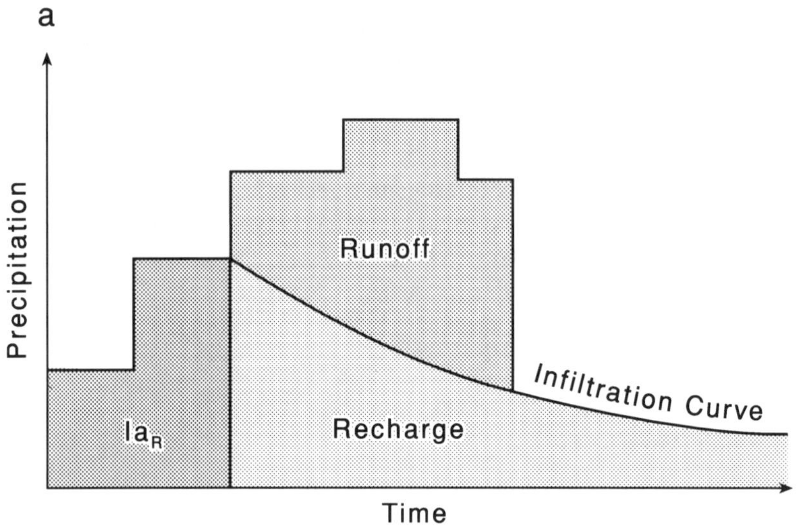

Ia_R= Initial Abstraction, mm of rainfall needed to saturate and pond the ground surface *before* runoff can occur *for runoff computation*

Fig. 1a. Graphical representation of USSCS concepts.

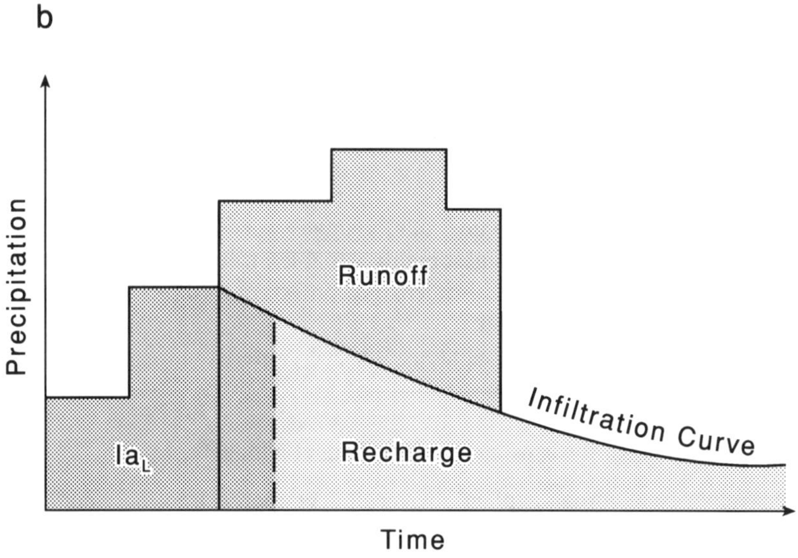

Ia_L = Initial Abstraction, mm of rainfall needed to saturate and pond the ground surface *before* runoff can occur *for recharge computation.*

Fig. 1b. Graphical representation of adjusted USSCS concepts.

On the basis of rainfall simulation tests, the range of the amount of simulated rainfall needed before runoff occurs on initially dry soil is 16-22mm. The regularity of runoff produced using a value for Ia of 16mm is consistent with observation, however, recharge values are 30-40 per cent of total rainfall which is unrealistically large. Higher Ia's produce no runoff at all and this is not consistent with observation. To solve this apparent contradiction two different values of Ia as defined above.
In other words, runoff is initially assumed to occur while the soil is continuing to absorb extra 'initial abstraction'. This is designed to reflect the intense nature of the rainfall which causes runoff to occur before the soil is wet enough to allow infiltration and recharge.

The model comprises the following basic steps:

For each rainfall gauge:

1. partition the basin into Thiessen Polygons around a rainfall gauge ensuring the land within each polygon is of a similar altitude and therefore rainfall;
2. use rainfall simulation test data to obtain the initial abstraction value used to calculate runoff (Ia_R);

For each storm at each rainfall gauge:

1. compute the previous five days rainfall total and adjust Ia_R to reflect the antecedent moisture conditions (AMCs), the greater the previous rainfall total the lower the value for Ia_R;
2. compare Rainfall (rf) and Ia_R , if rf > Ia_R then compute runoff using USSCS formulae;
3. select Ia_L ($Ia_L > Ia_R$), adjust for AMCs and sum Ia_L and runoff to compute total losses;
4. recharge = (rainfall - total losses) x (area of polygon for the rainfall gauge).

It was considered necessary to make the following assumptions to run the model:

- recharge is assumed to occur rapidly in the winter months, therefore evaporation is ignored;
- Ia values for runoff and total losses (hence recharge) were distinguished (see Fig. 1) in order to be consistent with observation and reflect the intense nature of the storms;
- Ia values were within the range of values derived from rainfall simulation tests with the exception of the high altitude Syrian rainfall gauges where a higher Ia_R was used to reflect higher vegetation cover in these areas (Ia values are usually derived from USSCS 'Catchment Curve Numbers' which reflect infiltration rate, storage, texture, composition, and vegetative cover of a soil);
- the daily rainfall value is assumed to be equivalent to a single 'storm';
- the model only considers direct recharge;

The average annual recharge predicted by the model for high altitude stations is 16.5±2 million m^3/yr, which would represent a total for the basin if the interpretation of stable isotope data by Drury (1993) is correct.

Errors relate to the assumptions listed above and the fact that extrapolation of 13 years of total annual rainfall to total annual recharge was undertaken for all rainfall gauges on the basis of the model outputs (the model was run for 6 years of storm by storm analysis).

The predicted recharge value for lower altitude stations is 21±4 million m^3/yr but this figure is thought to be an overestimate due to computation of excessively large recharge volumes for exceptionally intense storms and very low rainfall gauge density.

Analysis of the model output data shows that predicted recharge depends both upon rainfall magnitude and temporal distribution.

3. HYDROGEOLOGICAL APPROACHES

3.1. Hydrograph Analysis

'Recharge spikes' on continuous chart recorder hydrographs at observation boreholes beneath wadis were invariably found to match with rainfall events if a delay of 2-10 days was allowed. It should be noted that to perform an accurate analysis of this type it is necessary to assess the pumping regime of any proximal abstraction boreholes.

Modern indirect recharge appears to occur through wadi surface water runoff transmission losses. It is not possible to quantify these data.

3.2. Transient Groundwater Modelling

Simple transient modelling using the most favourable parameters for persistence of fossil groundwater gradients suggest the fossil recharge mound theory of Lloyd and Farag (1978) is inappropriate to the upper unconfined aquifer complex of the Azraq basin. In other words, using a steady state simulation of today's aquifer conditions calibrated with current water levels and estimates of aquifer parameters it is not possible to simulate present day hydraulic gradients without modern recharge.

4. CONCLUSIONS

Table 2 presents a tentative new water balance for the basin based on the latest findings:

Table 2. Tentative Azraq basin water balance

	Out (MCM/Yr)	In (MCM/Yr)	Balance (MCM/Yr)
Public Supply	28.84		
Agriculture	47.71	22.93[1]	
Leakage	93.3[2]	29.3[2]	
Recharge		37	
Total (with leakage)	169.85	89.23	- 80.6
Total (without leakage)	76.55	59.93	- 16.62

[1] Irrigation return, see Waddingham, 1994

[2] See Noble 1994

The key conclusion of this project is that on the basis of an improved water budget model for the northern part of the Azraq Basin, a minimum of 20 million m^3/yr of modern average recharge is occurring.

Even if upper estimates of total recharge such as 40 million m^3/yr are correct, it is clear from the above summary table that the upper aquifer of the Azraq Basin is being exploited beyond its safe yield and as such storage depletion and water quality deterioration will continue to occur.

ACKNOWLEDGEMENTS

My thanks to Mohammed Shahbaz and the staff at Safawi, to Roderic Dutton and CORD and to my supervisor, Jane Dottridge at UCL.

REFERENCES

Arsalan, F.A. (1976) Geologie und hydrogeologie der Azraq depression (Ost-Jordanien) Doctoral Thesis, Rheinisch - West Falischen Techniscen Hochschule, Aachen.

Barber, W. and Carr, D.P. (1973) *Digital Model of the Shallow Aquifer Complex of the Azraq Basin, East Jordan.*

BGR (Bundesanstalt fur Geowissenschaften und Rohstoffe (Federal Institute for Geosciences and Natural Resources)) (1987) *Geoelectrical Survey in the Azraq Area of Northeast Jordan.* Technical Cooperation Project No. 86.2278.9, Hanover.

Drury, D.M. (1993) The hydrochemistry of the Azraq Basin, N.E. Badia. Unpublished M.Sc. thesis, University College London.

Gibbs, B. (1993) The hydrogeology of the Azraq Basin, NE Badia, Jordan, Unpublished M.Sc. thesis, University College London.

HCST (1993) *The Badia, Rural Development...Investing in our common future..* Higher Council for Science and Technology, Jordan.

Howard Humphreys (1982) *Amman Water and Sewerage Authority Azraq wellfield evaluation.*

Lloyd, J.W. and Farag, M.H. (1978) Fossil groundwater gradients in arid regional sedimentary basins, *Groundwater* 16, p 388-93.

Mudallal, U.H. (1968) *Water balance study in Azraq area,* Groundwater division, Dept. of Research and Investigation, Natural Resources Authority, Jordan.

Noble, P. (1994) Quantification of recharge to the Azraq Basin, north-eastern Badia, Jordan. Unpublished M.Sc. thesis, University College London.

Parker, D.H. (1971) The hydrogeology of the mesozoic-cainozoic aquifers of the Western Highlands and plateau of East Jordan. Unpublished Ph.D. thesis, University of Nottingham.

SCS (1964) *A Method of Estimating Volume and Rate of Runoff in Small Watersheds.* Technical Publication 149, U.S. Department of Agriculture, Soil Conservation Service, Washington, D.C.

UNDP (1966) *General report on the ground water investigation of the Azraq Basin.*

Waddingham, J. (1994) Water demand in the north-east Badia, Jordan. Unpublished M.Sc. thesis, University of Newcastle-upon-Tyne.

WAJ (1989) *Azraq Basin Water Resources Study Draft Final Report.* Water Authority of Jordan, Amman.

CHAPTER 9

Water for Sustainable Development

JANE DOTTRIDGE
(University College London)
BEATRICE GIBBS
(British Geological Survey, Wallingford)

1. INTRODUCTION

In arid areas, water scarcity has traditionally restricted development; sustainable development relies on adequate, reliable and lasting supplies of fresh water. In 1992, the Jordan Badia Research and Development Programme (BRDP) was established to promote

> 'the sustainable development of the desertified Badia environment and the improvement of the standard of living of the inhabitants... under management systems which conserve the natural resources so that production levels will be sustainable in the long term.' (RGS 1992, HCST 1993)

Despite the apparently unfavourable geology and climate, the pilot area has substantial groundwater resources and regular winter flows in the wadis, originating from Jabal al-Arab in Syria. But the Badia programme area forms part of the Azraq basin, as illustrated by Figure 1, and the water issues cannot be considered in isolation from the overall basin resources; in particular the environmental deterioration at Azraq oasis and the adjoining wetlands, which are suffering from a reduction in spring flows (GEF 1993; Jones 1990). On the basin scale, groundwater is over-abstracted, with large-scale pumping for public water supply to Amman, Zarqa and Mafraq. This desert area is therefore an exporter of water to the wetter but more densely populated parts of Jordan. The local population relies on groundwater for domestic supplies, livestock and irrigation. The dilemma for the Badia programme and others working in the area is to provide sufficient water for local needs, without aggravating the problems of over-abstraction and increasing derogation at Azraq, which lies downstream of the programme area.

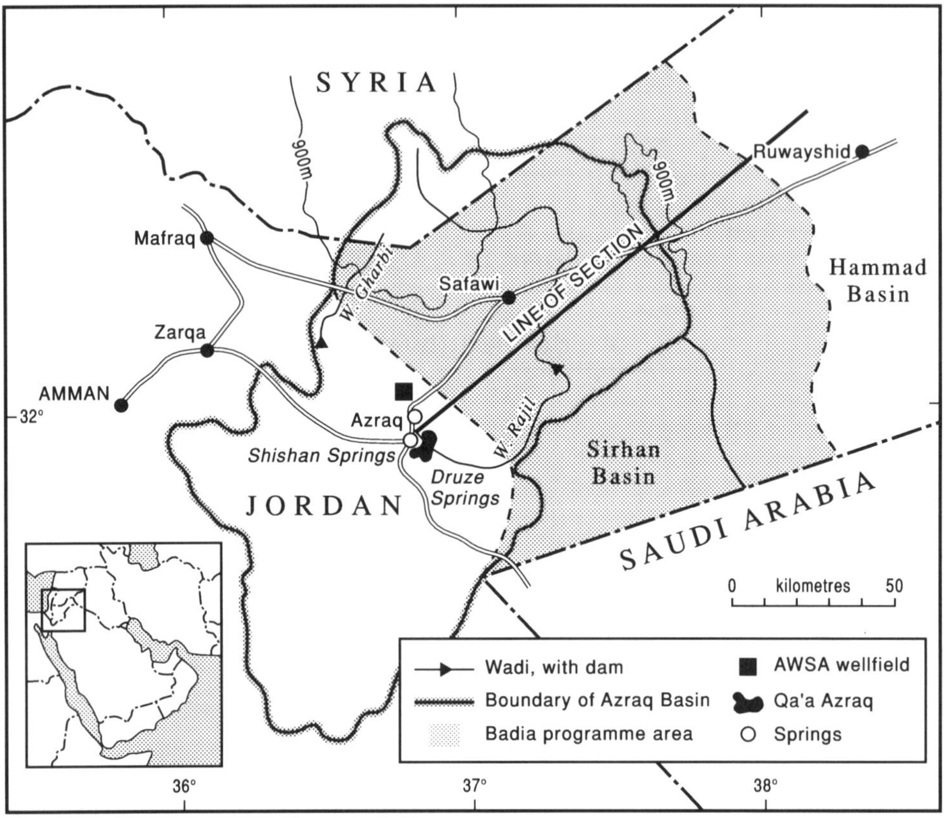

Fig. 1. Location map.

2. WATER RESOURCES

2.1. Groundwater

Groundwater in north-eastern Jordan occurs in three aquifer complexes (Parker 1971; Core Lab 1987; NRA 1993), as shown in the section in Figure 2:

- upper aquifer, comprised of Quaternary basalts and the underlying Rijam cherty limestone;
- middle aquifer, Cretaceous limestones of the Amman-Wadi Sir and Hummar aquifers, with very variable productivity and water quality, at depths of 300 to 1,000m;
- Lower aquifer, Cretaceous and older sandstones (Kurnub and Disi formations), at depths of 1.3 to 3.4km.

Although a recent project has explored the middle aquifer (Salzgitter 1994), only the upper aquifer is currently used. In this system, flow in the fissured basalt and Rijam limestone is approximately radial towards Azraq, with a gentle piezometric gradient of 0.001. Both depth to water and saturated thickness of the aquifers decrease southwards towards Azraq. The transmissivity of the basalts is highly variable, ranging from 5 to 30,000m^2/d (WAJ 1987) and shows a correlation with major faults. The major fault zones

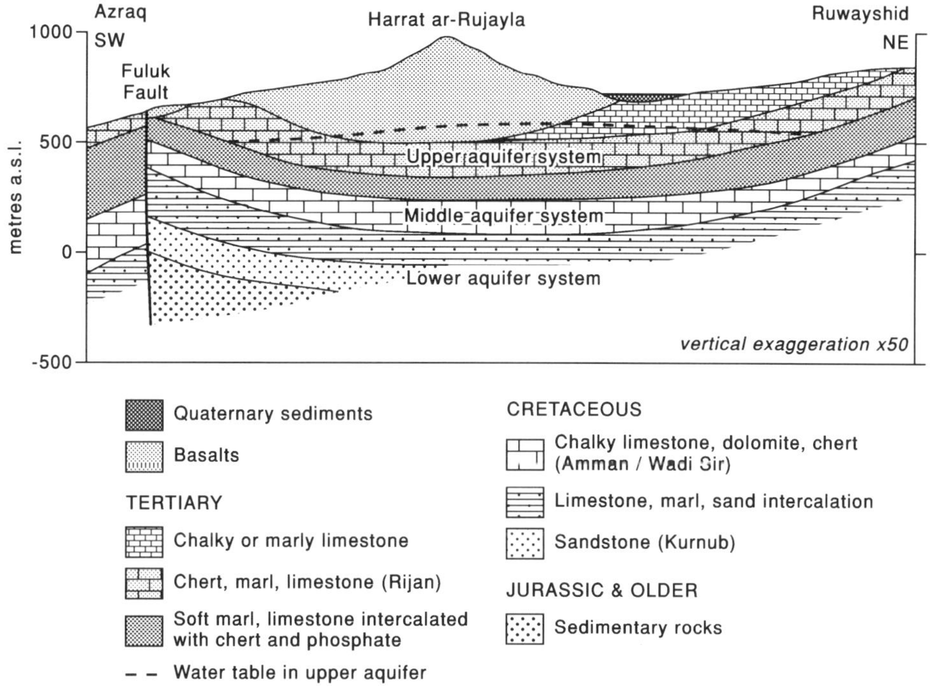

Fig. 2. Schematic section showing the main aquifers.

provide conduits for rapid flow of very low salinity water, while the Fuluk fault acts as a partial barrier to groundwater flow (Holden 1995). The transmissivity of the Rijam limestones is lower, 6 to 230m^2/d. The specific yield is estimated as 1 to 3 per cent (WAJ 1989), although test pumping at Azraq gave higher values of 12 to 22 per cent (Howard Humphreys 1982).

The water quality in the upper aquifer is extremely good, suitable for drinking or irrigation. The distribution of total dissolved solids confirms the radial flow pattern with elongated zones of low salinity, from 190 to 400mg/l, coinciding with the permeable fault zones (Drury 1993). Water with high salinity, up to 300,000mg/l, occurs locally in a zone of shallow water table around the playa of Qa'a Azraq.

Recharge to the upper aquifer occurs mainly through direct infiltration in the wetter, northern part of the basin, above 1,000m elevation, and by infiltration of wadi flows. In wet winters, some recharge may also occur in the larger, more arid area below 1,000m. The estimates of annual recharge range from 0 to 67 million m^3/year, mostly averaging between 22 and 36 million m^3/year. A credible minimum value is 16 million m^3/year, which equals both the value calculated for recharge above 1,000m (Noble 1994) and lowest estimate of historical flow from the springs at Azraq (Gibbs 1993).

The main natural outlets from the basin are the springs at Azraq, the Druze springs fed by water from the north and Azraq Shishan supplied from the south. In the mid 1950s, the discharge from the Druze springs was measured as 3.5 million m^3/year, with 23 million m^3/year from the lower altitude Shishan springs (Parker 1971; WAJ 1986). The spring flows have declined over the last forty years, until the Druze springs ceased to flow at the end of 1987 and the Shishan springs dried up at the end of 1992. The decline of spring flows and the accompanying decline in piezometric head, at rates between 0.35

and 0.6m/year, in the central part of the basin were matched by a continual increase in pumped abstraction since the 1940s. The trends with time are illustrated in Figure 3.

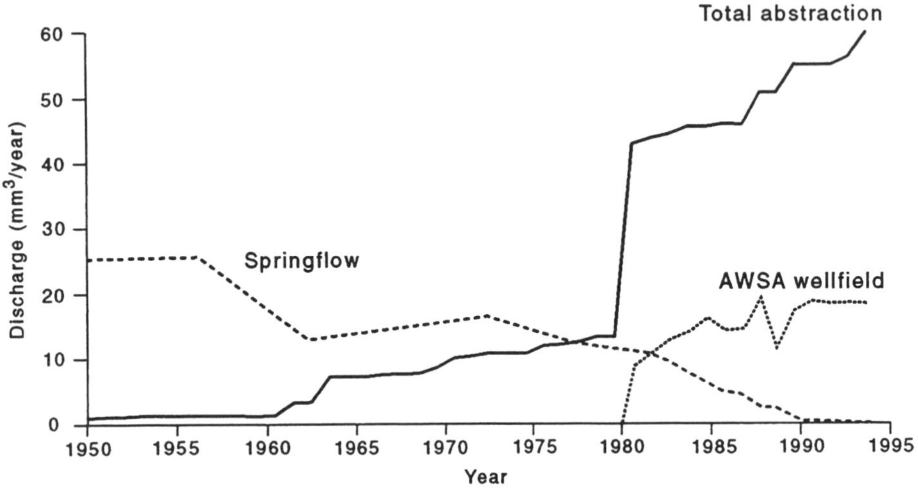

Fig. 3. Changes in abstraction, springflow and groundwater levels with time.

2.2. Surface Water

Although there are regular, short duration floods during winter in the wadis flowing south from Jabal al-Arab, there were no measurements of flow until six new gauges were installed in 1995. Total flows have been estimated from rainfall data to average 27 million m^3/year. It is reported that flood frequency and magnitude have declined markedly in recent years, due to the construction of dams on the upper reaches of the wadis in Syria. A Jordanian dam was constructed on the lower reach of Wadi Rajil in the 1980s, but the reservoir has only filled once, in the winter of 1994/95.

3. CURRENT WATER USE

Groundwater is used for:

- large scale public water supply to the Amman area, and on a smaller scale to Mafraq;
- local small supplies for domestic use and livestock;
- industrial users, with low water demands of only 56,000m^3/year at present;
- intensive irrigation concentrated around Azraq;
- scattered smaller scale irrigation in the north of the project area.

Abstraction rates for all sectors have increased rapidly since 1980. The single largest abstractor (18 million m^3/year since 1988, 19.2 million m^3/year in 1994) is the Amman Water and Sewerage Authority (AWSA) wellfield, north of Azraq. In 1994, only 0.87 million m^3 of this total was used for local supply, with the remainder exported from the area. A further 5.15 million m^3 from wells along the Mafraq-Safawi road was supplied to Mafraq. In the northern part of the basin, local supplies in 1994 accounted for 2.6 million

m^3, 18 per cent of total abstraction, but drilling of new boreholes for local public supply continued in 1994 and 1995. Losses in the distribution system are very high, estimated to be 61 per cent (Waddingham 1994).

Irrigated agriculture, with pumping from over 600 boreholes, is concentrated in the agricultural areas to the east of Azraq and the north of the basin. The abstraction for irrigation in the northern agricultural area was calculated as 6.6 million m^3 in 1994. Although drip irrigation is used, poor design and maintenance results in moderate distribution efficiencies, represented by a measured value of 68 per cent. This suggests that the crops' peak requirements are only just satisfied by the present systems. In the Azraq area, abstraction for irrigation was estimated to be between 25.35 million m^3 (Dottridge 1995) and 32.55 million m^3 (Waddingham 1994), with much lower efficiency, below 50 per cent, due to the coarser soil and continuing use of channel and basin techniques.

Combining all the components leads to estimates of total basin abstraction in 1994 between 58.9 and 65.1 million m^3. These figures are derived from detailed investigation of each activity, are consistent with the larger of the earlier estimates (Gibbs 1993) and are therefore considered to be reliable and realistic. Comparison of recharge and discharge figures shows that the annual abstraction is at least two or three times the quantity replenished.

4. WATER POLICY

In Jordan, any well or borehole abstracting more than $5m^3$/hour requires a licence, but licences allow pumping in perpetuity. Licence conditions are strict, unpopular, difficult to enforce and often flouted. The controls include a minimum distance between wells, typically at least 1km in the Badia, maximum abstraction of $50,000m^3$/year, pump testing supervised by the Water Authority of Jordan (WAJ), installation of a flowmeter, restriction of irrigation to the owner's land by efficient methods and prohibition on the sale of water.

Since 1971, no new licences have been allowed in the central Azraq area; this was later extended to cover most of the northern part of the Azraq basin. In May 1992, a three year ban on the issuing of new well licences throughout Jordan was imposed, with the exception of special circumstances. Direct action to close unlicensed wells followed a government decree in November 1988, issued in response to the 1987 RAMSAR Convention's recommendations on the conservation of the Azraq wetland. The campaign resulted in plugging of 151 wells and 35 boreholes (Jones 1990), but was abandoned due to problems with well owners. The decree formed part of a comprehensive set of recommendations from a Cabinet committee formed to consider Azraq's water situation, which also included restricting total abstraction to 20 million m^3/year, with 14 million m^3 allotted to the Amman area, only 3.5 million m^3 for irrigation and minimum spring discharge of 2.5 million m^3. These figures are in marked contrast with the recent estimates in section 3. Restoration of the Azraq wetland is the subject of a $3 million project, funded by UNDP and due for completion in 1996.

Despite the regulations, there are still large numbers of unlicensed wells and boreholes, with 203 boreholes and 164 dug wells recorded in a census in 1993 (Dottridge 1995). Most of these are in the Azraq area where the water table is shallow and drilling costs low; in the north, the high cost of boreholes deters farmers from drilling without a licence (Waddingham 1994).

The Department of the Environment is responsible for the administration of the RAMSAR Convention and for enforcing environmental protection measures, but the Ministry of Water and Irrigation has responsibility for both regulation and development of water. This dichotomy of interest in the institutional framework leads to lack of implementation of the regulations.

5. DEVELOPMENT OPTIONS

In view of the extent of over-abstraction in the Azraq basin, planning of water use for future development is problematic, and abstraction must be reduced if spring flows are to be restored. In contrast to the central part of the basin, the Badia Programme's project area still has ample supplies of groundwater, recharge exceeds local abstraction and the water quality is excellent. However, additional development of groundwater would contravene government policy and contradict the Programme's aim of sustainable development, by increasing volumes of 'mined' groundwater in the Azraq basin. The Programme has therefore concentrated on maximization of water availability by evaluating alternative sources of water, including surface flows, groundwater from the middle aquifer and the adjoining Sirhan and Hammad basins, artificial recharge of the upper aquifer, reuse of waste water and leakage control.

The most substantial and under-used resource is surface flow, which is capable of providing significant additional supplies through water harvesting. There will be no significant impact downstream because, at present most of the surface water flows onto the fine silty surface of the playas (*qa'as*) and then evaporates. Only a small proportion of the water is used, mostly by natural vegetation on the *marabs* (playas with outflow), which provide pasture, and in a few localities water is diverted and stored for local use. The largest of these systems is at a farm on Wadi Gharbi, to the north-west of Azraq, where the wadi flows are captured by a small dam and pumped into a deep storage reservoir. The extensive fields of olive trees and presence of water in the reservoir in October 1995 demonstrate that surface water harvesting can provide a year round supply of water, providing the winter rainfall is sufficient, but a borehole is used as a reserve supply.

The feasibility and optimum methods of utilizing the surface water resources form the main objectives of the water theme for the second phase of the BRDP (Dottridge 1995). However, groundwater will continue to form the main source of water for the area, due to its reliability and widespread availability.

6. CONCLUSIONS FOR FUTURE RESOURCE MANAGEMENT

When considering use of finite water resources, sustainable development policies cannot be implemented without considering the influences of water users both upstream and downstream of the project area. For the BRDP's pilot area, water planning must consider the whole of the Azraq basin, including the high rainfall area in Syria and the main discharge area centred on Azraq. However, over-abstraction on the basin scale need not prevent future development, provided that alternative sources of water can be utilized. In the Badia area, the surface flows represent an additional water resource, as now, almost all the water is lost through evaporation.

Although some uncertainty remains about the scale of over-abstraction in the Azraq basin, due to lack of precision in values of both recharge and abstraction, there is no doubt about the conclusion that current pumping rates are at least twice the recharge rate. The uncertainty should not therefore be a reason for lack of action to control abstraction.

The Jordanian government has tough policies on licensing and controlling groundwater use, but these are not always enforced when other considerations receive a higher priority. Enforcement of controls and firm resource management are essential to define the strategy and priorities for water use, co-ordinate different groups with conflicting aims and to implement the management plans to safeguard the resource in the long term.

ACKNOWLEDGEMENTS

The views expressed in this paper are those of the authors alone.

The authors acknowledge the contribution of the staff and researchers of the BRDP, and numerous others in the water sector in Jordan. The work was made possible by the co-operation and generous provision of data by WAJ, Ministry of Water and Irrigation, Natural Resources Authority (NRA) and Ministry of Agriculture. Special thanks to the water resources staff of WAJ for their continuing expert assistance. Financial support from the RGS, HCST, Royal Society, Natural Environment Research Council and Royal Jordanian are gratefully acknowledged.

REFERENCES

Core Lab (1987) *Azraq basin study*. Volume IIID, Reservoir Geology Hydrodynamics. For NRA, Amman, Jordan.

Dottridge, J. (1995) Water resources: progress and priorities, Jordan Badia Research and Development Programme.

Drury, D.M. (1993) The hydrochemistry of the Azraq Basin, NE Badia. Unpublished M.Sc. thesis, University College London.

GEF (1993) Arid region nature and natural resources conservation and rehabilitation - Azraq component - Project Document. For Global Environmental Facility UNDP and Government of Jordan.

Gibbs, B. (1993) The hydrogeology of the Azraq Basin, NE Badia, Jordan. Unpublished M.Sc. thesis, University College London.

HCST (1993) The Badia, Rural Development...Investing in Our Common Future. Higher Council for Science and Technology, Amman, Jordan.

Holden, W. (1995) Modelling and isotopc study of the Upper Aquifer of the Azraq Basin. Unpublished M.Sc. thesis, University College London.

Howard Humphreys (1982) *AWSA Azraq wellfield evaluation.*

Jones, T.A. (1990) *Azraq Oasis, Jordan.* RAMSAR Convention Monitoring Procedure Report No 16.

Noble, P. (1994) Quantification of recharge to the Azraq Basin, NE Badia. Jordan, Unpublished M.Sc. thesis, University College London.

NRA (1993) Al Azraq 3353 1. 1:50000 Geological map.

Parker, D.H. (1971) The hydrogeology of the mesozoic-cainozoic aquifers of the western highlands and plateau of East Jordan. Unpublished Ph.D. thesis, University of Nottingham.

RGS (1992) *Framework document*, The Jordan Badia Research and Development Programme 1992-1995 (Stage One: Safawi).

Salzgitter (1994) *Middle Aquifer Exploration Project*, Azraq basin. Final report.

Waddingham, J. (1994) Water demand in the north-east Badia, Jordan. Unpublished M.Sc. thesis, University of Newcastle-upon-Tyne.

WAJ (1986) *Springflow data in Jordan prior to 1985*, Department of Water Resources Development, Technical Paper no. 51. Water Authority of Jordan.

WAJ (1987) *North Jordan Groundwater Project*. Water Authority of Jordan.

WAJ (1989) Azraq basin water resources study, Draft Final Report. Water Authority of Jordan.

CHAPTER 10

Water Demand

JOHN WADDINGHAM
(Environment Agency, Ipswich)

1. INTRODUCTION

Virtually all groundwater abstraction in the Jordan Badia Research and Development Programme area is from the Azraq Basin (Fig. 1), and abstraction from the Azraq Basin is concentrated in the three demand centres: Azraq, Umm al-Quttayn and southern Syria.

Management of water resources is best done at the catchment level. Unfortunately, it was not possible to obtain an estimate of water demand from the Syrian part of the Azraq Basin. However, the results of this study, which concentrated on the area around Umm al-Quttayn, when combined with previous work in the vicinity of Azraq should give an approximation of the total water demand for the Jordanian part of Azraq Basin.

The town of Azraq is in the centre of the basin and historically was built around an oasis formed by the emergence of groundwater spring. Over recent decades the expansion of irrigated agriculture the increase in population and raised standards of living have forced greater groundwater abstraction, resulting in depression of the regional groundwater table (Ch. 4) and the consequent degradation of most of the oasis (GEF 1993). This has concentrated water demand management research on the Azraq area.

Umm al-Quttayn is in the north-western part of the basin and early settlements in the locality obtained their water by storing flash flood runoff in purpose built reservoirs (Al-Ayyash 1993), since at this elevation the groundwater is around 400m beneath ground level. This method of obtaining water has now been replaced by groundwater abstraction from deep boreholes. Although abstraction of water from any part of the basin could have repercussions for the Azraq oasis, this study is the first comprehensive estimation of water demand for the area around Umm al-Quttayn.

Umm al-Quttayn and surrounding villages have seen a dramatic increase in groundwater abstraction over the last decade. Because of the problems at Azraq, farmers have been given financial incentives to practise drip irrigation in the northern part of the basin where the climate is less arid and the effect on water levels beneath the oasis is not so acute. In addition, recent legislation has allowed Syrians to sell land they own on the Jordanian side of the border. This is being bought by Jordanians who have moved out of a nomadic lifestyle due, amongst other reasons, to overgrazing of the Badia and the restrictions of living so close to the border. Through the process of settling down they increase both domestic demand and agricultural demand when they invest the proceeds of selling their flocks in the construction of irrigation boreholes. This increase in groundwater abstraction is set to continue, both for the above reasons and because orchards already planted in the area will gradually require more water as they get older.

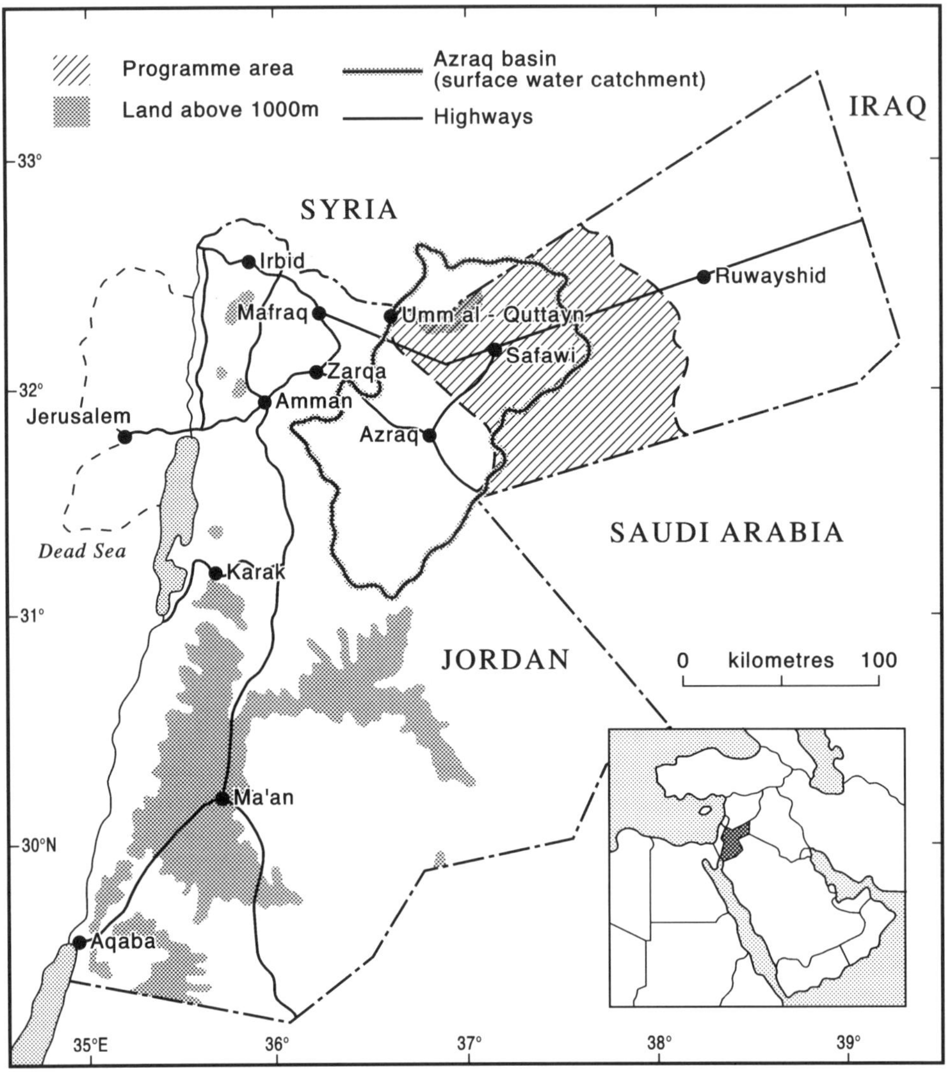

Fig. 1. Location map, showing Umm al-Quttayn.

During the five weeks of fieldwork in the summer of 1994, every farm in the region of Umm al-Quttayn was visited and questions were asked that could enable estimation of water demand throughout the year. Domestic water demand was also investigated. Some visits were made to Azraq and nearby farms in order to correlate this study with previous work.

2. DOMESTIC WATER DEMAND

Jordan is divided into eight governorates, each is responsible for its own water supply. In the Azraq Basin the northern villages in the region between Umm al-Quttayn and

Safawi, fall within the Mafraq governorate while the town of Azraq to the south is part of the Zarqa governorate. The population of the Azraq Basin is low compared to neighbouring areas and most of the water abstracted for municipal supply is exported from the basin. Most significant is the 14-well AWSA wellfield immediately to the north-west of Azraq which since 1986 has been abstracting about 19 million m^3/year, largely to supply the high population density region of Amman to the west. In 1992 the AWSA wellfield supplied 876 thousand m^3 for domestic consumption to the town of Azraq (WAJ 1994). Given that during this year 4,500 people had access to a mains supply in Azraq, the per capita daily demand was 533 litres. This is not an accurate reflection of the actual consumption by each individual, as it is thought that unaccounted losses (e.g. leaking pipes) are significant and also a lot of mains water is used for agricultural purposes (e.g. stock watering and small-scale irrigation).

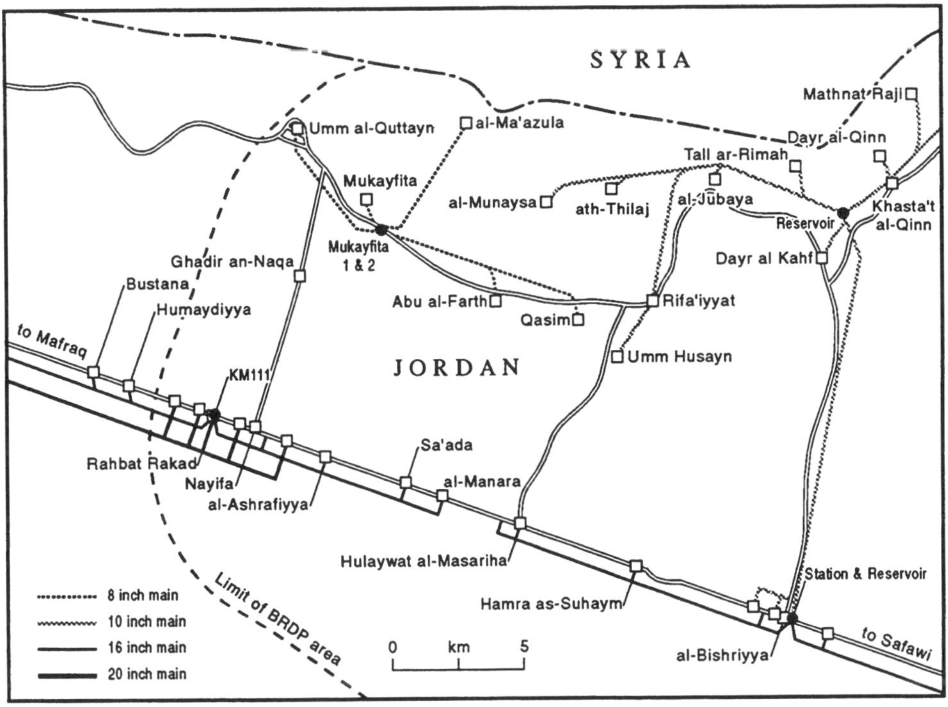

Fig. 2. Water distribution network, north-west Badia Programme area.

The Umm al-Quttayn region is different from Azraq in that it covers a larger area (34 villages), has an older and less sophisticated distribution system (Fig. 2) and the water is abstracted using a widely spaced network of 11 municipal wells. In 1993 the total abstraction was 6.7 million m^3 (WAJ 1994), 71 per cent of which was exported from the region. Given that the population of the region was roughly 20,000, the apparent per capita daily demand was 266 litres. A case study of what was taken to be a typical household showed that the actual per capita daily demand was 115 litres and so unaccounted losses and agricultural purposes are thought to be significant here also. It is likely however that compared to Azraq unaccounted losses play a more important part in domestic water demand than agricultural purposes, given the less arid climate and the older water distribution network in the Umm al-Quttayn region. Unaccounted losses may be as high as 60 per cent.

The regional water authority of Mafraq plan to modernize the Umm al-Quttayn distribution system by drilling more public water supply wells, increasing the mains pressure and frequency of availability of mains water to outlying villages, connecting the remaining 18 per cent of the population and replacing leaking surface-lain pipes with new buried mains. The extra consumption resulting from connecting the remaining population could be accommodated by reducing leakage. However, if a shortage of funds means that leakage is not significantly reduced then these proposals could increase the total annual domestic abstraction from the Umm al-Quttayn region of 2.5 million m^3.

3. IRRIGATION WATER DEMAND

In 1994 there were 26 farm boreholes near the villages east and south of Umm al-Quttayn (Fig. 3). Drip irrigation was the only method of distributing water to the crops.

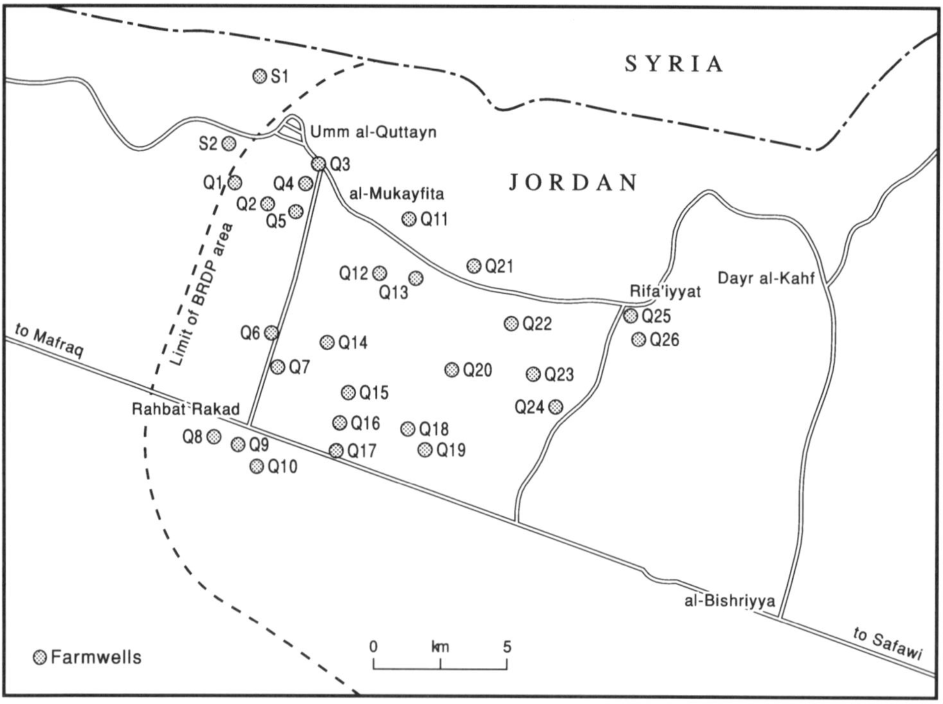

Fig. 3. Location of wells used for drip irrigation.

There were very few working water meters, though it is understood that the Water Authority of Jordan (WAJ) is now stricter on this, so estimates of water consumption had to be made by a variety of direct methods. The farmers were also asked for the total areas of each type of crop and their irrigation scheduling so that an estimate could be made of irrigation water demand throughout the year. It was anticipated that there might be irrigation in excess of the crop water requirement to allow for leaching and poor system design and maintenance. A combination of questioning and field experiments was therefore conducted. These identified the extent of over-watering (Fig. 4) and the reasons for it.

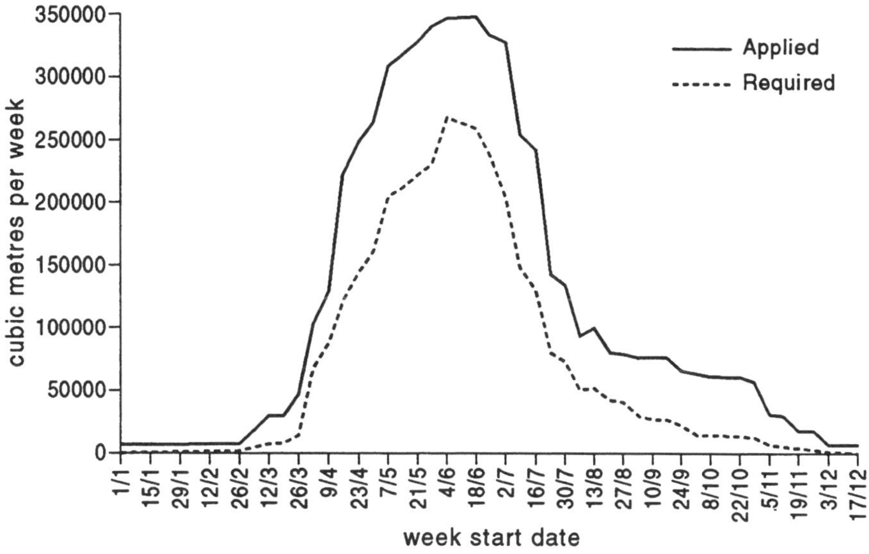

Fig. 4. Applied and required irrigation rates.

The total area under irrigation in the Umm al-Quttayn region was 482 hectares. Some 70 per cent of this area was split evenly between tomatoes and water melons, 17 per cent was fruit trees and the remaining 13 per cent in decreasing area included cauliflower, yellow melons, olives, grapes, cabbages, peppers, dwarf beans, eggplants, potatoes and peas. The projected amount of water abstracted during 1994 for irrigation in the Umm al-Quttayn region was 6.3 million m^3. This is similar to that abstracted for public water supply, although none of the water abstracted for irrigation was exported from the basin. A third of the total amount applied for irrigation could be expected to pass into the unsaturated zone, though given the great depth of the water table here it is unlikely that this would benefit the available resource.

The total crop water requirement and leaching requirement for the Umm al-Quttayn region was 3.8 million m^3 assuming an optimal efficiency of 0.9, which is feasible for drip irrigation systems (James 1988). Investigations revealed that the over-watering was mainly to compensate for poor system design and poor maintenance of the older systems (resulting in poor emission uniformity) rather than overzealous scheduling.

The relatively low technology surface irrigation systems around Azraq were found to be working much closer to their optimal efficiency and so although drip irrigation systems should theoretically be more water efficient their use was hardly saving any water and probably not justifying their extra cost.

The water near Azraq has a much higher salinity than in the region of Umm al-Quttayn, probably due to the dissolution of intraformational evaporites within the limestone aquifer, therefore the leaching requirement is higher. Mature orchards predominate and their presence is an indication that the leaching requirement is being met. By incorporating data from a 1992/93 survey by WAJ the total agricultural abstraction on the farms near Azraq in 1993 could have been 32.6 million m^3, almost twice that abstracted for public water supply. Half of the total amount applied could be expected to return to the aquifer.

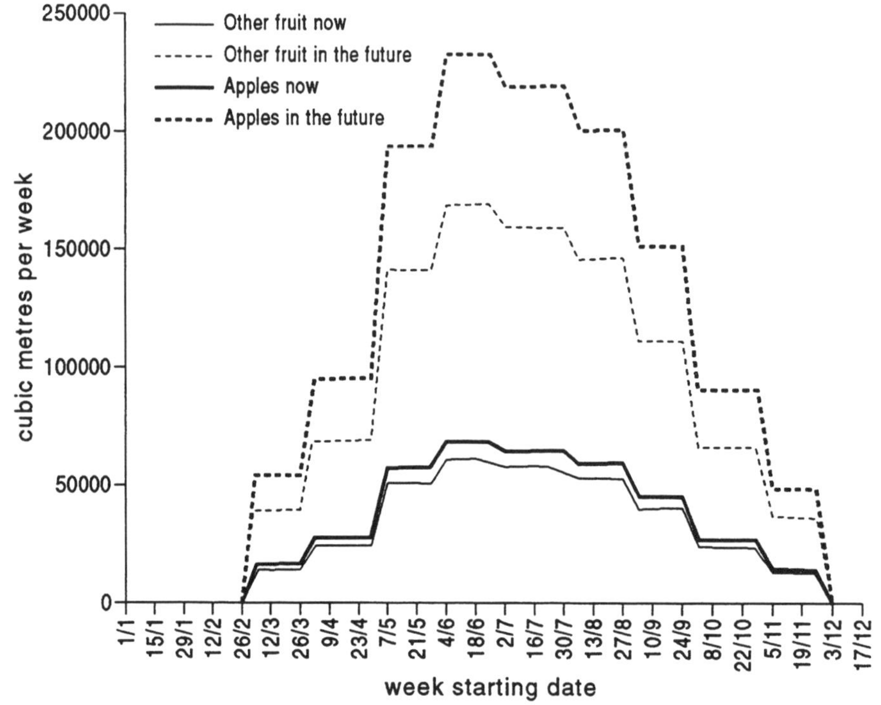

Fig. 5. Potential growth in demand for irrigation.

It is now illegal to drill any new wells in the Azraq Basin and though this rule might be an effective deterrent in the region of Umm al-Quttayn, where deep boreholes are expensive, it is unlikely to stop the digging of new wells near Azraq where the water table is shallow. Many farmers expressed an intention to expand their farms by planting orchards and this could increase annual irrigation water demand around Umm al-Quttayn by 650,000 m^3 at first, rising to 2 million m^3 when the trees are 10 years old (Fig. 5).

4. CONCLUSIONS

Almost as much water is lost from the local domestic distribution networks as the volume that is over-applied during irrigation. Investment is needed to reduce leaks and to enforce licence conditions for private abstractors. Follow-up advice for farmers who have invested in complex irrigation equipment should be readily available to ensure that the systems remain water efficient. The management of water resources needs to be co-ordinated at the catchment level and cost-benefit analysis should be carried out in order to distribute the scarce resource efficiently between competing needs.

ACKNOWLEDGEMENTS

I would like to thank the Royal Geographical Society, CORD at Durham University and the Higher Council in Jordan for creating this opportunity for field research. The fieldwork would not have been possible without the logistic support provided by the Centre at Safawi and the local knowledge, translating skills and enthusiasm of the young Jordanian research staff based there. I would also like to thank my supervisor at Newcastle University, John Gowing, for his interest and advice.

REFERENCES

Al-Ayyash, S.M. (1993) A development of water harvesting systems at Safawi area in northern Jordan Badia, Unpublished M.Sc. thesis, Jordan University of Science and Technology.

GEF (1993) Conservation of the Azraq oasis. Unpublished project document of the arid region nature and natural resources conservation and rehabilitation project - Azraq component. Project of the Government of the Hashemite Kingdom of Jordan, UNDP.

James, L.G. (1988) *Principles of Farm Irrigation Design*, Wiley, New York.

WAJ (1994) Untitled and unpublished report on the water resources of the Azraq basin by Ra'ed Daoud. Draft internal report prepared for the updated national water master plan, Water Authority of Jordan, Amman, Jordan.

CHAPTER 11

The Effect of Intensive Irrigated Agriculture upon Soil Degradation: A Case Study from Ashrafiyya

ALASTAIR KIRK
(University of Durham)

1. INTRODUCTION

The north-eastern area of the Badia includes the 'pan handle' of Jordan which separates Syria from Saudi Arabia and the land which stretches westwards to the major town of Mafraq. The area, *the Harra*, is distinct from the rest of the Badia further south primarily because of its geology, but also due to its surface and groundwater characteristics. Much of the north-eastern Badia lies on a tertiary basalt pavement which gives rise to a very specific geographical environment. The area is a complicated collage of basalt flows which have, at different times, been extruded from several series of volcanic cones lying along large fault and dyke sequences (Ibrahim 1993). Basalt boulders of different sizes currently dominate the ground surface. The age and mineralogy of the basalts are the prime factors influencing clast size, drainage density, wadi morphologies and slope angles (Allison and Higgitt in press). In addition, it is this basalt plateau which controls the groundwater resources and the Jabal al-Arab, which is almost entirely derived from the basalt, is vitally important in determining the rainfall and subsequently the surface hydrology of the surrounding footslopes.

During various periods of history this land has been populated by a mixture of both nomadic and sedentarized peoples. Their livelihoods have been, and to a large extent still are, based upon speculative farming and herding. However, during the last two decades, there has been a moderate but perceptible change towards intensive agriculture. Such organized agriculture has probably not been seen since the Roman periods of occupation fourteen hundred years ago. It is important to monitor this growth and identify possible detrimental effects it could be having upon the fragile physical resource base.

2. PATTERNS OF FARMING IN THE PAST

The north-western part of the Badia Programme area, which is the main focus for the research carried out, lies close to the Syrian border and marks the approximate eastern limit of both current irrigated agriculture and settled population. Evidence for human groups has been found as early as the Neolithic and indicates that a highly mobile and diversified nomadic population lived in north-eastern Jordan (Homès-Fredericq and Hennessy 1989). One of the oldest known walled cities in the world, Jawa, is situated on

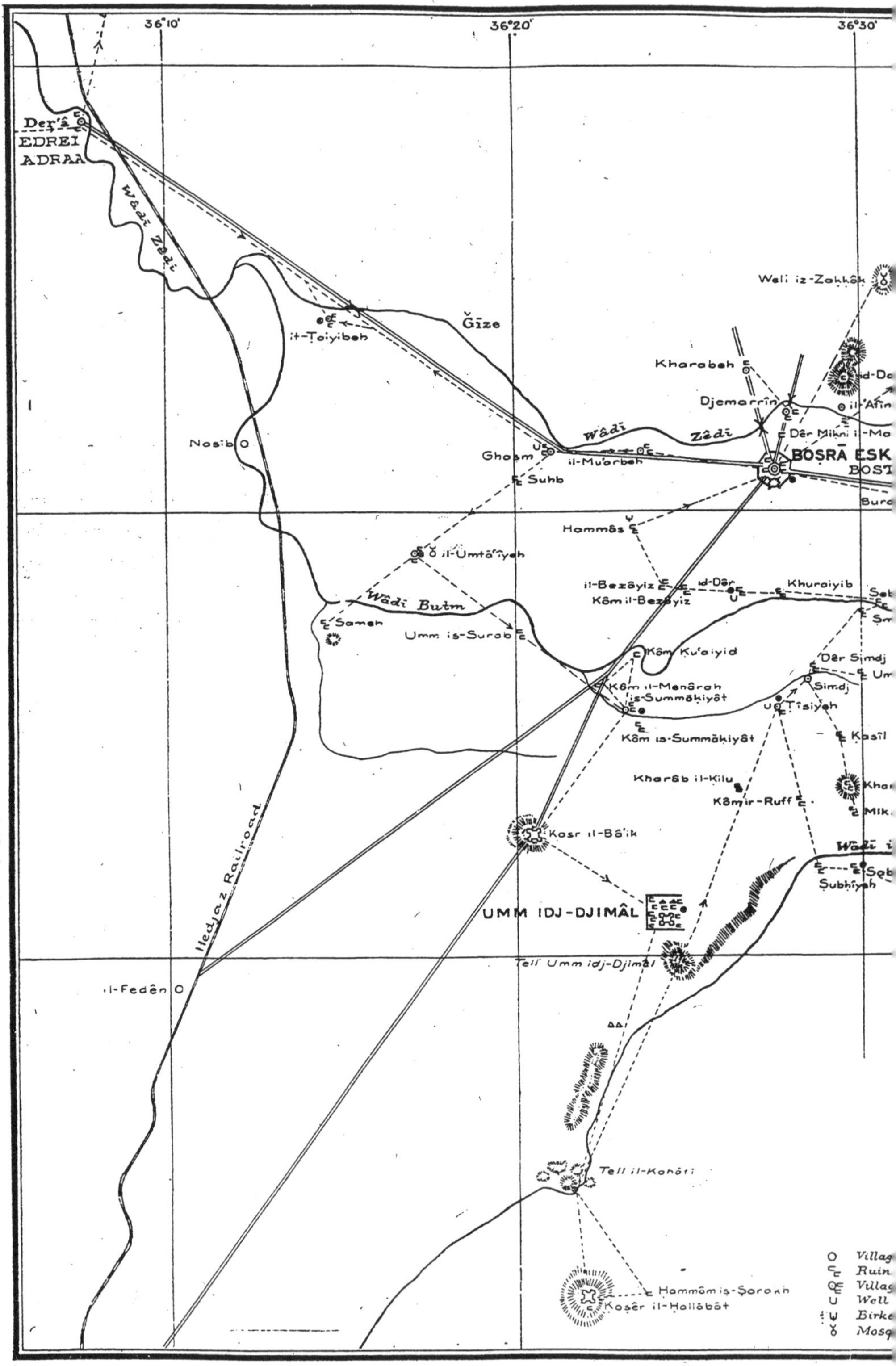

Source: Butler (1919)

Fig. 1. The Southern Haurân.

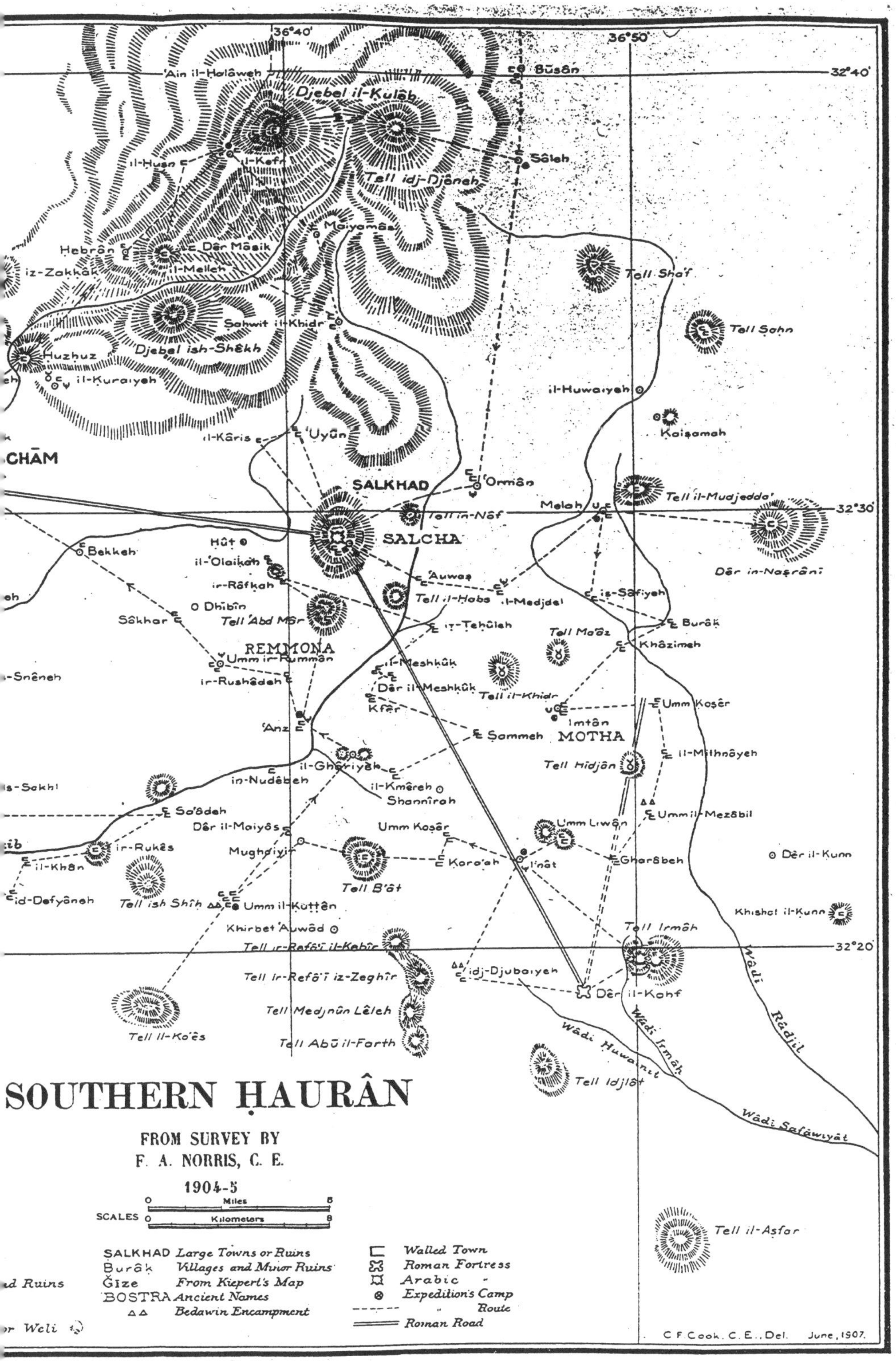
36°40'
36°50'
32°40'
32°30'
32°20'
'Ain il-Halâweh
Djebel il-Kulêb
Bûsân
Sâleh
il-Husn
il-Kefr
Tell idj-Djâneh
Maiyamâs
Hebrân
Dêr Mâsik
iz-Zakkâk
il-Melleh
Tell Sha'f
Tell Sahn
Sahwit il-Khidr
Djebel ish-Shêkh
Huzhuz
il-Kuraiyeh
il-Huwaiyeh
Kaişamah
il-Kâris
'Uyûn
CHĀM
SALKHAD
'Ormân
Tell il-Mudjedda'
Melah
Tell in-Nôf
SALCHA
Hût
Bekkeh
il-'Olaikah
Dêr in-Naşrânî
ir-Râfkah
'Auwaş
Tell il-Habs
il-Medjdel
is-Sâfiyeh
Dhibîn
Sâkhar
Tell 'Abd Mâr
it-Tehûleh
Burâk
Tell Ma'âz
Khâzimeh
REMMONA
Umm ir-Rummân
il-Meshkûk
is-Snêneh
ir-Rushêdeh
Dêr il-Meshkûk
Tell il-Khidr
Kfêr
Umm Koşêr
Imtân
'Anz
Şammeh
MOTHA
il-Mithnâyeh
il-Gharîyeh
Tell Hidjân
is-Sakhl
in-Nudêbeh
il-Kmêreh
Shannîrah
Sa'âdeh
Dêr il-Maiyâs
Umm Koşêr
Umm Liwân
Umm il-Mezâbil
ir-Rukês
Mughaiyir
Koraîeh
il-'Anât
Gharâbeh
Dêr il-Kunn
il-Khân
Tell B'ât
id-Defyâneh
Tell ish Shîh
Umm il-Kuttên
Khishat il-Kunn
Khirbet 'Auwâd
Tell Irmâh
Tell ir-Refâ'î il-Kebîr
Tell ir-Refâ'î iz-Zeghîr
idj-Djubaiyeh
Dêr il-Kahf
Tell Medjnûn Lêleh
Tell il-Ko'ês
Tell Abû il-Forth
Wâdi Huwânît
Wâdi Irmâh
Wâdi Râdjil
Tell Idjlât
Wâdi Safâwiyât
SOUTHERN ḤAURÂN
FROM SURVEY BY
F. A. NORRIS, C. E.
1904-5
Miles
Kilometers
SCALES
SALKHAD Large Towns or Ruins
Burâk Villages and Minor Ruins
Gize From Kiepert's Map
BOSTRA Ancient Names
Bedawin Encampment
Walled Town
Roman Fortress
Arabic "
Expedition's Camp
" Route
Roman Road
Tell il-Asfar
C.F. Cook. C.E., Del. June, 1907.

an island in the Wadi Rajil (Helms 1981; Betts 1991) on the eastern slopes of the Jabal Haurân, better known today as Jabal al-Arab. It is here that water catchment and harvesting schemes dating back to 3200 BC have been discovered. Moving on in history, the Holy Bible talks of the Giant cities of Bashan, the capital of which was the large underground city of Edrei (Num. 21: 33; Deut. 3; Josh. 13: 31), close to modern-day Der'a in Syria.

The greatest period of building was that of the Byzantine, stretching from the second quarter of the fourth century until the seventh (see Fig. 1). The Romans had always been keen to protect their realm and road networks from Bedouin attack (Issac 1992) and so the regular scattering of forts such as the one at Dayr al Kahf is not so surprising. During this time a considerable population was living in the northern Badia if estimates of cities such as Umm al-Jimal having a population 10,000 are close to the truth. Intensive farming must have been in operation during the period in order to service such a prosperous community (Lankester Harding 1959). It is clear from archaeological studies that water harvesting systems were in force to bring water into the settlements (Butler 1919) and therefore it is possible to propose that there was some irrigated agriculture in progress at that time. The decline of the region during the eighth century had little to do with the degradation of the physical resources due to overpopulation in a fragile environment, but because of the catastrophic nature of an earthquake which hit the region in AD 747 (De Vries 1981).

Butler's expedition early in this century came to the conclusion that there had been a considerable change in climate over the last fifteen hundred years. He says in his description of Dayr al-Kahf:

> "It is quite certain that the region was not desert in ancient times; for all the way from I^c nât there are remains of ancient walls that divided fields and the traces of furrows still remain as evidence of former cultivation. These signs of former fertility extend farther to the south and south-east, as far as the eye can reach" (Butler 1919, 145)

Butzer (1955) confirms that there were periods between AD 1-180, AD 390-415 and AD 670-925 when the whole of Jordan was wetter than it is today, and then agriculture was widely practised as far east as Dayr al-Kahf. The basalt covered ground would have been cleared and the blocks used to enclose fields around the various settlements. The removal of the basalt blocks would have reduced infiltration, encouraged the erosion of fine material by the wind and as a consequence, reduced the fertility of the top soil. In conclusion, the modern geomorphological landscape and pedological condition has been sculpted by anthropogenic behaviour, not of the recent past, but of the ancient past.

3. CONTEMPORARY FARMING: THE MOVEMENT TOWARDS HIGHLY INTENSIVE IRRIGATED AGRICULTURE

The majority of today's farming, in the north-eastern Badia is small-scale speculative rain-fed growing of wheat and barley and dominated by pastoral keeping of livestock, especially sheep and goats. In the north, around villages like Jubaya, Dayr al-Kahf and al-Rifa'iyyat, there is often enough rainfall for there to be a reasonable crop, but further south the risks of rain-fed farming increase. One factor which has made rain-fed agriculture less viable in recent years is the construction of Syrian dams on the Jabal al-

Arab. Formerly, there was regular wadi flow down from the mountains and through the foothills of the southern Haurân, but now the only flow in the wadis is generated by the much lower rainfall totals which fall in the Jordanian foothills. However, in the last decade, irrigation agriculture has been encouraged by loans from the government to drill private wells. As the population has settled, land cleared by earlier generations has been utilized and as the new generation of entrepreneurial farmers has moved eastward, new land has been brought into use by clearing the basalt. Boulder removal is almost complete near Mafraq and gradually reduces until it fades out about 65km east of Mafraq.

The choosing of Low Farm in the village of Ashrafiyya as a site of research was because it represents the approximate current eastern fringe of irrigated agriculture over the area as a whole. The farm was given a loan to drill a deep well which reaches down into the upper basalt aquifer, lying 292m below the surface (Drury 1993). Such a figure represents a considerable financial investment, in the region of JD60,000. The rate of water extraction is approximately $60m^3$/second which is fed onto the fields using a series of pipes. Irrigation lines called *laterals* draw off the main pipes and submerge under plastic sheeting. Water is delivered to the plants by means of drippers or emitters at a rate of between 2-8l/h. The emitters are placed at selected intervals (approximately 0.5-1m) along the irrigation laterals depending on the planting distance, the type of crop to be irrigated and the type of soil (Goldberg *et al.* 1976). The soil in the immediate vicinity becomes saturated during irrigation, but the moisture level drops away from the emitter.

Irrigation begins in April and continues throughout the summer until October and the usual crops are tomatoes, chillies, okra and water melon. Other methods of irrigation are also used to irrigate forage crops and fruit trees, but these are less abundant because of the lower short-term return that they give to the farmer. At the end of the season, the plastic and the irrigation pipes are all rolled up and stored and the plant residue is left as forage until the farmer decides to plough and plant wheat, barley, shallots or cauliflowers as rain-fed crops for the winter season. Once the winter crops have been gathered in the following March, the fields are left bare while another set of fields on the other side of the well are prepared for the next year's irrigation. A three or four-year rotation is practised so that after three years of being fallow a field is once again ploughed and prepared for another season of irrigated crops. In preparation for irrigation, the land is cleared of basalt if it is the first year of cultivation and ploughed before the winter rains. The land is then left for four or five months, during which time a crust will have formed. In late March or early February the soil is ploughed once again and then rolled before laying the irrigation pipes and planting. Urea fertilizer is then applied in strips parallel to the irrigation lines, which are spaced at two metre intervals, and covered in black plastic. Soil is then heaped on to the edge of the plastic from the intervening area which causes the morphology of the ground surface to take on a ridge and furrow effect (see Fig. 4).

4. PEDOLOGICAL DEVELOPMENT IN THE NORTH-EASTERN BADIA

Although it is suggested that the soils of the northern Badia are almost entirely formed as a result of the weathering out of the basalt (Al-Homoud *et al.* 1995; Ch. 2), a small amount of analysis can show that such a view is untenable. From elementary thin section analysis, it can be seen that the fabric of the soil is dominated by a fine grained quartz matrix which could not have been derived from the basalt. Research into the origin of soils in the Negev and Sinai has shown that much of the land surface is covered in a ca. 14m layer of loess called the Neviot Series which is thought to be pluvio-aeolian in origin. The composition of the sediment is similar to that of contemporary dust carried by the Arabian Hamseen winds during early and late summer. There are alternating silty and clayey bands which Bruins and Yaalon (1979) have attributed to moister and drier conditions in the Upper Pleistocene and Holocene. These differing sedimentological layers have probably been produced because clay is more likely to be deposited during pluvial conditions (Ganor 1975). For the soils of eastern Jordan, where the Hamseen winds also occur, it is much more likely that the source of much of the mineralogical matrix is aeolian.

The analysis which follows has several purposes: first, it seeks to elucidate the way in which the present soil has evolved in terms of the two main processes of *in situ* weathering and aeolian processes. The rest of the chapter picks up these questions and tries to answer them at different scales. Soils which have recently been brought into agricultural usage are compared with their state before. The whole question of the scale of erosion and depositional processes within a hillslope that is undergoing agricultural activity is tackled.

5. FINGERPRINTING LOCAL BASALT SIGNATURES TO COMPARE WITH THE SOIL

In order to understand the origin of the soil, it is important to find some way of analysing it. X-ray fluorescence (XRF) and X-ray defraction (XRD) are methods by which the mineral chemistry and refractive properties of a material can be examined. By comparing the signature of the natural basalt rock with that of the soil it can be seen how or if they are related.

The <5μm fraction of powdered basaltic glass material taken from Jabal Aritayn was examined using X-ray fluorescence (XRF) and X-ray defraction (XRD) by Dwairi (1987) who observed two main clay peaks at 6° 2θ and 12.3° 2θ. The first of these peaks corresponded with the diagnostic d-spacing of 17Å for smectite clay which was confirmed by its collapse after heating. The other peak was initially identified as phillipsite, but after heating it was considered to be kaolinite. Dwairi analysed the mineralogy of samples taken from Aritayn, which make up the most recent volcanic facies in the Harrat Ash Shaam basalts and found that the most predominant clay was smectite (see Table 1).

In both marine and non-marine palagonitic environments, it has been suggested that the formation of smectite clays is in fact an essential pre-requisite for phillipsite crystallization (Arrhenius and Bonatti 1965; Hay 1966; Kastner 1976). Similarly in zeolite terrains, it has been found that magnesium-rich clays, such as smectite, saponite, chlorite etc., precede the formation of zeolite in the alteration sequence.

The XRD signatures from samples taken at Ashrafiyya suggest some interesting similarities with those obtained by Dwairi (1987). The kaolinitic peak at 12.3° 2θ is not particularly surprising although in both cases the peak is relatively small which is concomitant with the low energy weathering environment found in the Jordanian Badia. The smectite peak, which Dwairi describes as an extended peak, is very similar in nature to a mixed-layered clay comprising chlorite and illite-type clays found in the soil. It would be foolish to make too much of this apparent correlation because palagonitic alteration would have only taken place close to the cones, however, there are grounds to propose that minerals and their associated authigenic counterparts have had a role to play in soil formation in the region.

Table 1. Occurrence of minerals after the alteration of basaltic glass

Site	pyroxene	olivine	phillipsite	chabazite	faujasite	smectite	illite	kaolinite	quartz
S-3	x	x	x	tr	-	x	-	x	x
S-11	x	x	x	-	x	x	tr	x	x
S-16	x	tr	x	tr	tr	x	x	tr	tr
S-17b	x	x	x	-	tr	x	tr	tr	tr
S-26	x	-	x	-	x	x	tr	x	x
S-27	x	tr	x	-	x	x	x	tr	x
S-30	x	tr	x	-	tr	x	tr	x	x
S-46	x	tr	x	-	-	x	tr	x	x
S-54	x	tr	x	-	-	x	tr	x	x
S-57	x	x	x	tr	tr	x	tr	tr	x

adapted from Dwairi (1987) x: Occurrence of minerals tr: Trace -: None found

6. X-RAY EXAMINATION OF SOIL AND ROCK MATERIALS AT LOW FARM

The samples of basalt, weathered basaltic material and the soil were tested to understand the mineralogical make-up of the three materials and compared in order to elucidate the evolution of the soil from the basalt; the results are given in Table 2.

The basalt exhibits a regular chemical composition although the total is less than would be expected suggesting about 4 per cent of CO_2 and 3 per cent structural and adsorbed water. The differences between the basalts can be explained by the fact that they are derived from different flows and that the smaller percentages of sodium, potassium and titanium relate to the acidity of the basalt. The low proportion of magnesium oxides and higher aluminium oxides suggest a basalt which is high in aluminosilicates indicating a relatively late evolution. The high levels of calcium correlate well with the high strontium levels and come from two possible sources. First, basalts can contain much calcium-rich alumino-silicate plagioclase feldspar and this is shown by the large number of peaks of anorthite in the XRD especially at 23.8° 2θ and between 27.6 and 28.1° 2θ. Secondly, it may be taken up as calcium carbonate acting as an impurity in the form of veins. In addition, the XRD results in Table 3 show that there is no quartz and that the other significant mineral present is diopside, a pyroxene made up of a magnesium-calcium silicate, which occurs around 30 and 35.5° 2θ.

Table 2. X-ray fluorescence data for Low Farm

Major oxides	Basalt (%)	Basalt (%) †	Weathered basalt (%)	Soil (LF3) (%)
SiO_2	40.02	44.6	18.49	43.90
Al_2O_3	15.16	16	5.56	13.59
Fe_2O_3	12.25	11.2	2.19	6.67
MgO	6.04	4.6	1.91	4.40
CaO	11.76	11.0	40.86	11.46
Na_2O	2.44	3.9	0.31	0.74
K_2O	1.153	2.1	1.040	1.972
TiO_2	2.309	3.6	0.293	1.188
MnO	0.214	0.1	0.051	0.180
P_2O_5	0.655	0.5	0.282	0.263
CO_2 estimate	4		25	
H_20 estimate	3		4	
Total	92.00%	97.6%	70.99%	83.18%
Trace elements(in ppm)				
Nb	44.1		7.3	21.6
Zr	182.5		75.5	340.1
Y	21.2		14.3	31.7
Sr	654.8		219.2	246.1
Th	5.8		2.3	5.7
Pb	8.5		13.6	15.9
Ba	21.4		8.4	13.8
Zn	96.6		43.4	76.1
Cu	56.3		14.3	18.4
Ni	191.8		19.5	43.7
Cr	229.3		14.5	122.2
V	205.4		61.8	149.1
Rb	11.1		22.3	45.3
Ba	459.2		108.7	429.3
Ce	31.9		17.4	39.4
Nd	27.3		27.5	32.2
La	17.9		25.8	30.9

† Basalt XRF data from Dwairi (1987)

A cursory glance at the weathered basalt data, which was material taken from nodules within the basalt, would suggest large differences in the mineralogy from that of the parent basalt. The large deficit in the total means that there is a large percentage of CO_2 assumed to be linked with the large calcium value creating over 73 per cent calcite which is evident in the XRD. The XRD suggests that there is about 10 per cent quartz leaving approximately 8 per cent for clay minerals. The fact that the strontium value has decreased significantly despite the increase in calcium indicates that the calcium is being derived from elsewhere other than the basalt and hence diluting the strontium. On much of the basalt in the Badia there are often significant deposits of caleche on the underside of the blocks. This may suggest two possibilities: on the one hand that the large amounts of calcium within the basalt are relatively mobile and that upon chemical weathering the basalt releases the calcium quickly, and on the other hand, it can be postulated that the calcium is derived from elsewhere. The basalt overlies large formations of limestone and it may outcrop occasionally over the Jabal al-Arab. The evidence from the soil thin sections, in the form of calcium carbonate nodules, would substantiate the implication that there is a limestone influence. Maybe in an environment which has a very high evaporative flux for most of the year, groundwater is forced to the surface through capillary action and that as it passes through bands of limestone the water is heavily charged with calcium carbonate.

From the XRF data for the soil sample, it can be seen that there has been a dramatic increase in SiO_2, but that the iron, magnesium and titanium oxides have decreased substantially. The combination of a low total value and a CaO content of 11.46 per cent give rise to an estimate of 25 per cent calcite. The rest of the total value can be explained by the presence of a greater volume of structural and adsorbed water within the soil fabric. It is evident that the influence of the basalt fraction has been vastly reduced; the anorthite and diopside have been weathered to form some of the clay minerals and are otherwise virtually unrecognizable. The dominant mineral is now quartz which almost completely dominates the silt fraction. The quartz has obviously been incorporated into the soil profile from elsewhere because there is no source in the vicinity as neither basalt of limestone has quartz as a weathering product. It is likely that it is derived from an aeolian process probably originating in the Arabian Peninsula if the evidence of Bruins and Yaalon (1979) and Ganor (1975) is taken into account.

In terms of the clay content found in the soil sample (see Table 3), the XRD results show that about 15 per cent of the clay is made up of kaolinite confirmed by the peak at 12.4°2θ which is removed upon heating. In addition there is evidence for a somewhat diluted peak between 8 and 9°2θ which corresponds to approximately 30 per cent illite mica clays which can be seen in the XRD graph. The majority of the clays (about 55 per cent) can only be classified as mixed-layered clays randomly orientated and containing units of chlorite and illite. There is a possibility that the mixed-layered clay contains some smectite, because there is a peak between 6.1 and 6.3°2θ corresponding to the Dwairi results. However, there are no clear signs of change upon glycolation giving rise to the idea that there is a majority of chlorite and illite. Whatever the exact composition of the mixed-layered clays, it is likely that they do exhibit a swelling nature similar to that of the smectites because they are very fine grained crystalline amorphous clays.

Table 3. X-ray diffraction data for Low Farm

Mineralogy	Basalt %	Weathered basalt %	Soil (LF3) %
Quartz	0	10	45
Calcite	0	75	25
Clay	0	10	30
Anorthite	60	0	0
Diopside	40	0	0

7. SEDIMENT DYNAMICS AT DIFFERENT SPATIAL SCALES

It is clear that the soil is complex in nature and is related to the underlying basalt, but also is heavily influenced by deposition of fine particles of aeolian origin. Wind erosion is one of the most important variables which determine sediment movement in the contemporary landscape. Fine material is constantly being deflated from the upper layers of the soil profile, especially during and after ploughing when there is much human disturbance and an absence of a seal to reduce erosion and protect the soil.

In order to investigate these processes in the field, the field site was chosen at a farm in Ashrafiyya. The fields were surveyed using a *Leica TC1010* Electronic Distance Measurer (EDM) in order to construct an accurate digital elevation model (DEM) (Fig. 2) of the farm so that slopes could be measured and to allow small-scale changes in topography to be taken into account which were not noticeable to the human eye. The sampling took place along the transect a-a', normal to the ridge and furrow sequence and allowed comparison between cropped and uncropped areas as well as giving a downslope component to the study.

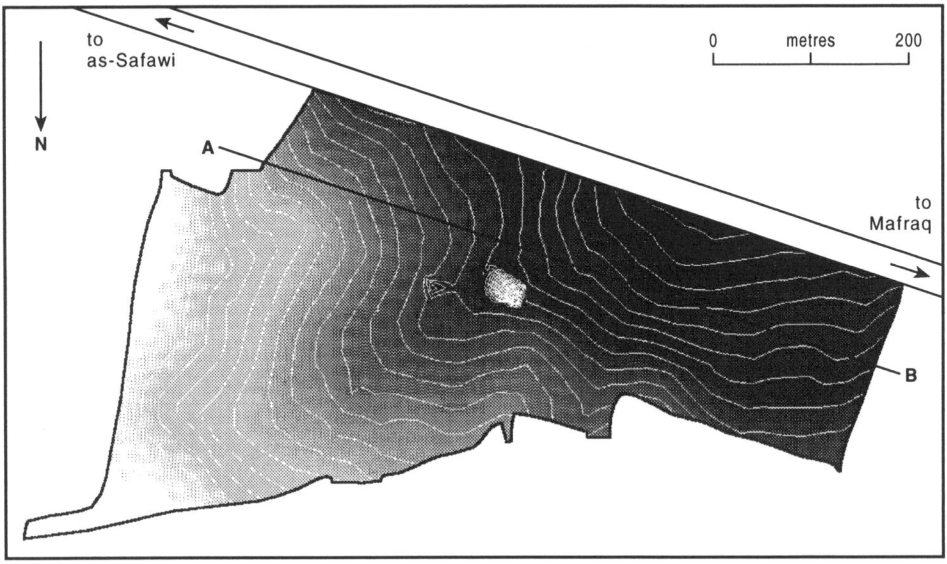

contours at 0.5m intervals

Fig. 2. Digital Elevation Model of the farm at Ashrafiyya constructed using Tins in ARC-INFO.

8. AN INTRODUCTION TO THE SAMPLING STRATEGY EMPLOYED

The methodology of sample collection seeks to embrace changes in soil characteristics at different temporal and spatial scales. Frequently in geomorphological investigation a deductive approach is taken too far in that only one unit of the land surface is studied to the detriment of others which actually have important influences upon the system under study. In the landscape of the Badia, changes in soil characteristics are often subtle, due to the very low relief, and therefore it is important to take a more inductive systematic approach. With little prior knowledge of the landscape and the processes acting within it, it is important to sample in such a way that looks at large-scale changes occurring over hundreds of metres of hillslope, taking in areas that are cultivated and non-cultivated. This was done by sampling down a transect at 50m intervals at the farm. Such an approach if taken in isolation can mean that small-scale changes actually mask changes at the larger scale. In order to find out what was happening at the two scales, it was seen to be essential to focus on changes occurring within a furrow to investigate the relationships between local and regional signals of sediment transfer. This meant that for every point on the large-scale transect, between three and six additional samples were taken depending on their position within the ridge-furrow sequence.

In addition to the spatial sampling, there was an in-built temporal framework. First, it can be postulated that by looking at cropped and uncropped areas there is a degree of temporal change which can be observed albeit in a different landscape position. Work was carried out to compare how previously uncropped land had changed once it had been brought into cultivation, by comparing areas over two to three years. Lastly, the spatial sampling strategy was repeated three times in different years to try to obtain information on changes over time, especially at a ridge-furrow scale.

9. PROCESSES ACTING WITHIN THE SCALE OF THE HILLSLOPE

It is important to look at the sediment fluxes at different scales as they change at different rates over various temporal scales. These attributes are controlled by the climatic variables as well as the soil characteristics, but underpinning the natural factors, the attributes are constrained by the anthropogenic activity active within the system. It is important to consider the changes in soil chemistry and soil physical attributes within the framework of irrigation management, crop rotation and cultivation practices.

From the data from Low Farm showing downslope changes in each size fraction, there are no significant trends within the data which are immediately clear, except to say that there is more variation in the crust samples in comparison with the auger samples and that there is more variability in the furrow crust samples than the crust ridge samples. This fact is perhaps not surprising considering that the average slope angle is about 1°. Neither of these conclusions gives a consistent view of trends downslope as they result from processes acting at a smaller scale. It would be expected that only one year after ploughing there would be very little change in the subsurface soil structure and this is born out by the stability in the data for furrow auger samples at Low Farm. The anomaly at the base of the transect of the same data for the ridge auger samples probably corresponds to the wadi-bed conditions at the base of the slope.

When the individual size fractions are displayed as part of a larger size fraction, it is possible to identify changes downslope more readily as can be seen in Figure 3a-d. As the coarse silt fraction is the dominant fraction, it is treated on its own in order to compare it with the fines which have been defined as material below 20μm and the sand which lies above 63μm. The auger samples retain their characteristics downslope and there is no significant change. The furrow crust (FC) shows some variation, but it is not slope-related and can therefore be accounted for by local differences in the depositional regime. The ridge crust (RC) shows the only distinct reduction in fines and increase in sand along the transect. The RC is the only part of the ridge and furrow sequence which is in any way exposed and therefore it is possible to postulate that the removal of fines and an increase in sand is a natural buffering mechanism. Although there is no visual differences in the FC, Table 4 shows that both the RC and the FC have lost on average 16 per cent of the fines and increased a similar amount in the sand fraction when compared with the auger samples.

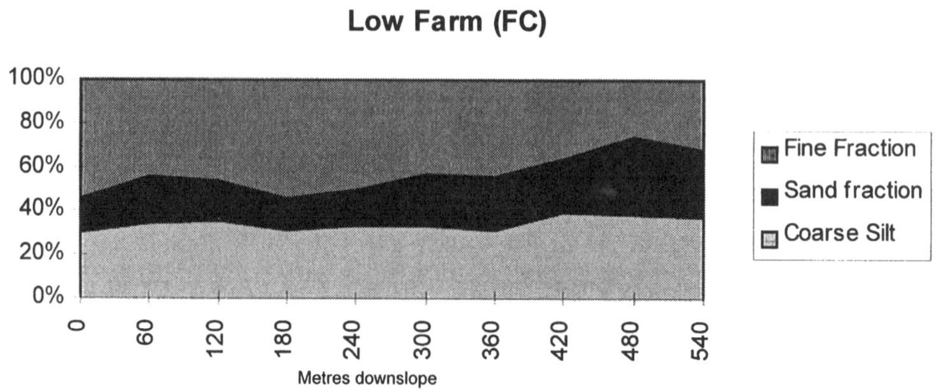

Fig. 3a. Downslope changes in the three major particle size fractions FC = Furrow Crust.

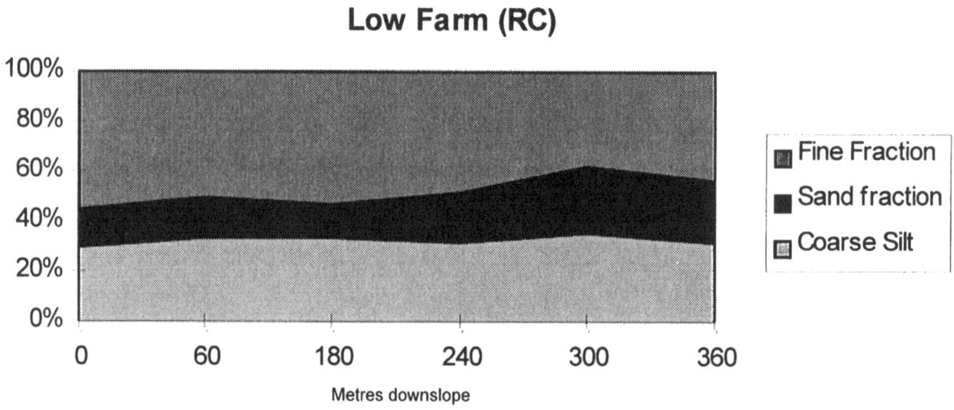

Fig. 3b. Downslope changes in the three major particle size fractions RC = Ridge Crust.

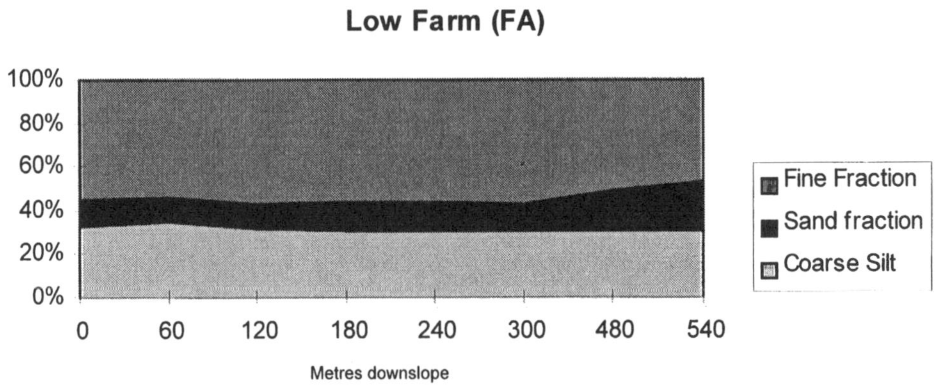

Fig. 3c. Downslope changes in the three major particle size fractions FA = Furrow Auger.

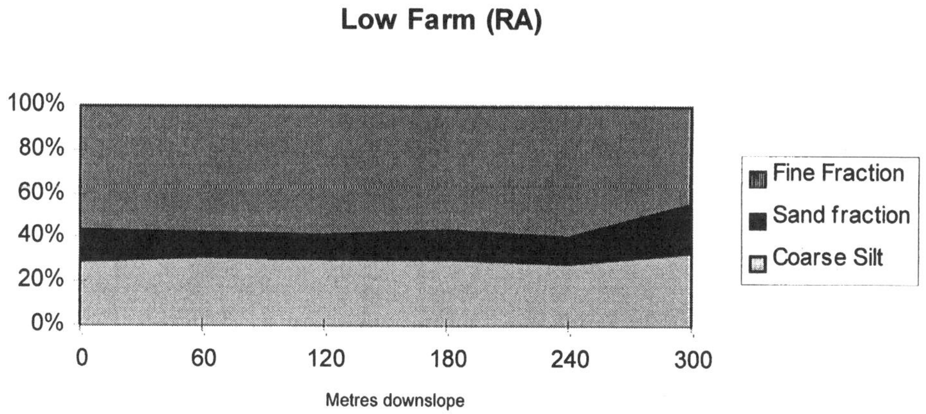

Fig. 3d. Downslope changes in the three major particle size fractions RA = Ridge Auger.

Table 4. Average values for particle size variation downslope at Low Farm

Site	Sand %	Coarse silt %	Fines %
Low Farm (FC)	18.97	32.6	48.5
Low Farm (FA)	12.25	31.8	56.1
Low Farm (RC)	19.3	32.5	48.2
Low Farm (RA)	14.25	30.33	55.5

Table 5 shows clearly that there is a distinct difference in the soil surface conditions at sites which have been cleared of basalt, but are not undergoing cultivation. For the auger samples there is very little change compared with auger samples under the land which has been ploughed and cultivated. However the crust displays a totally different fabric. All of the sand fractions have increased so that the total sand lies around 30 per cent compared with about 19 per cent and 18 per cent for the average furrow and ridge crusts respectively. The coarse silt is likewise high, with a percentage of approximately 18 per cent greater than the newly formed crusts. The medium silt shows no apparent change, but both the fine silt and clays have drastically reduced. If it is postulated that the auger samples represent the real particle size distribution then there is a 32 per cent and 62 per cent reduction in the fine silt and clays respectively. This is a very important fact because it means that the natural soil crust will develop, as long as there is no cultivation taking place, to an equilibrium stage. The finest material is relatively quickly removed (see above) and the sediment making up the crust becomes much more dominated by the sand and coarse silt. These act as a buffer zone against further wind erosion and even rainsplash erosion during precipitation events.

This represents a considerable change in focus from most soil crust research. Whereas much of the research sees a soil crust as a sign of degradation, reduction in infiltration rate and an encourager of soil erosion, the present data seem to suggest that desert soils use the formation of a soil crust much more as a defensive mechanism. In observing the current landscape, the major wind erosion occurs as a result of ploughing at the wrong time allowing much more fine material to be lost as the soil is turned over.

Table 5. Changes in particle size distribution between cropped and uncropped soil at Low Farm

Site		Coarse sand %	Medium sand %	Fine sand %	Coarse silt %	Medium silt %	Fine silt %	Clay %
Average Furrow	Crust	6.89	3.66	8.43	32.56	15.3	13.1	20.1
Average Furrow	Auger	4.19	3.17	6.67	31.79	15.21	14.13	24.93
Average Ridge	Crust	6.06	3.59	8.41	33.26	14.39	13.9	20.44
Average Ridge	Auger	6.92	3.96	6.93	31.57	14.7	12.93	23.04
Average Uncropped	Crust	11.8	6.6	11.43	38.53	14.2	9.33	8.13
Average Uncropped	Auger	4.95	4.6	10.5	30.55	14.05	13.8	21.5

10. SPATIAL VARIATIONS WITHIN THE RIDGE-FURROW SEQUENCES

The ridge and furrow sequence used in the contemporary farming techniques of the Badia represent a complex set of slopes albeit at a relatively small-scale in comparison with the general slope characteristics. For the purposes of the research, each position within the sequence needs to be given a specific label. The first position is the outer ridge which is artificially formed by the farmer as he attempts to cover the plastic irrigation sheeting; the crops often lie in a slight depression on the far side of the ridge having been encouraged to grow up through holes within the plastic. The second position is the outer furrow flank which is the slope dividing the outer ridge from the depression of the furrow. The third position is the base of the furrow and the fourth position is the centre flank. The final position is the smaller ridge which lies approximately half way between the rows of crops. As an example, Figure 4 shows two transects across the sequences LF3-3 and LF3-4 respectively showing various positions which have been described above. However, it must be noted that at some sites only three or four samples were taken and that not every sequence showed exactly the same characteristics. The following Table 6 attempts to consolidate all the possible sequence positions with the actual samples taken at each site.

It can be seen that the flanks of the main furrow at points **B** and **E - F** have a much higher relief than the ambient slope angle, often with local slopes exceeding 30° and regularly between 15° and 30°. Although erosion processes do relate to slope length, there is a case for arguing that there can be considerable erosion and deposition in progress. Rainsplash erosion is least likely to affect the sequence during the first year of cultivation, because the sequence is only created in late March or early April, by which time rain is unlikely. It would be a process that would act over subsequent seasons, however, if the soil is not reploughed. In the short-term, the more likely forms of erosion would be compaction and aggregate break-up from human activity on the one hand, and on the other, sheet erosion caused be over-irrigation.

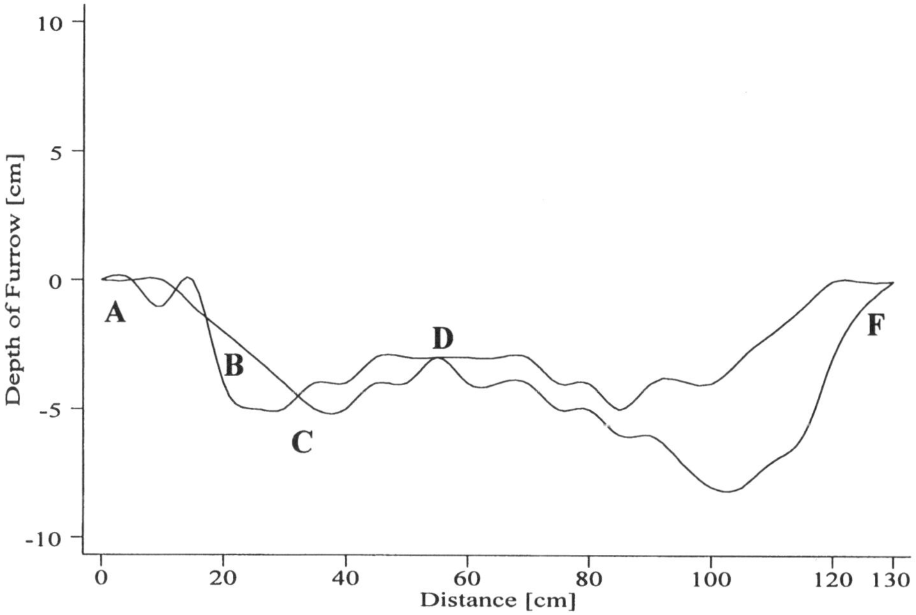

Fig. 4. Two transects of the ridge and furrow sequence at LF3-3 and LF3-4.

Table 6. Compilation of sample details taken at each site

Site	Ridge under plastic	Outer furrow flank	Furrow base	Inner/central furrow flank	Central ridge
LF3-1	D & E	C	B	A	
LF3-2	A	B & F	C & E		D
LF3-3	A		B & D		C
LF3-4	A	B & F	C & E		D
LF3-5	D	C	B	A	

It is clear from the following graphs of inter-furrow variation (Figs. 5 to 10) that there is a considerable amount of variability. As each graph is analysed, it is important to keep in mind the different positions each sample occupies within the sequence.

In the first example (LF3-1), samples **C** and **A** have a significant reduction in the clay and fine silt fraction in comparison with the other three locations. This would suggest that the fines are being preferentially removed from the flanks, but not the furrow base. If wind erosion was seen as the only potential erosive agent, it is unlikely that such a difference would occur. The control has to be the set of auger samples in Fig. 5 which suggest that clay content should be approximately 21 to 24 per cent as there is no possibility of the auger soil being affected by wind erosion except indirectly over a long period of time. The furrow base has to have had some deposition from the surrounding flanks which has compensated for any wind erosion.

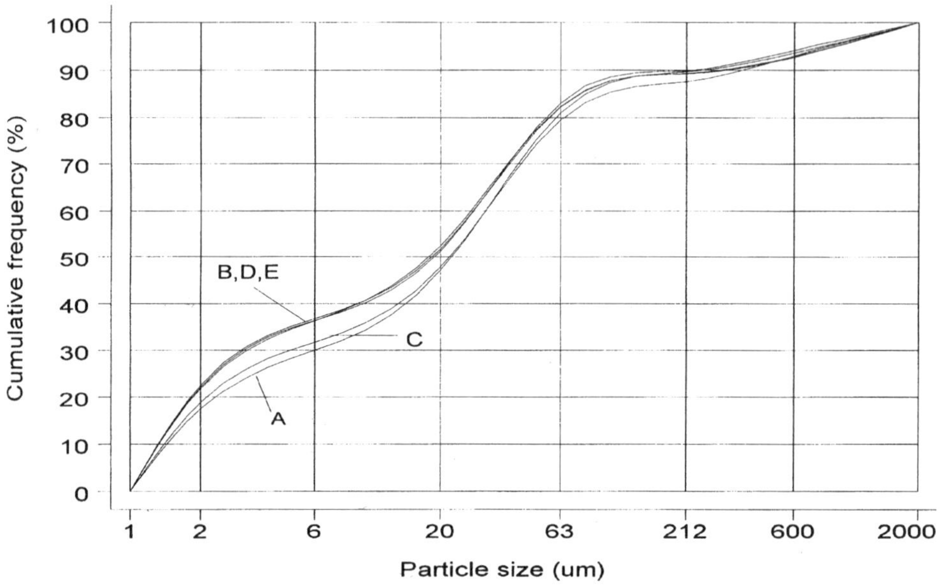

Fig. 5. Inter-furrow variation in particle size - Furrow 1 Crust (LF3-1).

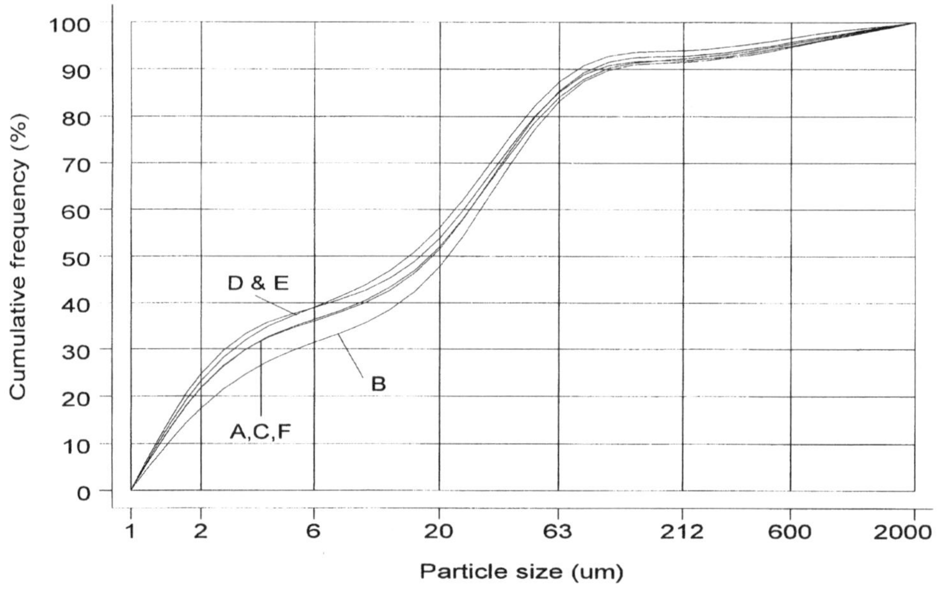

Fig. 6. Inter-furrow variation in particle size - Furrow 2 Auger (LF3-2).

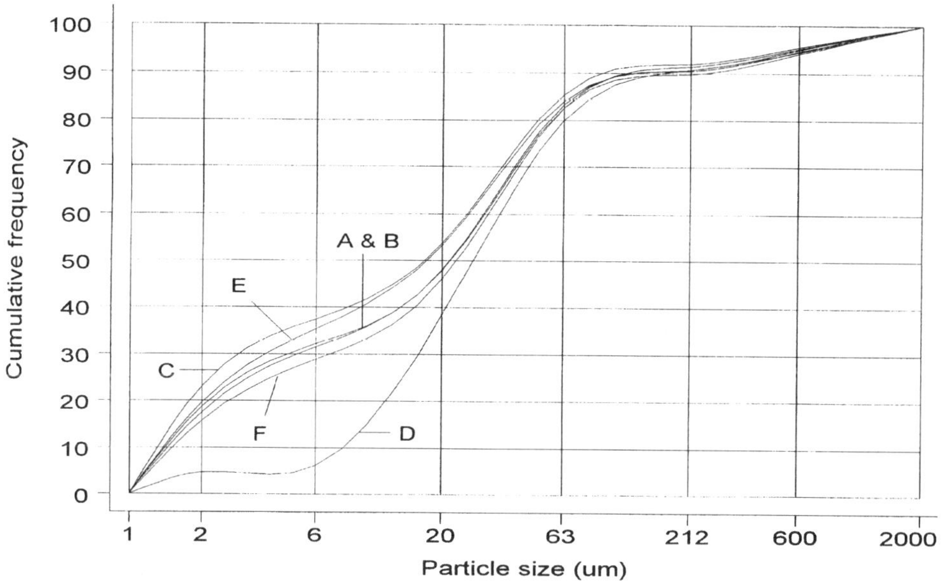

Fig. 7. Inter-furrow variation in particle size - Furrow 2 Crust (LF3-2).

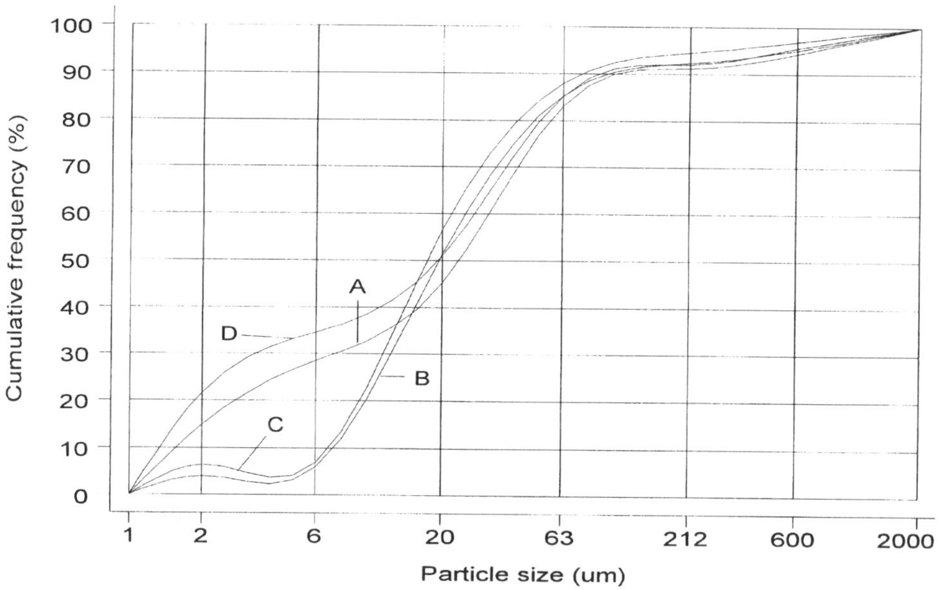

Fig. 8. Inter-furrow variation in particle size - Furrow 3 Crust (LF3-3).

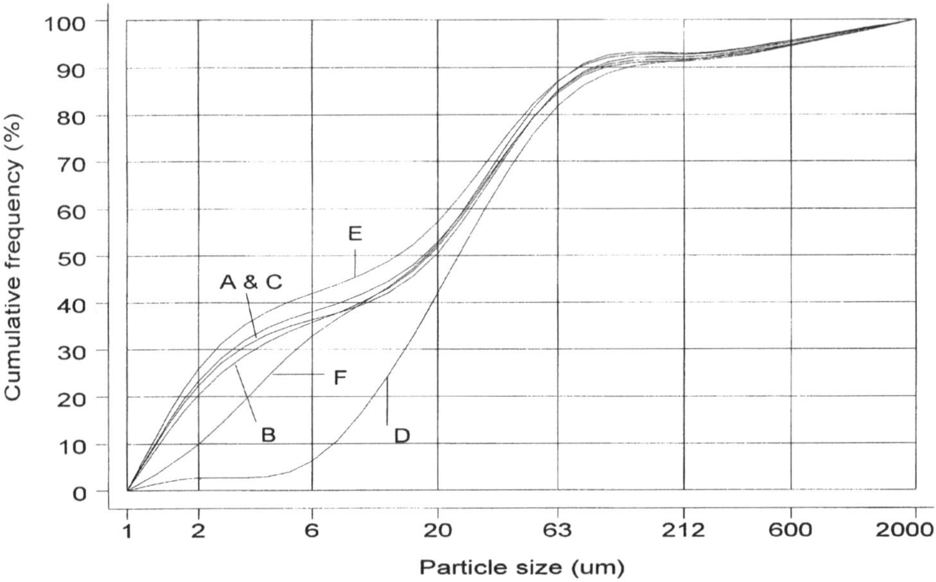

Fig. 9. Inter-furrow variation in particle size - Furrow 4 Crust (LF3-4).

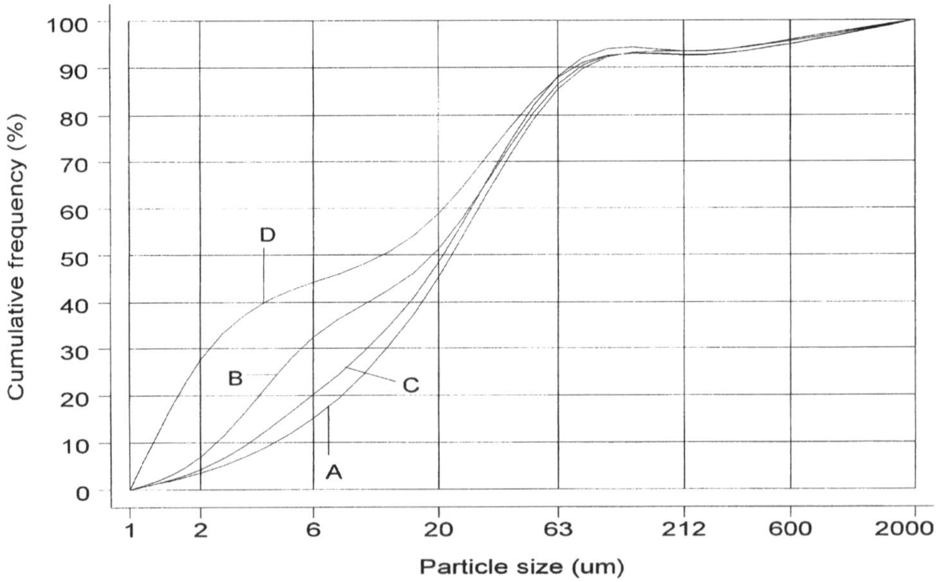

Fig. 10. Inter-furrow variation in particle size - Furrow 5 Crust (LF3-5).

The two figures representing the auger and crust samples taken at site 2 should be taken as a single entity as direct comparison can be made with the control. At site 2, **B** and **C** are the only samples which display little change between the auger and crust samples, the latter having a similar disposition to site 1. **A** is reduced despite being sampled under the plastic but this manifests itself to the order of 2 per cent which is probably not significant in relation to likely errors in sampling and analysis. Both **F** and **E** have a reduction of clay content of approximately 5 per cent in the crust while **D** experiences a drastic reduction in both clay and fine silt to the order of 20 and 10 per cent respectively. Despite a reduction of clay in **E**, both **E** and **C** remain the crust samples with the highest fines content which correlates well with the furrow bottom sample findings from site 1. Sample **D** is the sample which would seem to give the most problems in interpretation as there is so rapid a decrease in all the fine material below 6μm. However, it would seem prudent at this point to look at the other distributions to see if there are similar anomalies.

It is immediately noticeable that not only does the phenomenon of sample **D** at site 2 exist elsewhere, but that its occurrence is more abundant in Figs. 8 to 10. Moreover, in the field the samples which had low clay and fine silt were stained darker brown than the surrounding soil. Such a feature of the data, with very low clay and fine silt contents and a comparatively large medium silt fraction cannot be fully understood within the confines of the particle size data and a simple model of erosional processes. Instead it is necessary to study the chemical data (Fig. 11) for the same samples to see if they can shed light on it.

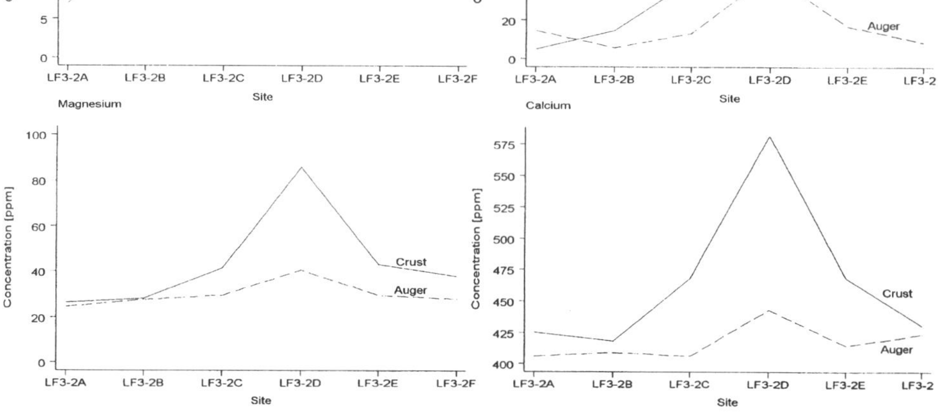

Fig. 11. Soil chemical characteristics at site 2.

Table 7. Correlations between physical and chemical characteristics at site 2

	sodium	magnesium	calcium	clay	fine silt
sodium	1.000				
magnesium	0.9631	1.000			
calcium	0.9493	0.9858	1.000		
clay	0.7067	-0.8241	0.7582	1.000	
fine silt	0.7616	-0.903	-0.8763	0.9403	1.000
medium silt	0.8898	0.9690	0.9481	-0.9238	-0.9645

Looking at Figure 11 and the accompanying correlation data (Table 7), it can be seen that there is a general increase of the three cations and especially sodium as well as medium silt towards sample **D** and then away again. The four variables in this case have an almost perfect negative correlation with the fine silt and clay. The exact nature of the relationships between the < 63μm particles and the soil chemistry will be investigated later in the chapter. For the purposes of analysing the particle size data it is important to recognize the link between the two sets of data and to postulate that the increasing salinity in the middle of the furrow is vital to the understanding of the soil data.

As for site 3 (Fig. 8), it was noticed during field investigation that samples **B** and **C** were stained dark brown. Furrow base **D** is again high for the fine silt and clay fractions although a little low in the medium silt fraction. All the samples are high in the coarse silt fraction. **A** is slightly lower than expected, but only by a matter of about 2 per cent when compared with the previous sites which is not significant. The central ridge sample **C** is displaying similar tendencies to sample **D** in the previous example, although here the medium silt goes up to over 40 per cent of the total sample. **B** is behaving in a similar way which suggests that the actual position in the sequence is not primarily responsible, but rather there is a complex relationship between the irrigation input, the associated wetting bulb and their effect on the soil surface conditions.

Six samples were taken at site 4 (Fig. 9). Here **C** and **E** remain high with **E** showing signs that there has been some possible deposition of fines. **A** is likewise unchanged and has similar characteristics to **C**. The outer furrow flanks, **B** and **F** show differing amounts of clay loss with B showing very little, but **F** displaying a considerable reduction in clay which is quite unlike sample **D** because there is a relatively high amount of fine silt although chemically there is a similarly high amount of calcium, magnesium and sodium. **D** displays similarities with central ridge samples from previous sites.

Site 5 (Fig. 10) presents the largest interpretative problems which stem from the fact that they come from a part of the field which had been replanted with wheat and had not been left fallow like sites 1 to 4 and therefore **D** is sampled amongst the wheat. The ridge-furrow sequence is partially destroyed as the soil has been reploughed before planting and yet there is a residue of the former soil chemical and physical attributes. **A**, **B** and **C** are all affected by high quantities of sodium, calcium and magnesium which manifests itself once again by low clay contents and for the flanks, low fine silt contents as well. **B** seems to have a much higher level of fine silt and comparatively less medium silt which would suggest that there has been some redeposition in the furrow base. Conversely, sample **D** shows a rather high level of clay which is likely to have been influenced by replanting and will have had some protection by the plants from wind erosion.

11. LINKAGES BETWEEN SOIL PHYSICAL AND CHEMICAL CHARACTERISTICS

It has been emphasized in the previous section that the physical and chemical attributes of the soil cannot be taken in isolation of each other. Table 8 shows correlation results between each of the particle size fractions for the whole of Low Farm.

It is clear from the Tables 8-10 that the sand fraction plays a very neutral role within the soil. There is rarely a large amount of sand except in crust samples where all the fines have been blown away, which suggests that they create stability in the soil by being a buffer to the soil below. It is the <63μm fraction which is important to study in the environment of the Badia as it remains the largest proportion of the soil fraction at all times and it is the stability of the fine material which is vital in the understanding of the crust formation processes and the likelihood of wind erosion.

Table 8. Correlations of soil physical characteristics (n=103)

Site	Coarse sand %	Medium sand %	Fine sand %	Coarse silt %	Medium silt %	Fine silt %	Clay %
C. sand	1.0000						
M. sand	0.7335	1.0000					
F. sand	0.4546	0.6265	1.0000				
C. silt	0.0919	0.1373	0.4762	1.0000			
M. silt	-0.1891	-0.2131	0.0788	0.4117	1.0000		
F. silt	-0.2148	-0.2749	-0.4293	-0.3454	-0.6498	1.0000	
Clay	-0.2629	-0.2282	-0.5209	-0.8175	-0.6742	0.4136	1.0000

Table 9. Correlations of selected physical and chemical characteristics, LF3 (n=103)

Site	Coarse silt	Medium silt	Fine silt	Clay	Ca	Mg	Na	K
C. silt	1.0000							
M. silt	0.4665	1.0000						
F. silt	-0.2685	-0.7304	1.0000					
Clay	-0.8041	-0.8071	0.4037	1.0000				
Ca	0.3989	0.7947	-0.4737	-0.7037	1.0000			
Mg	0.4620	0.9110	-0.6401	-0.7591	0.8832	1.0000		
Na	0.4126	0.7630	-0.3789	-0.7102	0.8769	0.8429	1.0000	
K	0.3843	0.5707	-0.3482	-0.5763	0.4526	0.5656	0.6092	1.0000

From the correlations in Tables 9 and 10 it can be seen that the medium silt is the controlling physical variable because it has high correlations with every other attribute except potassium. Indeed if Figure 12 is taken into account, it can be seen that the relationship is particularly close in reference to magnesium, which means that it could be used as a predictive tool to identify areas which have high electrical conductivities because magnesium is in turn closely related to calcium and magnesium (see Figs. 12 to 15).

Table 10. Correlation and regression coefficients between the fine fraction and soil chemical attributes at LF3

Variables	Correlation	r-squared	r-squared for transformation
Coarse silt - Clay	-0.8	0.65	n/a
Medium silt - Fine silt	-0.73	0.42	n/a
Medium silt - Clay	-0.81	0.65	0.77 (log10 of medium silt)
Medium silt - Calcium	0.79	0.63	n/a
Medium silt - Magnesium	0.91	0.83	0.84(or ^{3} of medium silt)
Medium silt - Sodium	0.76	0.58	0.61(log10 or ^{3} of medium silt)
Fine silt - Magnesium	-0.64	0.41	n/a
Clay - Calcium	-0.7	0.5	n/a
Clay - Magnesium	-0.76	0.58	n/a
Clay - Sodium	-0.71	0.5	n/a

It has been postulated earlier in the chapter that irrigation water is profoundly linked with the salinity and sodicity of the soil crust. Evidently there is an increase in sodium, calcium and magnesium in the middle of furrows and this leads to an increase in the proportion of medium silt and decreases in the percentages of clay and fine silt. This leads to the following questions. First, why is there a special case in the middle of furrows rather than closer to the irrigation water sources? Second, why is there no evidence of movement of potassium to the surface as there is with the other cations measured? Third, how can the apparent paradox of increasing salinity and sodicity lead to an overall weakening of the crust be solved?

The idea of salinity increasing towards the middle of the furrow is not a foreign one as there is plenty of research to show that this happens. Salinities increase towards the boundaries of the wetting bulb (Hillel 1971, Bresler *et al.* 1982). Evaporation is reduced closer to the plants by the layers of black plastic which cover the irrigation pipes. As the wetting front extends laterally there is an increasing rate of evaporation and so as the water moves outwards it dissolves salts. When it comes to its equilibrium position, the lateral movement stops and upward movement takes place by capillary action (Rose 1996). As the water reaches the surface, the water evaporates into the atmosphere leaving a high quantity of salts in the surface layers of the soil.

Potassium salts are the only ones which do not seem to be affected by this lateral movement of irrigation water through the soil. The previous graphs and correlations show that potassium is reasonably stable within the soil profile. Potassium is much less likely to be taken up by water as it remains in the silicate structures of the clays and is therefore less available for cation exchange purposes (Sparks and Huang 1985).

The third issue concerns the apparent paradox between stability and instability due to the contrasting effects of salinity and sodicity. It is well documented that higher EC values of water allow soil to flocculate and their tendency to crust is vastly reduced (Shainberg 1985). However, if the SAR or ESP is increased then the opposite process occurs, that of dispersion which causes crusting to occur more rapidly (Abu-Sharar 1988). Looking back to Figures 11, 13 and 15 it would seem that, relatively speaking, sodium becomes the most mobile, which would anticipate higher dispersion rates. This allows the freeing up of the finest fractions of the soil which then can be entrained by the wind. The structure of the crust is actually weakened rather than strengthened because there is no mechanical impact (Le Bissonais 1990). In the field, the darker brown

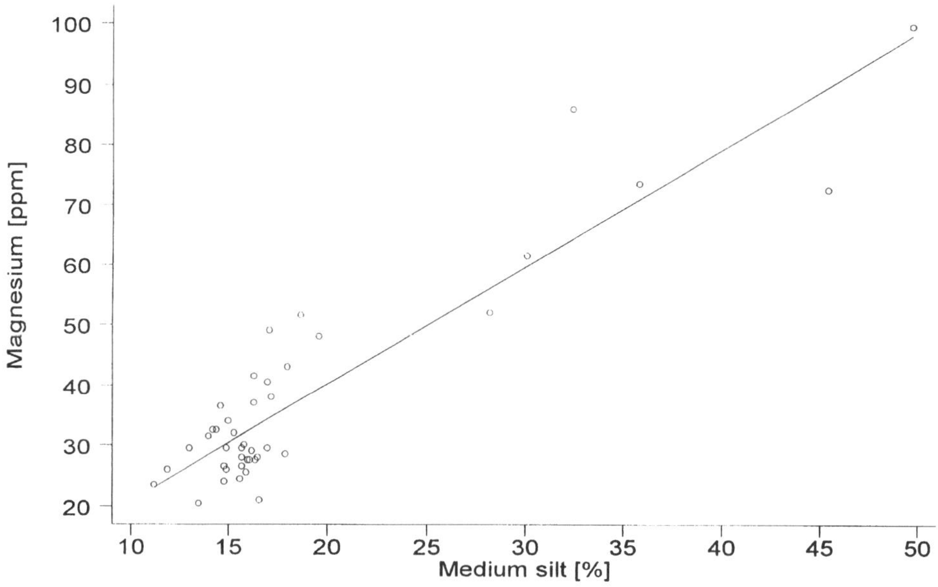

Fig. 12. Relationship between medium silt and magnesium at LF3.

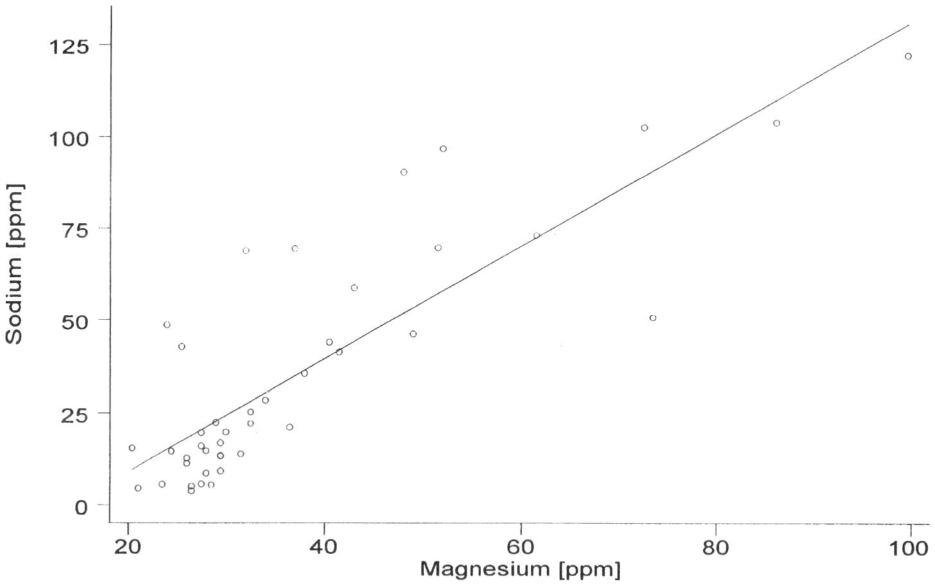

Fig. 13. The relationship between magnesium and sodium at LF3.

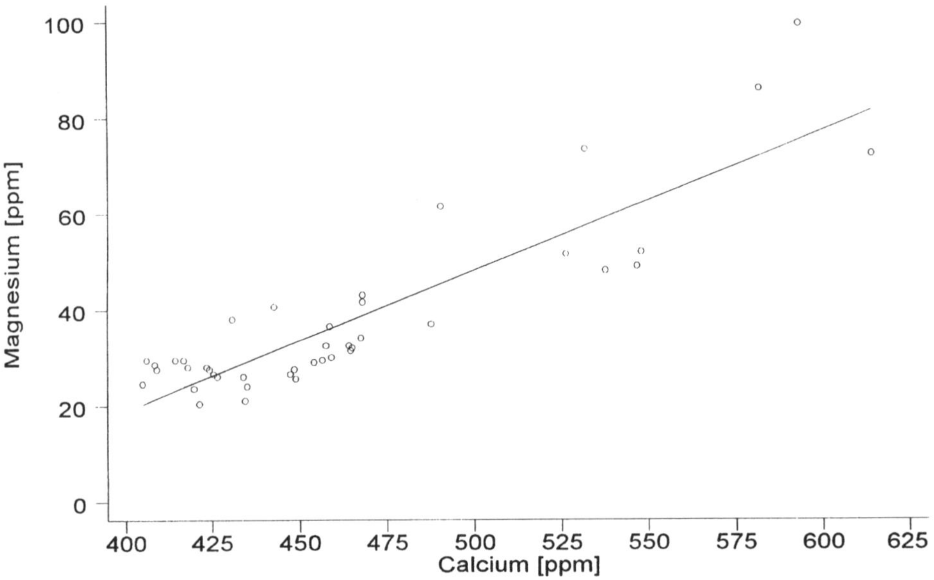

Fig. 14. The relationship between calcium and magnesium at LF3.

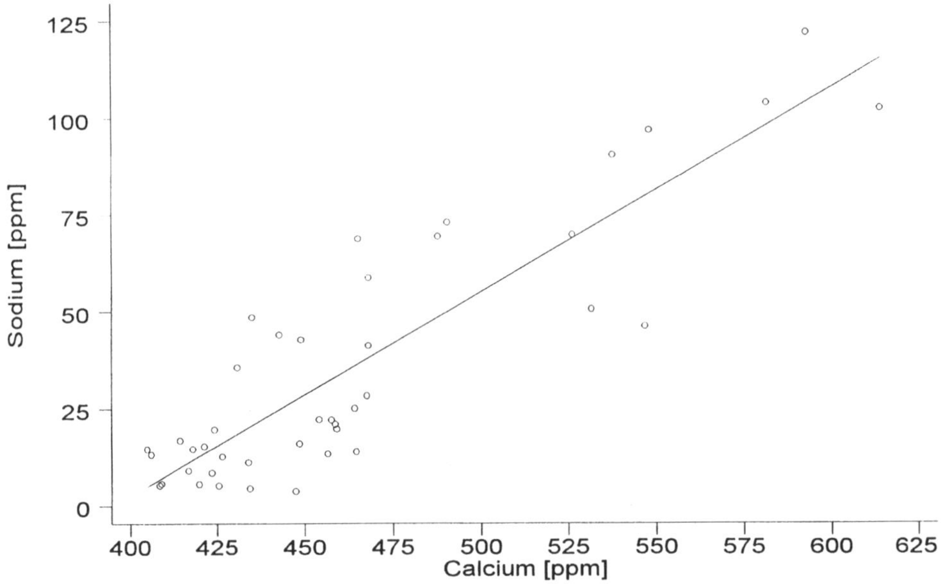

Fig. 15. The relationship between calcium and sodium at LF3.

patches represent staining from the sodium salts and their structure is crumbly because the sodium has dispersed most of the clay which has then been eroded by the wind. Such circumstances allow a relationship to be made between the medium silt and clay, Figures 16 and 17, because the former increases relatively speaking as the latter is eroded.

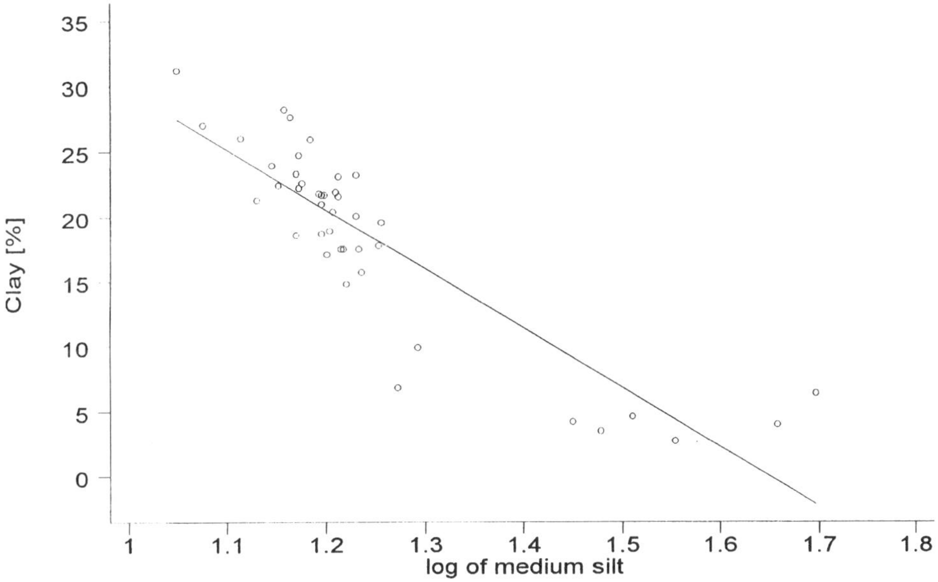

Fig. 16. Relationship between medium silt and clay in LF3.

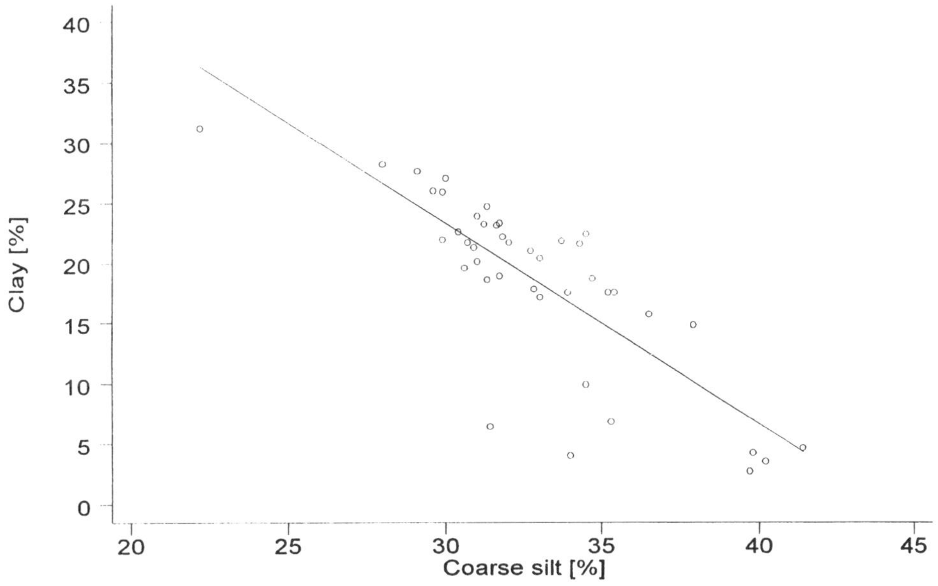

Fig. 17. Relationship between coarse silt and clay at LF3.

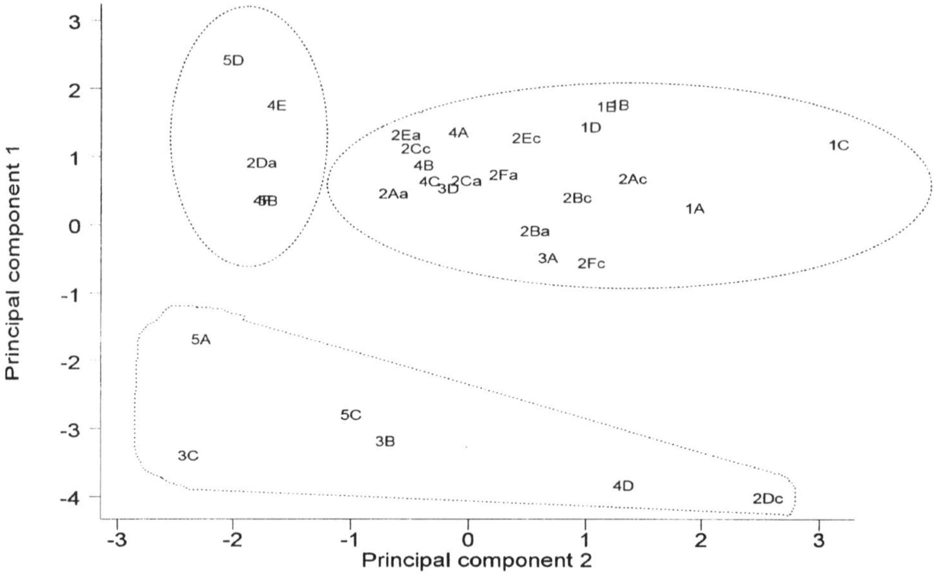

Fig. 18. A summary of soil chemical data for LF3 using principal components.

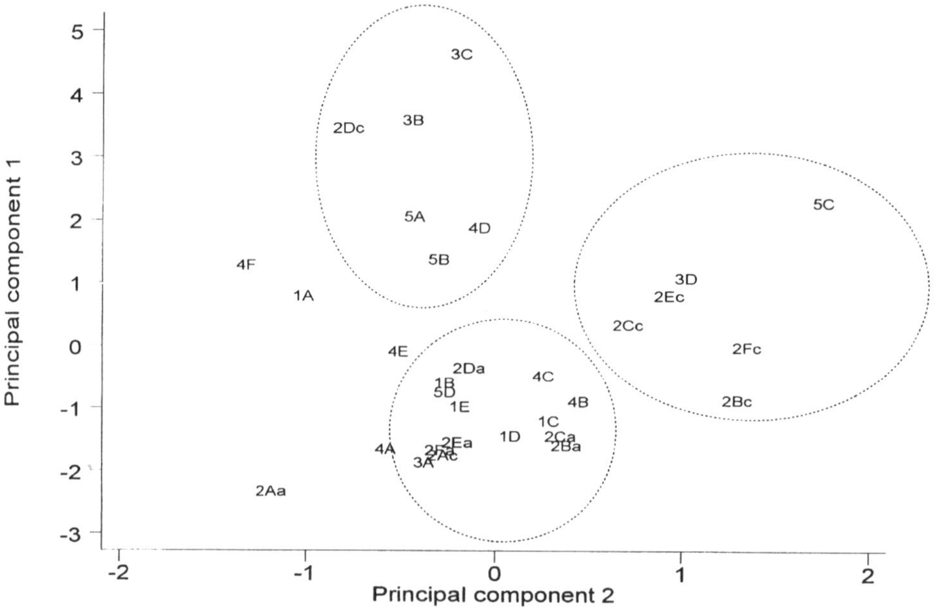

Fig. 19. A summary of particle size data for LF3 using principal component.

Finally, a summary of the soil research can be undertaken by using Principal Components Analysis (PCA) in order to use the variability in the dataset to come up with certain groups of samples which can be identified by their relative positions within the ridge and furrow sequence. The top group in Figure 18 and the bottom group in Figure 19 represent all the samples which lie in the middle of the furrow sequence and are affected by the increasing salinity and sodicity. The samples on the right of Figure 19 identify the furrow base samples which are probably a mixture of erosional and depositional features.

12. CONCLUSION

Historical and archaeological accounts suggest that relatively intensive farming has taken place, at various times, since the beginning of the Roman Empire. Clearance of the boulder-strewn landscape has always been a necessary preparation for cultivation and it is likely that such an activity has had profound effects on the landscape evolution, especially with regard to wind erosion.

Modern-day intensive farming techniques have increased the amount of land which can be brought into production, whilst rainfall shortage is overcome by the pumping of water from the basalt aquifer. However, new machinery causes considerable increases in the propensity for wind erosion. Fine material is rapidly deflated from newly ploughed soil, but after a small amount of precipitation, a soil crust forms which protects the soil directly beneath it. Over time the crust becomes dominated by the coarser fractions of silt and sand as the fine silts and clays are eroded and acts as a barrier to further erosion.

The research presented here has concentrated upon the origin of the present soil in the north western Badia and the contemporary processes which are operating on it. Whilst there is some evidence to suggest that the soil is partially derived from the basalt in terms of the presence of heavy minerals and decaying feldspars, there is rather more evidence which would point to limestone and aeolian sources for much of the sedimentary matrix. Analysis of weathering nodules within the basalt show a great preponderance of calcite which would seem to derive from a dissolved form of the free calcium within the basalt itself. Calcium is definitely the major cation dominating the soil profile and along with salts of sodium and magnesium is readily taken into a dissolved state during irrigation. These three salts concentrate around the edge of the wetting profile and as evaporation takes place, they are precipitated out on the surface of the soil crust. This has a considerable influence on the micro-scale processes acting at a ridge and furrow scale. There is an increase in salts building up in the crusts with distance from the irrigation laterals. Where they are at a maximum, it is possible to find thirtyfold increases in sodium. Such an increase causes large amounts of dispersion in the crust which, as a result, partially breaks down and the fine silts and clays either filter into the soil profile or more likely get eroded by the wind.

These results point to the fact that over-irrigation will lead to considerable mobility of salts and areas of high ESP will develop in the long-term creating problems of fertility. In addition, the research highlights the importance of the formation of the soil crust in combating soil erosion. Farmers need to become managers of the soil crust where they attempt to maximize infiltration from rainfall during the winter, but minimize the loss of fines by wind erosion. In conjunction with such soil conservation measures, there needs to be more investment in reducing wind erosion by using the basalt blocks to enclose fields and to encourage the planting of trees around the farm boundaries.

REFERENCES

Abu-Sharar, T.M. (1988) Stability of soil aggregates as inferred from optical transmission of soil suspensions. *Soil Sci. Soc. Am. J.*, 52, 951-4.

Al-Homoud, A.S., Allison, R.J., Sunna, B.F., White, K. (1995) Geology, geomorphology, hydrology, groundwater and physical resources of the desertified Badia environment in Jordan. *GeoJournal, 37*, 51-67.

Allison, R.J. and Higgitt, D.L. (in press) Slope form and associations with ground surface boulder cover in the Eastern Badia, Jordan. *Geomorphology*.

Arrhenius, G. and Bonatti, E. (1965) Neptunism and volcanism in the ocean. *Progress in Oceanography* 3, 7-22.

Betts, A.V.G. (ed.) (1991) *Excavations at Jawa 1972-1986*. Edinburgh University Press, Edinburgh.

Bresler, E., McNeal, B.L. and Carter, D.L. (1982) *Saline and sodic soils*. Springer-Verlag, New York.

Bruins, H.J. and Yaalon, D.H. (1979) Stratigraphy of the Neviot section in the desert loess of the Negev. *Act. Geol. Acad. Sci. Hung.* 22, 161-9.

Butler, H.C. (1919) *Publication of the Princeton University archaeological expedition to Syria 1904-5 and 1909*. Division II ed. Vol. Section A, Architecture, South Syria. Leiden.

Butzer, K.W. (1955) Some aspects of postglacial climatic variation in the Near East considered in relation to movements of population. M.Sc. Thesis, McGill University, Montreal.

De Vries, B. (1981) The Umm el-Jimal project 1972-1977. Bulletin of the American Schools of Oriental Research 244, 53-72.

Drury, D.M. (1993) Hydrogeology of the Azraq Basin. M.Sc. Thesis, University College, London.

Dwairi, I.M.A. (1987) A chemical study of the palagonitic tuffs of the Aritain area of Jordan with special reference to nature, origin and industrial potential of the associated zeolite deposits. Ph.D. Dissertation, University of Hull.

Ganor, E. (1975) Atmospheric dust in Israel: sedimentological and meteorological analysis of dust deposition (in Hebrew with English Abstract). Ph.D. Dissertation, The Hebrew University, Jerusalem, Israel.

Goldberg, D., Gornat, B. and Rimon, D. (1976) *Drip irrigation: principles, design and agricultural practices*. Drip Irrigation Scientific Publications, Kfar Shmaryahu, Israel.

Hay, R.L. (1966) *Zeolites and zeolitic reactions in sedimentary rocks*. Special Paper of the Geological Society of America, No. 85.

Helms, S.W. (1981) *Jawa: lost city of the black desert*. Methuen, London.

Hillel, D. (1971) *Soil and water: physical principles and processes*. Academic Press Inc., San Diego.

Homès-Fredericq, D. and Hennessy, J.B. (eds.) (1989) *Archaeology of Jordan*. Vol. II Field Reports. Peeters, Leuven.

Ibrahim, K. (1993) *The geological framework for the Harrat Ash-Shaam basaltic super-group and its volcanotectonic evolution.* Geology Directorate, Natural Resources Authority, Amman.

Issac, B. (1992) *The limits of empire: the roman army in the east*. Revised Edition ed. Clarendon, Oxford.

Kastner, M. (1976) *Diagenesis of basal sediments and basalts of sites 322 and 323 Leg 35: Bellinghausen Abyssal Plain.* Initial Report of the Deep Sea Drilling Program, Vol. 35. United States Government Printing Office, Washington D.C. pp. 513-18.

Lankester Harding, G. (1959) *The antiquities of Jordan.* Jordan Distribution Agency, Amman.

Le Bissonais, Y. (1990) Experimental study and modelling of soil surface crusting processes. *Catena Supplement* 17, 13-28.

Rose, D.A. (1996) The Dynamics of soil water following single surface wettings. *European Journal of Soil Science* 47, 21-32.

Shainberg, I. (1985) The effect of exchangeable sodium and electrolyte concentration on crust formation. *Advances in Soil Science* 1, 101-22.

Sparks, D.L. and Huang, P.M. (1985) Physical chemistry of soil potassium. In: *Potassium in Agriculture.* (ed: Munson, R.D.) American Society of Agronomy, Madison, 201-276.

CHAPTER 12

A Reconnaissance Survey of Surface Water Infiltration Characteristics

JEFF WARBURTON
(University of Durham)

1. INTRODUCTION

Increasing development pressure in arid and semi-arid regions of the world has focused attention on soil and water conservation issues (Hudson 1987). A key aspect of this topic is the need to understand how rainfall received at the surface is redistributed. Quantitative understanding of the partitioning of rainfall into components of direct runoff, surface storage, infiltration, soil moisture storage, percolation and groundwater storage is essential before water resources in arid areas can be effectively managed. Of these components infiltration is arguably the most important hydrological process because, by definition, it is the movement of water through the soil surface into the ground and therefore controls the proportions of surface and subsurface water. However, as Betson (1964) points out, 'The concept of infiltration is easily understood. Measuring and understanding the physical factors controlling the infiltration process, however, is exceedingly difficult.' This is especially true of arid environments where rainfall is sparse, erratic and irregular; runoff from ephemeral drainage systems tends to be irregular and short-lived; and infiltration in to channel alluvium and desert sediments is highly variable (Agnew and Anderson 1992).

Many factors control the rate of infiltration: soil texture, soil structure, vegetation cover, biological activity, soil moisture content and surface conditions (e.g. pavement formation). Although knowledge of infiltration in arid areas is still limited, it is acknowledged that it is important in relation to the following:

1. Surface runoff and sediment production on hillslopes. In general, fine-textured soils generate surface runoff faster than coarse-textured soils and are therefore more susceptible to surface soil loss (Abrahams *et al.* 1988). Runoff is produced when the rate of rainfall is greater than the infiltration capacity of the surface soil.
2. Soil profile development and subsurface enrichment. Infiltration influences rates of illluviation and the mobilization of soluble compounds and organic matter in the soil.
3. Transmission losses from the bed of ephemeral wadi systems. Transmission losses of up to 57 per cent have been reported (Cooke *et al.* 1993). Thornes (1977) identified four conditions resulting in rapid transmission losses of wadi discharge: inputs are relatively small; transmission losses are high where the alluvium is thick; water tables are relatively deep and sediments have high hydraulic conductivities,

therefore outflow discharge is high; and channels have high width-depth ratios and large bed areas which promote rapid infiltration through the bed.

4. Groundwater recharge. Lloyd (1986) identifies four possible means of groundwater recharge in arid areas: direct recharge from precipitation through the soil profile; indirect recharge from runoff through joints; indirect recharge from runoff ponding; and indirect recharge through transmission losses during river runoff.

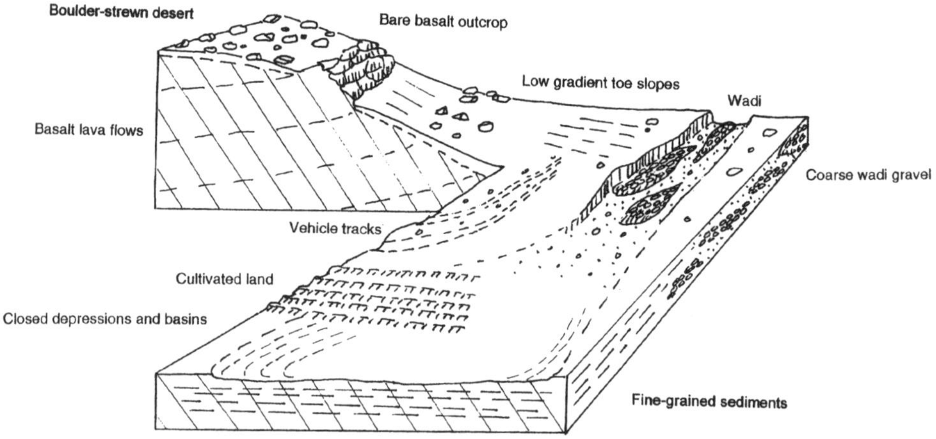

Fig. 1. Block diagram illustrating the main land systems components and their interrelationships within the study area.

5. Applied context. Estimates of infiltration rates are useful in determining infiltration in areas of surface or flood type irrigation; infiltration basins for groundwater recharge; seepage losses from streams; canals and reservoirs; and the effects of clay crusts and compaction (Bouwer 1986).

The aim of this study is to carry out a reconnaissance survey of surface water infiltration characteristics in the Northern Badia, Jordan. The objectives are: develop a methodology for the rapid assessment of infiltration characteristics; determine variability of infiltration rates in the Northern Badia land system of Jordan; establish the main factors controlling infiltration rates; and illustrate how estimates of infiltration can be used in water resources problems.

2. FIELD AREA

The infiltration investigations described in this report were carried out in the Northern Badia Region of the Hashemite Kingdom of Jordan, as part of the Jordan Badia Research and Development Programme for the sustainable development of the desertified Badia environment and the improvement of the standard of living of the inhabitants.

This area of the Badia has a rainfall of approximately 50 to 75mm/yr potential evaporation of 1,900mm/yr and an average annual temperature of 17° C. The area is within the arid climatological zone. Observations are carried out in the upper part of the Azraq Basin (area 12,700km^2). Al-Homoud *et al.* (see also Ch. 2) estimate that of the available rainfall 85-92 per cent is lost through evaporation, 5-11 per cent is infiltrated and 2-4 per cent generates runoff. These values are gross averages and highly speculative because the hydrological network and knowledge of local hydrological processes are very poor.

In deciding on a sampling strategy for the present survey a land systems approach provides a suitable framework. Land systems are subdivisions of regions into areas which have within them common physical attributes different from adjacent areas. A particular land system is usually characterized by a pattern of topography, soils and vegetation reflecting the geology, geomorphology and climate under which these processes function. This concept is particularly useful when examining the hydrology of arid lands. It can easily be applied in the present context to the infiltration characteristics of the wadi systems in Jordan's Northern Badia.

Figure 1 shows a block diagram illustrating the main land systems components and their interrelationships within the study area. Three main zones are distinguished: side slopes covering the land outside the wadis and including areas of boulder-strewn desert; wadis including all the deposits with the channel system and low-angle closed depressions and basins known locally as *marabs* and *qa'as* (Fig. 1). Within these zones surface soil characteristics are modified by human action such as cultivation and the use of vehicles. These land systems are used for stratifying sampling within the study area.

3. FIELD SAMPLING AND METHODS

The problems of measuring infiltration in arid lands are highlighted by Agnew and Anderson (1992, 128), who suggest that sophisticated infiltration apparatus tends to be large, difficult to transport and install: although infiltration rings offer a practical alternative, large amounts of water are required which in turn can be difficult to obtain and transport. In this study two main methods were used to examine infiltration in the field: a single ring flooding infiltrometer and a infiltration moat. The infiltration ring was used for quantitative determination of rates and the infiltration moats used for qualitative understanding of water movement into the soil. In total 40 tests were carried out: 32 ring tests and eight infiltration moats (Fig. 2).

The flooding infiltrometer used in the field consisted of a small metal ring (ring diameter 100mm) inserted into the soil to a depth of approximately 30mm (Fig. 2A). This is deep enough to prevent water leakage ('water blow-out') around the cylinder and shallow enough to minimize disturbance. Where a fine crust was present, a thin ring of clay was placed around the inner and outer walls of the cylinder to ensure good contact between the crust and soil. The area enclosed inside the ring was flooded with water and the rate at which water moved into the ground was measured. A small inverted cup was placed on the soil surface to prevent erosion when the water was being added. In order to reduce evaporation losses during the test, a loose lid was placed over the ring and the area around the test area covered by a plastic sheet. Water was kept at a constant head by adding water to a known level in the cylinder and recording the rate of fall. Using this method variations in pressure head were between 10 and 30mm. These measurements were continued until the rate became essentially constant, a time period of 2 hours had elapsed or all the available water for the test was used up. The use of small ring infiltrometers is justified because of the logistics involved (transport, large travel distances, lack of available water) and the need to collect reconnaissance measurements of relative differences in rates in a short period of time.

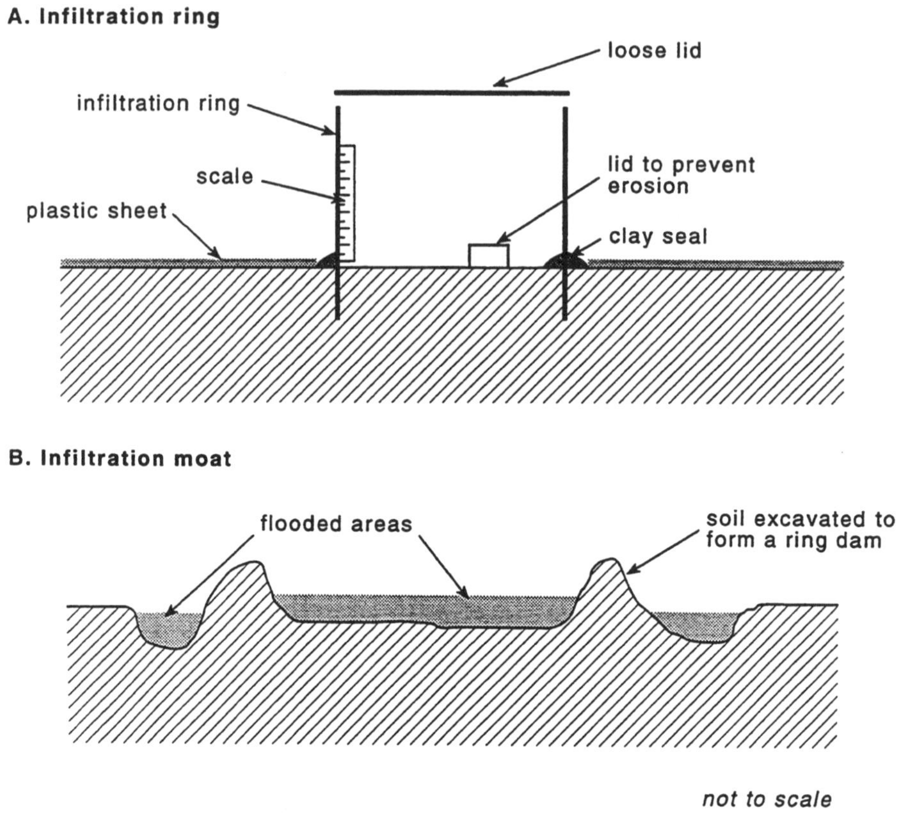

Fig. 2. Field measurement techniques. Two methods are shown: (A) infiltration ring used for quantitative determination of rates and (B) infiltration moats used for qualitative understanding of water movement into the soil.

This provided a simple and direct approach of determining the capacity at which the ground absorbs water under the prevailing environmental conditions. The results obtained only provide an estimate of the local hydraulic conductivity of the soil and are of value in determining relative rates of infiltration for a variety of surface cover and surface soil types. Errors arise in this method because: the influence of entrapped air is absent due to the capacity for lateral escape; placing the ring in the ground results in disturbance of the soil at the surface; and the effects of rainfall impacts on surface soil structure are absent.

In order to examine infiltration over larger areas, an infiltration moat was constructed. This consisted of a circular trench of approximately 10-15cm width and 1m diameter dug to a depth of approximately 10 to 15cm. Soil excavated from the trench was piled up as a rampart on the inner wall of the trench to a height of approximately 10cm producing an enclosed area of approximately 80cm diameter with a rampart and trench around the circumference (Fig. 2B). The outer trench was first flooded forming a moat which began to infiltrate; then the inner area was flooded to a depth of between 4 and 5cm. A plastic cup placed on the ground surface prevented erosion when the water was being added. This created a crude dual ring infiltrometer. The test was run until all the water in the central area had infiltrated and then a trench was rapidly excavated across the full width of the test area exposing the wetted profile. Field observations and measurements of wetting front depth were taken in the exposed profile.

Grain-size analysis was carried out on 28 samples collected from the infiltration test sites. Samples were air-dried and disaggregated with a rubber pestle and mortar. Samples were sieved through 16, 8, 4, 2, 1, 0.5, 0.25, 0.125 and 0.063mm mesh sizes. The fraction less than 0.063mm was retained for pipette analysis and moisture and organic content determination. The hygroscopic moisture content was determined gravimetrically by placing approximately 6g of the sample in an oven at 150° C overnight then reweighing the samples the following day after cooling in a desiccator for 30 minutes. Organic matter content was determined by loss on ignition. Samples were placed in a preheated muffle furnace at 375° C for 16 hours. Samples were cooled in a desiccator for 30 minutes, and then reweighed. Approximately 15g of the less than 0.063mm sample was used for pipette analysis of the silt and clay fractions. The sample was mixed with a dispersant of 50g $(NaPO_3)_6$ to 1 litre of distilled water before pipette samples were taken. These methods are appropriate for the granular, slightly cohesive samples collected in this study.

4. RESULTS

Infiltration rates varied greatly between sites. Maximum rates varied from 114,000 to 14mm/hr and minimum (stable) infiltration rates 74,500 to 3mm/hr (Fig. 3). This variability is associated with specific land system units and soil surface types. The highest infiltration rates occur in the coarse-grained wadi systems, the slowest rates with side slopes and small closed basins (*qa'as*). Figure 3 shows representative infiltration curves for seven different landscape elements. The highest rates are for the wadi gravels and wadi margins where coarse gravelly and sandy deposits dominate. The slowest rate is in the side-slope areas where fine-grained, often crusted desert soils tend to occur. The fine-grained closed basins and land surface depressions also have relatively low rates. In these areas surface characteristics such as compaction due to vehicles or ploughing for cultivation do not appear to alter the overall infiltration rates radically. Interestingly, maximum initial infiltration rates in many of the lowest stable infiltration rate soils are at least as high as the wadi areas (Fig. 3). Differences in the duration of the infiltration tests shown in Figure 5 reflects the different rates because in the wadi areas the water supply available for the test became rapidly exhausted.

From the preceding description it is clear that there is an association between infiltration rates and soil texture. Analysis of the textural properties of the infiltration test sites shows a large variation in grain sizes (Fig. 4). Median grain sizes vary from fine gravel down to silt with some samples containing up to 30 per cent silt and clay. Wadi gravels are clearly the coarsest deposits and sediments of the land surface depressions are the finest. Correlations of infiltration rates with grain-size characteristics are shown in Table 1.

The grain-size characteristics include D_{90} (grain-size for which 90 per cent of the sample is finer), D_{50} (grain-size for which 50 per cent of the sample is finer), the fraction of the sample less than 2mm (sand fraction and finer) and the fraction of the sample less than 0.063mm (silt and clay fraction). Using the raw data, correlation coefficients are low and insignificant. D_{90} and D_{50} values are positively correlated with grain-size, while the less than 2mm and less than 0.063mm parameters show slight negative correlations.

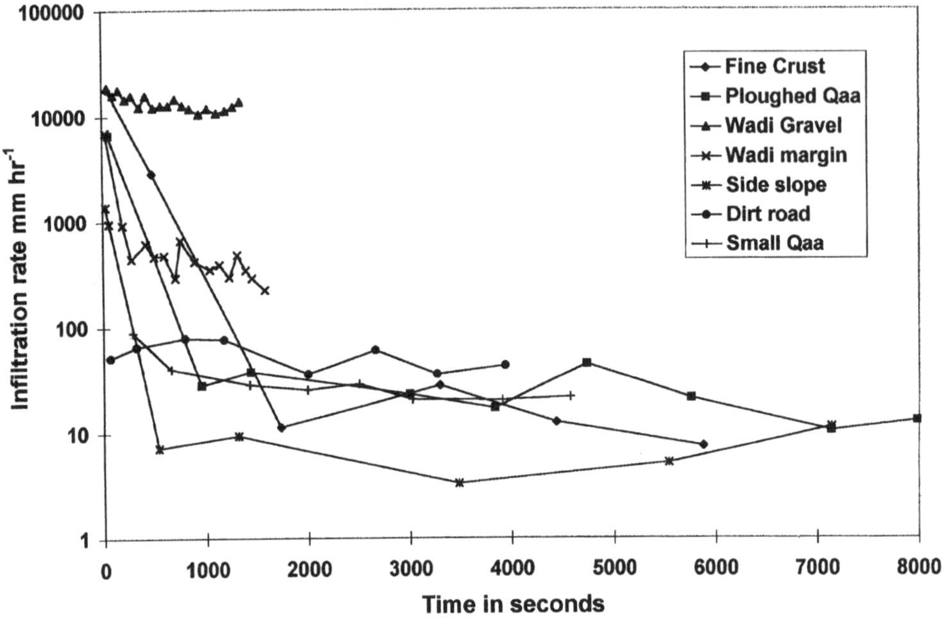

Fig. 3. Results of infiltration tests from seven sites representing differing surface conditions. Highest infiltration rates are associated with the wadis, lowest values correspond to sites on valley side slopes and in low closed depressions.

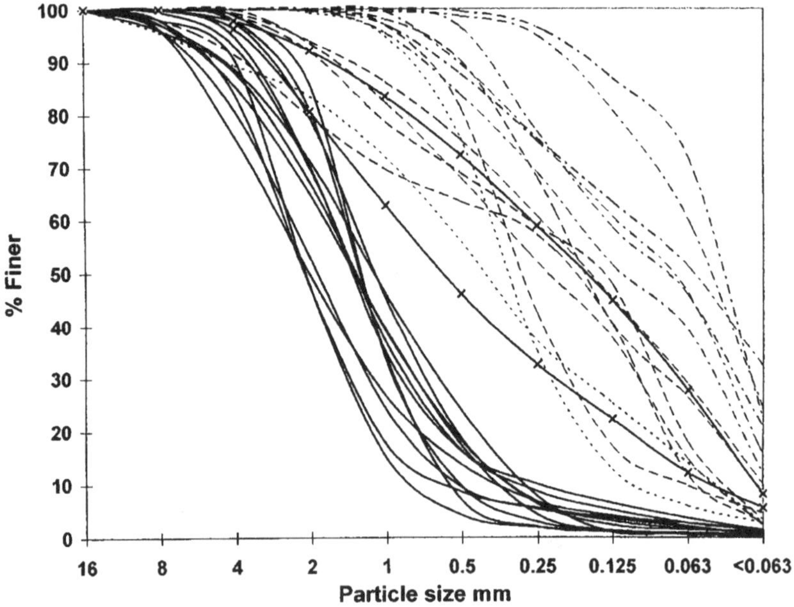

Fig. 4. Grain-size distribution curves for surface samples collected from 28 of the infiltration test sites. Five classes of curves are distinguished: wadi gravels (solid line), side slope areas (large dash line); fine sediment deposits in wadis (small dash line); closed depressions and pan areas (dash-dot line); and dirt roads (solid line with crosses). The diagram shows a large range of grain sizes. However, some clear trends are evident: wadi gravels are the coarsest deposits and deposits in closed depressions are the finest.

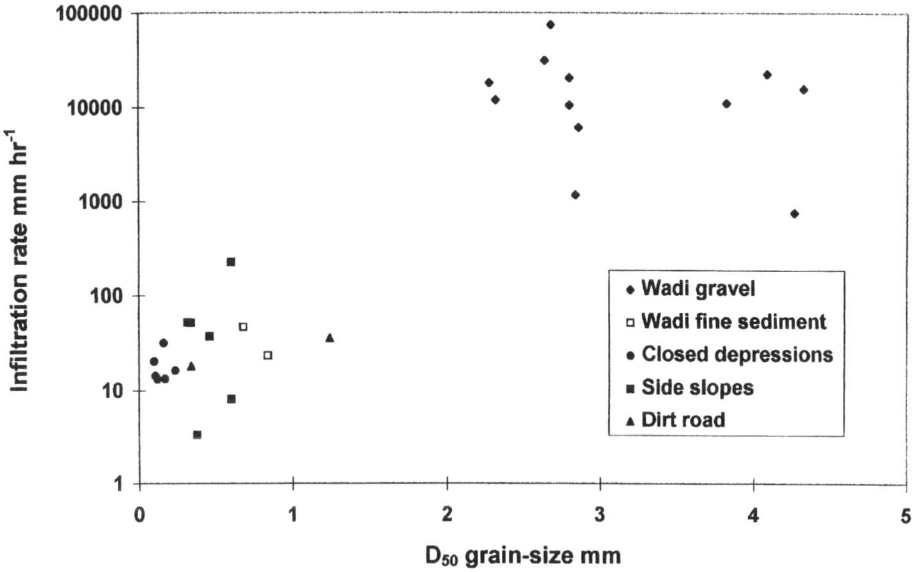

Fig. 5. Plot of D_{50} grain-size against infiltration rate in mm/hr. Five different surface soil types are distinguished. The correlation between these variables is seen in Table 2.

Transformation of the grain size/infiltration rate relationship to a log-linear relation allows the data to be clearly plotted and generates a significant linear relationship (Fig. 5). Although the scatter in this relationship remains high, the trends of the correlation coefficients are the same and the correlation coefficients are all significant. The highest correlation is between D_{50} and the stable infiltration rate (Fig. 5, Table 1). This plot clearly distinguishes the high infiltration rate of the coarse wadi gravels and the lower rates of the fine deposits of the closed depressions and side slopes.

Table 1. Correlation coefficients between maximum and stable infiltration rates and grain-size parameters. Results are shown for both raw and log infiltration rates.

	Maximum infiltration rate mm/hr		Stable infiltration rate mm/hr	
	Raw data	Log data*	Raw data	Log data*
D_{90}mm	0.32	0.61	0.32	0.67
D_{50}mm	0.46	0.75	0.49	0.86
< 2mm	-0.28	-0.62	-0.30	-0.69
<0.063mm	-0.38	-0.60	-0.37	-0.60

* All values significant at the 0.001 level.

This simple relationship between soil texture and infiltration rate disguises the complexity of how water moves into the soil. Figure 6 shows wetting front profiles for infiltration tests at six sites. The diagram shows vertical soil profiles for both infiltration ring and infiltration moat tests. The top two diagrams (A and B) show marked differences in the wetted profiles for two ring tests. In the coarse uniform soil a symmetrical wetting bulb develops while in example A water penetrates the soil through a series of fingers. In

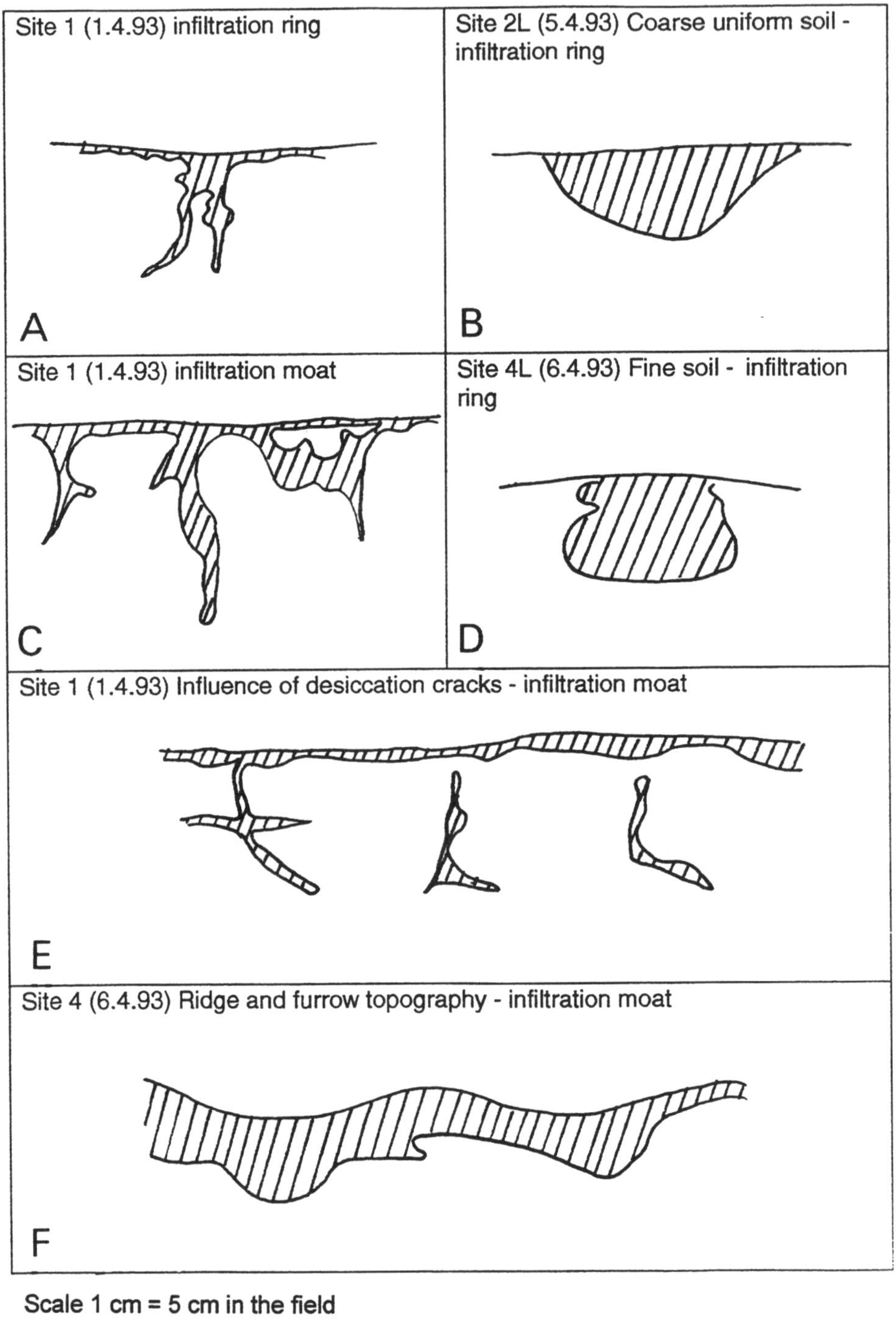

Fig. 6. Wetted profiles for infiltration tests at six sites. Results for both infiltration rings and infiltration moats are shown. Following the infiltration test a vertical face through the centre of the test areas was excavated and the extent of the wetting front determined.

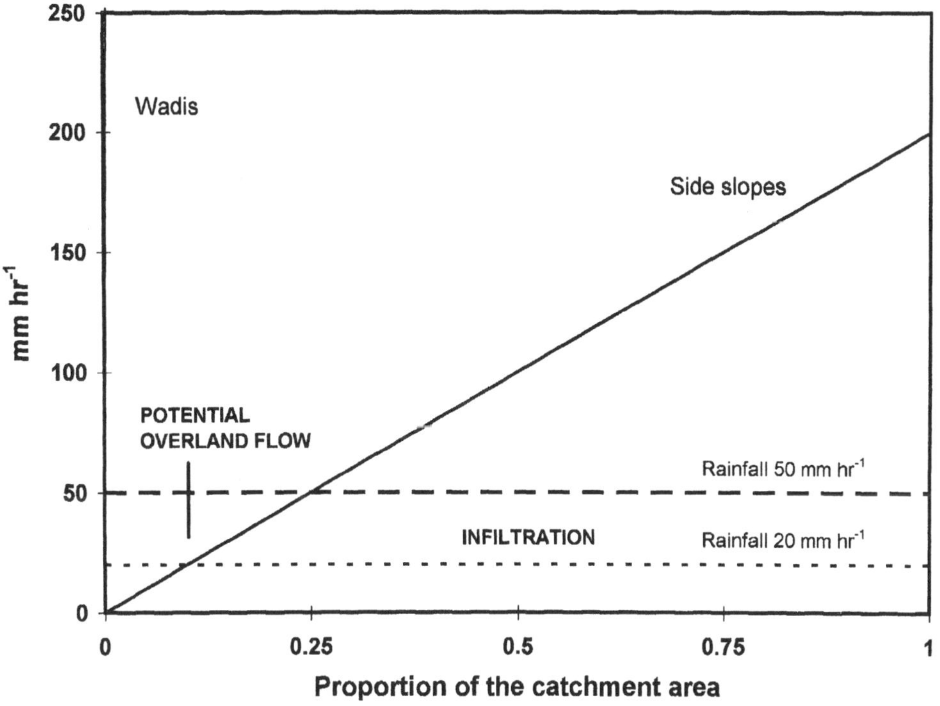

Fig. 7. Illustration of the Manley (1977) method of estimating catchment infiltration and potential overland flow. The inclined line represents an approximation of the areal distribution of infiltration based on the maximum infiltration capacity of the side slope areas. A second near vertical line represents an approximation of infiltration for the wadis. The two horizontal lines show rainfall rates of 20 and 50mm/hr. After Dingman (1994).

example D the infiltration ring test for the fine soil shows a well-defined wetting bulb. Lateral infiltration is far less than in the coarse soil (diagram B). Diagrams C and E show the wetting front over a larger area. This illustrates that the lines of maximum wetting front penetration are relatively evenly spaced (10 to 20cm). Lateral infiltration is high just below the surface crust and is directed along vertical fissures. These vertical fissures correspond to desiccation cracks within the surface soil horizons. This leads to very high local spatial variability in infiltration rates and could explain why the use of small infiltration rings may result in large errors when trying to estimate average infiltration rates. The final diagram in Figure 6 (F) shows the wetted profile for an area of ridge and furrow topography in an area of ploughed *qa'a.* Although infiltration is relatively uniform, there is greater wetting front penetration in the furrows. This is probably caused by convergence of flow lines and ponding of water in these depressions.

Infiltration into the soil profile is far from uniform. Lateral spreading involving a substantial horizontal infiltration rate was observed at many of the wadi gravel sites. Lateral infiltration six times the infiltration ring diameter was observed at one site. In finer grained deposits macropores, related to soil desiccation cracks, were seen to open up as water infiltrated. Macropores of 15cm length and 2cm width were observed just below the surface. Piping tends to occur in areas where high infiltration capacities are found in sediment of low intrinsic permeability. Pipes can enlarge or collapse in on themselves (Cooke *et al.* 1993) leading to surface water capture, increased pipeflow and the development of surface depressions. Excavations are therefore essential in understanding the process of infiltration and in interpreting measured rates.

5. APPLICATION OF THE RESULTS

Infiltration is a fundamental hydrological process and has major implications for runoff generation (flood hazard and erosion control) and water harvesting and agriculture practices in arid areas. This section attempts to apply some of the results from the reconnaissance survey to estimate runoff potential and determine best practice for sprinkling irrigation.

The physical response of a catchment to rainfall requires an understanding of the spatial variability of infiltration. High spatial variability in arid and semi-arid catchments requires careful estimation of infiltration properties (Yair and Lavee 1976; Berndtsson and Larson 1987). This implies that the partial area concept for runoff generation is important in catchments of this kind. Therefore point estimates of infiltration are difficult to use in estimating catchment runoff because of large variations in measured rates over short distances. This variability is difficult to incorporate in runoff models; however, a simple approach is to assume a linear cumulative distribution of infiltration capacity (i_c) over the entire catchment (Manley 1977; Dingman 1994). Applying this approach in the present study and differentiating between side slopes and wadis a simple rainfall-runoff relationship can be defined. In Figure 7 the inclined line represents the cumulative distribution of side-slope infiltration capacities reaching a maximum value at 200mm/hr (see Fig. 5, Table 1). A similar distribution is defined for the wadis but due to the extremely high infiltration capacities the slope on this line is so steep it is almost parallel to the vertical axis. Horizontal lines define rainfall rates (p) selected at 20 and 50mm/hr representing moderate and heavy intensities. The intersection of the horizontal line with the inclined lines defines an area proportional to the rate of runoff (R). Therefore for a rainfall intensity of 20mm/hr approximately 10 per cent of the side slopes may generate overland flow. At a rate of 50mm/hr this rises to almost 25 per cent. Direct precipitation on to the wadis will never generate runoff given the very high infiltration capacities. Runoff rates can also be calculated using the formula proposed by Manley (1977):

$$R = \frac{p^2}{2i_c}$$

For the rainfall intensities of 20 and 50mm/hr this yields runoff rates of 1 and 6.25mm/hr respectively from the side slopes. Therefore, although this analysis is very simple and makes gross assumptions about the hydrological response of this environment it nevertheless generates two important hypotheses. First, wadis do not contribute directly to runoff (through infiltration-excess overland flow) and second, side slopes only generate significant amounts of runoff via overland flow during periods of high intensity rainfall when a large part of the catchment is contributing.

It should, however, be recognized that these conclusions are sensitive to the estimation of the maximum infiltration capacity (set at 200mm/hr in this scenario). A value half this may be more realistic which would double the runoff rate and the proportion of the catchment contributing to runoff.

Estimates of infiltration are also important when developing irrigation schemes. Hillel (1980) provides a useful illustration of this using the Green and Ampt theory which has been shown to provide satisfactory results for infiltration into initially dry, relatively coarse soils of the type encountered in the study area. The infiltration rate is given by:

$$i = i_c + b / I$$

where i_c is the steady infiltration rate, I is cumulative infiltration and b is a constant. Using the values given in Table 2 for two of the test sites representing wadi and side slope conditions it is possible to solve for b and estimate infiltration rate for any value of cumulative infiltration. For example, using the data for the side slope site in Table 2 and the equation shown above, to infiltrate 250mm by steady sprinkling without exceeding the infiltration capacity would require a rate of 9.14mm/hr. These calculations are essential in planning irrigation so that flooding, runoff and erosion can be avoided.

Table 2. Infiltration data for two test sites representing side slope and wadi areas; used in calculation of infiltration characteristics for irrigation practices.

	Side slope	Wadi gravel
i (infiltration rate) mm/hr	24	40,600
i_c (steady infiltration rate) mm/hr	8	31,100
I (cumulative infiltration) mm	17.75	5,200
b	284	49,400

6. LIMITATIONS OF THE METHODOLOGY

An obvious limitation of the methodology used in this study is the small size of the infiltration rings. It has been demonstrated that the size of the infiltration ring is important if representative rates are to be obtained. Bouwer (1986) recommends a ring of at least 1m in diameter. The smaller the ring the more susceptible it is to local variability; however, few studies have been undertaken to estimate spatial variability in infiltration rates (see Sharma *et al.* 1980). Results of tests with different sizes of rings indicate that values can only be reliably reproduced with rings of greater than 150mm in diameter (Youngs 1991). With smaller rings local heterogeneity in soil structure produces significant errors. It has been argued that in order to estimate to within 30 per cent of the true mean rate, six replicate measurements are required (Knapp 1978). However this will vary with local soil conditions. Generally speaking the infiltration rate in smaller rings is higher than in larger rings and it reaches a uniform rate sooner (Youngs 1987).

Ideally estimates of infiltration using infiltrometers should be of the vertical infiltration rate. Unfortunately in most infiltration tests the water diverges laterally as it penetrates the soil. Therefore the infiltration rate within the infiltrometer will be higher than the true vertical rate. Lateral divergence can be caused by capillary forces, layers of reduced hydraulic conductivity low in the profile and by variations in water depth in the cylinder. The overestimation of vertical infiltration depends on the ratio of cylinder diameter and unsaturated flow capability of the soil. This means that the smaller the cylinder, the greater the overestimation of infiltration. Furthermore Bouwer (1961) demonstrated that infiltration rates within a cylinder infiltrometer are much greater than the true vertical rate. Therefore it is wrong to use a second ring to buffer for edge effects and lateral divergence (Swartzendruber and Olson 1961). The only way to reduce this is to make the infiltrometer as large as possible. Restricting layers also greatly enhances infiltration rates by causing the infiltrated water to spread laterally over an area much larger than the area of the infiltrometer. The pressure head of water at the surface is another factor: the larger the pressure head relative to the cylinder diameter the more the flow will diverge and the greater the divergence. The best way to overcome this is to reduce water depth in the cylinder.

When carrying out infiltrometer tests it is important to replicate actual field conditions as closely as possible (e.g. water quality, soil types and surface conditions). Errors arise due to soil disturbances such as compaction and/or crust destruction leading to reduced and enhanced infiltration rates respectively. This is reduced by using large diameter thin-walled cylinders. Water also needed to be added carefully so as not to disturb the soil surface. A very coarse dish cloth or similar fabric can be placed in the cylinder to protect the soil when adding water. Water quality and temperature should also be maintained at field conditions. However, the water used in the present test was clear water devoid of suspended sediments, which is very unlike the situation during hillslope runoff and flooding. Also, strictly speaking ring infiltrometers should not be used where water reaches the ground surface by rainfall: sprinkler-type infiltrometers are more appropriate.

The results of a comparison between duplicate infiltration tests of surfaces of similar soil type (Fig. 8) support this general approach. Values for replicate tests plot reasonably close to the line of perfect agreement. The log-log scale is used because of the large ranges of the infiltration values; however, this will tend to reduce the scatter in the data. Nevertheless replicates plot within the correct range and no major discrepancies occur. However, it is acknowledged that, because of the factors outlined above, infiltration rates measured by this method will be higher than the true vertical infiltration rate.

7. CONCLUSION

Surface water infiltration characteristics in this arid area of North-east Jordan show considerable variability. Infiltration rates vary over five orders of magnitude. Wadi systems have very high rates of infiltration but side slopes and closed depressions are characterized by low rates. This implies that under heavy rainfall, runoff can be readily generated from side slopes but wadis will never generate direct runoff. Transmission losses (and indirect groundwater recharge) are likely to be high along wadi courses but any water reaching the closed basins (*qa'as* and *marabs*) may exist temporarily as surface storage due to the low infiltration rates encountered in these areas. This hypothesis cannot be fully evaluated because measured rates of infiltration only apply to the upper soil layers, and the lateral and vertical extent of the various land system units has not been determined. Estimating the infiltration characteristics of these various land systems elements is nevertheless useful in preliminary water resource planning. Using measured infiltration rates, runoff coefficients can be estimated and rates of sprinkling irrigation determined.

The method of rapid assessment of infiltration characteristics used in this study provides acceptable estimates of infiltration rates and provides an insight into the main factors controlling infiltration. Measured rates are likely to be greater than the true vertical infiltration rate due to the small size of the infiltrometer rings used in the tests.

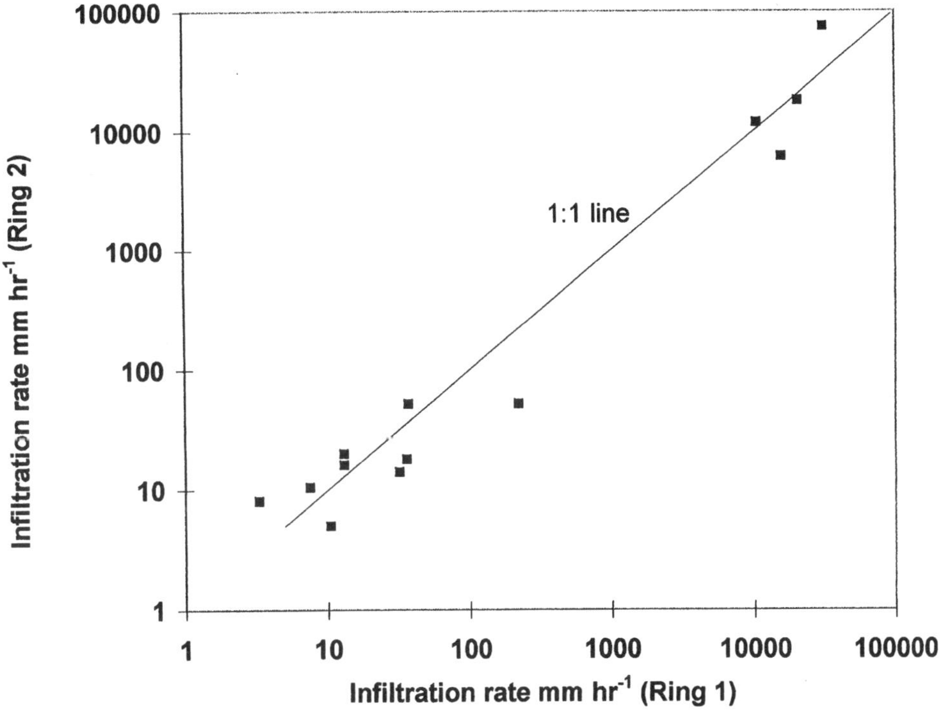

Fig. 8. Comparison of infiltration rates from paired samples. Tests were carried out on the same surface approximately 5 to 10 metres apart. Logarithmic scales are used in the diagram due to the large range in measured infiltration rates.

This method could be applied to other arid and semi-arid zones where vegetation is sparse and basic hydrological data are lacking.

REFERENCES

Abrahams, A.D., Parsons, A.J. and Luk, S. (1988) Hydrologic and sediment responses to simulated rainfall on desert hillslopes in southern Arizona. *Catena*, 15, 103-17.

Agnew, C. and Anderson, E. (1992) *Water Resources in the Arid Realm*. Routledge.

Al-Homoud, A.S., Allison, R.J., Sunna, B.F. and White, K. (1995) Geology, geomorphology, hydrology, groundwater and physical resources of the desertified Badia environment in Jordan. *GeoJournal*, 37, 51-67.

Berndtsson, R. and Larson, M. (1987) Spatial variability of infiltration in a semi-arid environment. *Journal of Hydrology*, 90, 117-33.

Betson, R.P. (1964) What is watershed runoff? *Journal of Geophysical Research*, 69, 8, 1541-52.

Bouwer, H. (1961) A study of final infiltration rates from cylinder infiltrometers and irrigation furrows with an electrical resistance network. *Transactions of the 7th International Congress of Soil Science*, 6, 448-56.

Bouwer, H. (1986) Intake rate: cylinder infiltrometer. (Chapter 32) In Klute, A. (ed.) *Methods of Soil Analysis. Part 1 Physical and Mineralogical Methods* (2nd Edition). Agronomy 9, 1. American Society of Agronomy/Soil Science Society of America, 825-44.

Cooke, R., Warren, A. and Goudie, A. (1993) *Desert Geomorphology*. UCL Press, London, 526p.

Dingman, S.L. (1994) *Physical Hydrology*. Macmillan Publishing Company, New York.

Hillel, D. (1980) *Applications of Soil Physics*. Academic Press.

Hudson, N.W. (1987) Soil and water conservation in semi-arid areas. *FAO Soils Bulletin 57*, UN, Rome, 172p.

Knapp, B.J. (1978) Infiltration and storage of soil water. In Kirkby, M.J. (ed.) *Hillslope Hydrology*. John Wiley and Sons, Chichester, 43-72.

Lloyd, J.W. (1986) A review of aridity and groundwater. *Hydrological Processes*, 1, 63-78.

Manley, R.E. (1977) The soil moisture component of mathematical catchment simulation models. *Journal of Hydrology*, 35, 341-56.

Sharma, M.L., Gander, G.A. and Hunt, C.G. (1980) Spatial variability of infiltration in a watershed. *Journal of Hydrology*, 45, 101-22.

Swartzendruber, D. and Olson, T.C. (1961) Model study of the double ring infiltrometer as affected by depth of wetting and particle size. *Soil Science*, 92, 219-25.

Thornes, J.B. (1977) Channel changes in ephemeral streams: observations, problems and models. In Gregory, K.J. (ed.) *River Channel Changes*, John Wiley, Chichester, 317-35.

Yair, A. and Lavee, H. (1976) Runoff generative process and runoff yield on arid debris mantled slopes. *Earth Surface Processes*, 1, 235-47.

Youngs, E.G. (1987) Estimating hydraulic conductivity values from ring infiltrometer measurements. *Journal of Soil Science*, 38, 623-32.

Youngs, E.G. (1991) Infiltration measurements - a review. *Hydrological Processes*, 5, 309-20.

CHAPTER 13

Characteristics of the Basalt Boulder Surfaces

DAVID L. HIGGITT
ROBERT J. ALLISON
(*University of Durham*)

1. FORMATION AND SIGNIFICANCE OF BOULDER COVERED TERRAIN

The majority of ground surfaces in the Eastern Badia are covered by basalt boulders. The size and spacing of clasts varies and associations with particular lava flow types are apparent. The clasts form a rock armour protecting the underlying orange-brown silty sediment from deflation and providing an enhanced surface roughness which promotes the deposition of airborne sediment deflated from either local or more remote locations, such as *qa'a* or *marab* surfaces. Disruption of the natural, boulder-strewn surfaces will reduce the resistance to erosion by wind and runoff afforded by the boulders. In the north-west of the Programme Area, where phases of agricultural development have taken place, the basalt boulder surfaces have been partly removed in order to create fields.

Stone surfaces are characteristic of many desert areas but their origin remains a matter of contention (Dixon 1994). The term *desert pavement* is generally applied to stone mantles in dryland regions but is usually reserved for relatively uniform mantles of one to two stones thickness, with relatively short axial dimensions, overlying finer material. The Arabic term, *hammada* is applied to boulder-strewn outcrops, pavements and rock surfaces with a residual scatter of stones (Cooke *et al.* 1993) and encompasses the variety in basalt boulder terrain found in the Badia. A quantitative analysis of the characteristics of basalt boulder surfaces has been conducted. The objectives were to identify surface characteristics associated with particular basalt types, evaluate evidence for the organization of boulder surfaces along slope profiles and to elucidate the formation of the surfaces and their interrelationships with slope evolution.

A number of evolutionary models have been advanced to explain desert pavement formation, which can be summarised as the deflation, water sorting, upward migration and aeolian infiltration hypotheses (Fig. 1). The first three of these provide explanations for mechanisms whereby clasts can be exhumed from a matrix of finer sediment. The finer sediment may be removed by the action of wind or runoff or the coarse particles may migrate towards the surface during freeze-thaw or wetting-drying cycles. The fourth hypothesis suggests that the clasts originate at the surface and that the underlying sediment increases in thickness over time as airborne dust is trapped between the clasts, gradually raising the elevation of the whole surface (McFadden *et al.* 1987).

Sections exposed in quarries and road cuttings in the Badia provide an opportunity to observe the vertical distribution of clasts. The upward migration hypothesis is supported

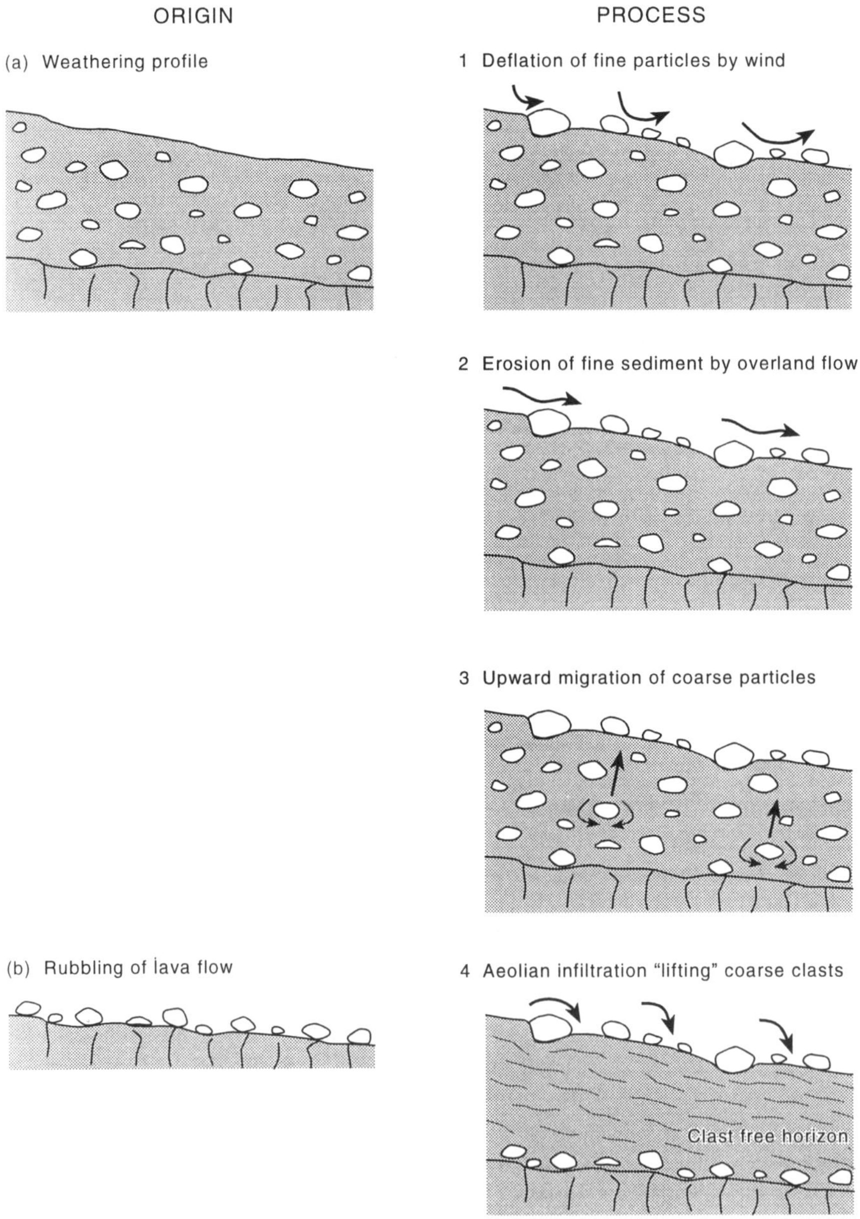

Fig. 1. Four possible models of boulder-strewn surface evolution.

by observations of spheroidal weathering of clasts throughout the profile. However, there appears to be a depletion of intact basalt clasts immediately below the surface, which may indicate the addition of aeolian sediment. The thickness of the sediment beneath the boulder cover is frequently observed to increase in a downslope direction, suggesting that sediment is transported by water during some precipitation events. Small rills, sediment splays and occasional scouring around the larger clasts provides confirmation of contemporary waterborne sediment transfer on boulder-covered slopes, although it should be noted that rates may have been higher during more humid periods in the past. Towards the crest of slopes, the sediment underlying the boulders is thin or sometimes absent and the boulders may be perched directly on bedrock. In these locations the majority of boulders stand proud of the substrate, such that a high proportion of the clast surface area is exposed. As a consequence of the increasing thickness of the sediment downslope, clasts become partially buried. A mechanism for the generation of the boulder covered surface through removal of fine sediment by runoff is implied.

Examination of the boulder surface characteristics alone cannot provide sufficient information to discriminate between the competing surface formation hypotheses. Observations suggest that an interplay of processes is likely to have occurred and further examination of the geochemistry of the sediments underlying the boulder surfaces and of the regolith of each of the basalt types is in progress to this end. It is likely that the balance of process operation will have shifted as climatic conditions changed. Little is known about the palaeoclimate of the Programme Area but it is evident from the widespread occurrence of caliche and from some documentary materials that wetter phases have occurred in the recent past. While the definitive set of processes for the formation of the boulder surfaces is not yet established, examination of their characteristics in relation to basalt type and slope position yields some useful information.

2. METHODS

A sampling framework for investigating boulder cover characteristics was undertaken in conjunction with slope profile survey (Allison and Higgitt, in press). At 13 of the 25 sites where profiles were measured, five points were located at the same relative location on each slope (crest, upper, mid, lower, toe). At each point a 2m by 2m quadrat was established on the ground and a number of measurements undertaken. The sampling programme included four different types of lithology and a variety of profile shapes and lengths (Table 1).

The quadrat sampling strategy was used in preference to a grid-based sampling framework where clasts are measured at the intersection of grid nodes over a larger area. Although such an approach avoids the problems of mis-representation by small-scale variation in boulder characteristics, it is time-consuming to establish and has problems of bias because larger clasts are more likely to fall under the grid nodes (Dunkerley 1996). Quadrats were established randomly at each point but their locations were rejected if affected by atypical microtopography. The quadrat measurement routine also had logistic advantages enabling measurements to be undertaken speedily by two field operators. Within each quadrat the long (a-axis) and intermediate (b-axis) diameter of each clast was measured. The thickness (c-axis) of each clast was ignored at this stage.

The a- and b- axis permit calculation of the two-dimensional surface area of each clast using a simple ellipse formula, where

$$A = \pi \, a \, b / 4$$

Table 1. Summary of site locations and properties

Basalt type	Slope form	Toe slope characteristic	Location
Abed	Convex-straight	Wadi	Wadi Safawi
Abed	Convex	*Marab*	Marab Suway'id
Abed	Convex	*Marab*	Marab Suway'id
Madhala	Convex	Wadi	Wadi Safawi
Madhala	Concave	Large *qa'a*	Qa'a al Buqei'wiya
Madhala	Concave	Large *qa'a*	Qa'a al Buqei'wiya
Bishriyya	Concave	Small *qa'a*	Shubayka road
Bishriyya	Concave-straight	Small *qa'a*	Shubayka road
Bishriyya - cleared	Concave	Field (arable)	al-Jidah
Bishriyya - cleared	Concave-straight	Field (grazing)	al-Jidah
Bishriyya - cleared	Concave-straight	Field (grazing)	al-Jidah
Scoracious	Concave	Small *qa'a*	Tulul ash Shahba
Scoracious	Concave	Small *qa'a*	Tulul ash Shahba

Although the area estimate assumes a regular plan form and clasts smaller than 5mm are ignored, the sum of the clast surface areas provide a measure of percentage surface cover of the quadrat. The number of clasts within the quadrat area varied according to relative slope position and basalt type, ranging from a minimum of 13 to a maximum of 151 (mean = 67.8). The size of clasts is conventionally represented by the mean intermediate b-axis length but the measurement of all clasts within the quadrat provides information on size distributions, dispersion and shape (represented by a/b axis ratio).

In addition to boulder dimension and cover statistics, information on boulder exposure was collected. A downslope increase in the thickness of the fine sediment underlying the basalt cover is apparent on many slopes, such that many clasts at lower elevations are partly buried. The extent to which boulders are exposed above or buried within the underlying sediment can be identified clearly in the field because of the marked staining where the rock has been in contact with the sediment. Ten randomly selected clasts within a 5m radius of the quadrat were exhumed and the proportions of the stained and unstained surfaces measured at each of the relevant sample locations. The computation of the volume of each boulder exposed above the contact line is by no means straightforward and an elliptical function was used to estimate this property. The average exposure at each sampling site was calculated as a volume weighted mean of the ten exhumed clasts.

The boulder sampling framework enables a number of properties and associations to be evaluated. These include clast dimensions (size distributions, degree of sorting and shape), ground cover and degree of clast exposure. Each of the variables can be examined in relation to lithological and slope properties. Analysis of Variance (ANOVA) is used to evaluate the statistical significance of variation between groups (basalt types, slope positions) relative to within group variation. The technique must be used with caution as a significant result may sometimes be attributed to one particular group being different from the rest. Where more detailed examination of variation between two samples is required, the t-test for difference of means can be employed.

3. LITHOLOGICAL VARIATION IN BOULDER COVER CHARACTERISTICS

The characteristics of the boulder-covered landscape varies considerably across the Badia. Indeed, changes in the nature of the surface cover provide the clearest indication in the field of a transition in the underlying geology. It is also the variation in boulder cover characteristics which helps to distinguish between different lava flows on satellite imagery.

The experimental framework enabled four types of basalt lithology to be examined - Abed, Madhala, Bishriyya and Scoracious material. The nature of the lithologies and their general landscape characteristics have been described elsewhere (Ch. 2). Together with the Salaman Flood Basalt, the Abed, Madhala and Bishriyya are the most important of the basalt types in the study area. Salaman outcrops are often found in association with the Abed and in this case no separate Salaman study plots were established. The fourth type of lithology considered was Scoracious volcanic deposits associated with recent volcanic cones. Their distribution is mainly linked to contemporary dyke systems and major structural lineaments. Because of their recent formation, the topography associated with the Scoracious deposits is the most dramatic in the study area and the slope gradients, lengths and relative relief tend to be longer than other lithologies. Three of the slopes located on Bishriyya lava flows have been subjected to some agricultural development in the past. Both slopes are located in the north-west of the Programme Area and are adjacent to locations where irrigation-fed agriculture has been introduced recently. Although these slopes are used for rough grazing at present, it appears that they have been exploited for rain-fed agriculture in the past, possibly during the Roman Period. As such the original basalt boulder cover will have been disrupted by human activity including clearance and trampling by grazing stock and are therefore considered as a separate category of 'cleared' slopes. In the drier regions of the Badia there is also evidence for disruption of the natural boulder surfaces for the construction of boundary cairns, burial mounds, corrals, tent settlement sites and a network of tracks. However, disruption is spatially limited and none of the other sites are considered to have been altered by human activity to any significant degree.

3.1. Boulder dimensions

The size of clasts is conventionally summarized by the mean intermediate diameter and a starting place for the analysis is to examine the difference in mean basalt clast size between different lithologies. Table 2 indicates the mean and standard deviation of mean intermediate diameter, b, from the individual quadrats. It is interesting to note that clast sizes are ranked in accordance with basalt age with the oldest basalt (Abed) having mean clast size of 16.2cm compared with 10.5cm on the recent Scoracious material. The size difference may owe more to the way in which the joint structure of the original lava flow has controlled weathering. Analysis of variance demonstrates that the variation between groups is significantly greater than the variation within groups and the majority of basalt types are distinctive from one another. A series of t-tests establishes that the Abed clast

Table 2. Mean intermediate diameter of clasts by basalt types

Basalt Type	Mean Intermediate diameter, *b*, (cm)	Standard Deviation	Significant difference (∝ = 0.05) with:
Abed	16.24	1.43	B, C, M, S
Bishriyya	11.98	1.17	A, C, M
Madhala	12.83	2.54	A, B, S
Scoracious	10.48	1.98	A, C, M
'Cleared sites'	14.11	2.07	A, B, S
ANOVA	df	F	Probability value
	63	11.92	0.000

sizes are significantly different from all other basalt types. The Bishriyya and cleared sites are also significantly different. Although the sites are located on different Bishriyya lava flows which may have inherent variation in surface characteristics, the implication is that human interference has led to a coarsening of the boulder surface.

The use of mean b-axis as a summary measurement from each quadrat raises some problems. As noted in many previous investigations of both desert surfaces (Abrahams and Parsons 1991) and fluvial gravels (Diplas and Fripp 1992), the majority of clast size distributions are positively skewed. In many ways the geometric mean might be considered a more reliable statistic:

$$\log b = \Sigma \log b_i / n$$

although the arithmetic mean is retained here because of convention. Recent work on the statistical properties of sediment mixtures (Dunkerley 1995) has indicated that a variety of weightings can be used to determine mean values and that the arithmetic mean, where the weighting component is number of clasts, may not be the most appropriate for hydrological modelling. The skewness and the standard deviation of individual quadrat measurements provide useful statistics in their own right. Because standard deviation, σ_b is partly dependent on clast size, the coefficient of variation, CV_b, provides a relative measure of dispersion:

$$CV_b = (\sigma_b / b) \times 100$$

The smaller the coefficient of variation, the more uniform the range of clast intermediate diameters in the sample which may indicate a higher degree of sorting by transport processes. Comparison of mean b-axis skewness and coefficient of variation from the individual quadrats again reveals some distinctions between the lithologies (Table 3). Once more, an age-related pattern is suggested with the Abed displaying the least skewed distributions and the lowest mean CV_b. Analysis of variance confirms the significant variation in skewness between basalt types but the equivalent test for Cv_b is not significant at $\propto = 0.05$. The evidence for preferential sorting is not strong but it is suggested that the older basalt flows are characterised by relatively few smaller clasts, which may have been preferentially removed over time. The shape of the clasts, characterized by the a/b axis, are highly scattered with Abed clasts being the most rounded (mean a/b = 1.46) and Scoracious the most angular (1.55) but there is no clear evidence for preferential shape sorting on any basalt type.

Table 3. Clast size distribution and sorting by basalt type

Basalt	Mean skewness	Standard deviation	Mean Cv_b (%)	Standard deviation
Abed	0.71	0.80	34.83	5.88
Bishriyya	1.41	0.50	37.81	10.99
Madhala	1.07	0.62	36.83	6.61
Scoracious	1.52	0.58	43.97	19.93
'Cleared sites'	1.11	0.59	43.77	6.61
ANOVA	df	F	Probability value	
skewness	63	3.01	0.025	
CV_b	63	2.15	0.086	

3.2. Ground Cover Characteristics

Ground cover characteristics include the percentage of the surface comprising basalt clasts and the degree of exposure of clasts from the underlying sediment. Not only are the Abed clasts larger than other basalt types, but they provide the highest degree of ground cover, comprising on average half the ground surface area. Ground cover afforded by Bishriyya and Madhala basalts are at similar levels but the cover on Scoracious materials is much lower (Table 4). The low cover may be misleading as the amount of sediment on the Scoracious slopes is limited and the clasts often overly intact bedrock or a gravely layer. Indeed the younger Bishriyya and Scoracious lithologies can be recognized from distance by their dark colour due to the lack of exposed orange-brown sediment.

Table 4. Ground surface cover and degree of clast exposure by basalt type

Basalt	Mean Ground Cover (%)	Standard deviation	Mean Exposure (%)	Standard deviation
Abed	50.77	14.27	62.80	16.44
Bishriyya	46.44	21.71	58.06	18.25
Madhala	39.61	15.30	61.22	10.41
Scoracious	19.68	13.11	61.72	16.29
'Cleared sites'	29.61	11.95	46.05	10.41
ANOVA	df	F	Probability value	
Ground Cover	63	7.54	0.000	
Exposure	63	3.40	0.014	
Exposure (excluding 'cleared')	48	0.20	0.900	

It might be expected that the variation in clast size and ground cover would lead to difference in clast exposure as sediment is transported downslope. Although, Analysis of Variance suggests a significant difference between the five categories, closer inspection shows that this is accounted for by the cleared sites where many clasts are embedded in the sediment. Recalculating ANOVA excluding the cleared category indicates a higher degree of variability within groups than between them. The high standard deviations are due to the importance of slope position as a control on the exhumation or burial of clasts and the variation in basalt boulder characteristics according to slope position is now considered.

4. SLOPE - BOULDER RELATIONSHIPS

4.1. Boulder Dimensions

Having established that some variation in the size distributions of clasts on the Badia surfaces are dependent on the underlying geology, the study also enables downslope changes in cover characteristics to be analysed. Consistent variation in the size, shape and degree of sorting would suggest an interaction between boulder surface formation and the transport of clasts. For example, slope variables such as gradient, profile curvature and length influence the hydraulics of overland flow which in turn may be capable of transporting part of the size range of the basalt surface cover. Although the profile data collected during the study permits the calculation of a variety of slope properties for each individual quadrat, this short review concentrates only on the relative position of quadrats as defined in the methods section.

Clast dimensions from many individual profiles tend to display a slight peak between the crest and toe slope locations, but Analysis of Variance on the aggregated data does not reveal any significant difference in mean intermediate diameter and relative slope position (Table 5). When the data is disaggregated according to basalt type clearer patterns begin to emerge. Mean intermediate diameter is highest at the upper and mid slope locations on Bishriyya and Scoracious slopes and then declines downslope. On the Abed, clast sizes increase towards the lower slope position, while the Madhala slopes show much dispersion. The first two lithological categories have concave slope form, the Abed slopes are convex and the Madhala are mixed. The peak in mean intermediate diameter therefore coincides with the locus of maximum steepness.

The b-axis skewness and the degree of sorting data are also highly scattered and do not indicate any consistent pattern with slope position. The lower and toe sites on Abed slopes appear to be better sorted and it is here that skewness is at a minimum. On Scoracious slopes the upper to lower segment of the slopes are better sorted and the mid-slope clast distributions are only slightly skewed. It can be tentatively suggested that on the steepest parts of the slopes there is some evidence for enhanced sorting with the preferential removal of the smaller clasts, which causes size distributions to be less skewed. However, the number of observations is limited when the data are disaggregated by lithology and slope position, and further analysis and/or measurement is required.

Table 5. Mean size distributions of clasts by slope position

Slope position	Mean intermediate diameter, *b*, (cm) (± s.d.)	Mean skewness (± s.d.)	Mean CV_b (%) (± s.d.)
Crest	13.31 (2.41)	1.15 (0.63)	39.32 (12.94)
Upper	14.29 (2.23)	1.37 (0.72)	41.53 (5.92)
Mid	12.94 (2.25)	1.11 (0.67)	37.66 (6.57)
Lower	13.10 (3.39)	1.12 (0.60)	37.09 (10.04)
Toe	12.09 (2.67)	0.82 (0.77)	40.80 (16.32)
ANOVA	df	F	Probability value
b	63	1.13	0.349
skewness	63	1.05	0.389
CV_b	63	0.40	0.811

4.2. Ground Cover Characteristics

On most individual slope profiles a peak in stone cover is observed between the crest and toe-slope position and the pattern is generally similar to the b-axis length data (Table 6). Analysis of variance indicates a significant difference in ground cover between slope positions, dominated by the limited cover at the toe-slope position. As the toe-slope position is defined as the point where the slope merges into a *qa'a*, *marab* or wadi feature, the lack of ground cover here is not unexpected. When toe-slopes are removed from the analysis, no significant difference is observed between the remaining slope positions. Again, disaggregation into lithological units assists interpretation. The older basalts (Abed and Madhala) have a gentle downslope decline in ground cover, while the younger lithologies (Bishriyya and Scoracious) have limited ground cover at crest sites and maximum cover at upper sites.

Table 6. Ground surface cover and degree of clast exposure by slope position

Slope position	Mean Ground Cover (%)	Standard deviation	Mean Exposure (%)	Standard deviation
Crest	42.88	17.08	69.48	11.67
Upper	48.11	14.94	65.75	10.74
Mid	40.98	16.02	58.19	11.86
Lower	38.65	18.42	55.79	12.54
Toe	16.96	9.00	37.35	6.53
ANOVA	df	F	Probability value	
Ground Cover	63	7.33	0.000	
Ground Cover (excluding toe-slope sites)	51	0.76	0.523	
Exposure	63	16.02	0.000	

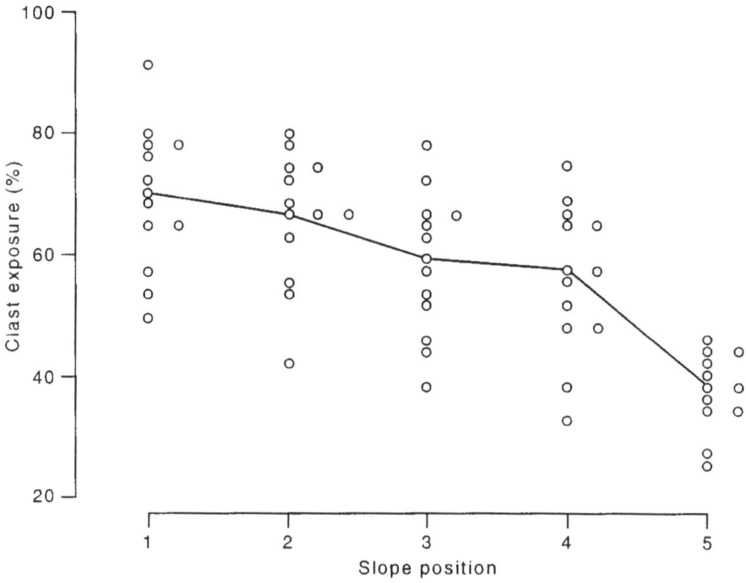

Fig. 2. Relationship between clast exposure against slope position.

The clast exposure data is perhaps the most convincing variable which changes with slope position. A dot-plot connecting median values shows a clear decline in the proportion of basalt clasts exposed above the surface of the fine sediment (Fig. 2). Exposure appears to decline downslope on concave profiles, whereas the Abed clasts remain at a constant degree of exposure at all sites except the toe-slope position. The clast exposure data provide an opportunity to examine interrelationships between sediment transport, slope evolution and boulder surface formation which are currently in progress.

5. DISCUSSION

The basalt boulder strewn surface of the Badia is a ubiquitous feature. However, the analysis of basalt clast size, cover and exposure distributions begins to demonstrate consistent variation according to geology and slope position. In general terms, geological controls are the primary influence on the size of clasts present at the surface and of the percentage ground cover which they occupy. The operation of hydrological processes on the slopes of the Badia has an important influence in modifying the nature of the basalt covered surface, through the erosion and deposition of the fine sediment underlying the basalt cover and by the size sorting of the clasts themselves. A consistent pattern between clast size and sorting by slope position on different basalt types highlights the importance of gradient as a control on surface cover characteristics.

The analysis of clast data has shown the importance of disaggregating the sample by geological and slope characteristics in order to elucidate patterns. A further issue which requires consideration is the extent to which the clast variables are interrelated. It might be expected, for example, that the size of the basalt clasts have an overriding influence on the degree of clast exposure and ground cover. A significant positive correlation (r =

0.528) is observed between b-axis length and stone cover, with significant relationships existing for each of the individual lithologies. Curiously the relationship for Abed is negative, such that the larger the mean b-axis length on Abed slopes, the further spaced the individual clasts. The Abed slopes are indeed characterized by coarse open-work boulder surfaces. There is no relationship between mean b-axis length and degree of exposure (r = 0.160), as might be expected since a greater depth of sediment is required to be transported to cover larger clasts. Exposure is positively correlated with ground cover (r = 0.483), but the relationship falls apart if the toe-slope locations are removed from the analysis.

6. CONCLUSION

The paper has described and accounted for variation in the characteristics of basalt boulder surfaces in the Programme Area. While understanding of the nature, evolution and process dynamics of desert boulder surfaces is of interest to geomorphologists in its own right, the applications of the investigation require some consideration. Characterization of the surface is primarily of importance for hydrological modelling, where the extent of stone cover and its relationship with the underlying sediment in terms of exposure or embedding is an important control on infiltration and runoff generation (Abrahams and Parsons 1991; Poeson and Lavee 1994; Dunkerley 1995). The decline in the volume exposure of boulders along slope profiles indicates the importance of downslope sediment transport as a geomorphological process in the Badia environment, even though the origin of the basalt boulder surfaces may involve aeolian activity. The removal of basalt clasts for agricultural development changes the hydrological processes of the surface drastically. The stone cover provides protection of the underlying sediment from soil erosion by wind and water and is susceptible to damage. Comparison of the natural and cleared Bishriyya slopes indicates numerous changes in the characteristics of the surface properties such as the preferential removal of the smaller clasts, reduction of surface protection and enhanced embedding of clasts within the fine sediment. Improved understanding of the relationships between the basalt boulder ground surface characteristics and the operation of geomorphological processes offers an opportunity for improved management of the Badia soils.

REFERENCES

Abrahams, A.D. and Parsons, A.J. (1991) Resistance to overland flow on desert pavement and its implications for sediment transport modeling. *Water Resources Research*, 27, 1827-36.

Allison, R.J. and Higgitt, D.L. (in press) Slope form and associations with ground surface boulder cover in the Eastern Badia, Jordan. *Geomorphology*

Cooke, R.U., Warren, A. and Goudie, A.S. (1993) *Desert geomorphology*, UCL Press, London. pp526.

Diplas, P. and Fripp, J.B. (1992) Properties of various sediment sampling procedures. *Journal of Hydraulic Engineering*, 118, 955-70.

Dixon, J.C. (1994) Aridic soils, patterned ground and desert pavements. In Abrahams, A.D. and Parsons. A.J. (eds) *Geomorphology of desert environments*, Chapman and Hall, London, 64-81.

Dunkerley, D.L. (1995) Surface stone cover on desert hillslopes: parameterising characteristics relevant to infiltration and surface runoff. *Earth Surface Processes and Landforms*, 20, 207-18.

Dunkerley, D.L. (1996) Stone cover on desert hillslopes: Extent of bias in diameters estimated from grid samples and procedures for bias correction. *Earth Surface Processes and Landforms*, 21, 573-80.

McFadden, L.D., Wells, S.G. and Jercinovich, M.J. (1987) Influences of eolian and pedogenic processes on origin and evolution of desert pavements. *Geology*, 15, 504-8.

Poeson, J. and Lavee, H. (1994) Rock fragments in top soils: significance and processes. *Catena*, 23, 1-28.

CHAPTER 14

Checklist of the Flora

T.A. COPE
(Royal Botanic Gardens, Kew)

DAWUD EL-EISAWI
(University of Jordan)

The following is an interim checklist of the known flora of the Badia region. To date, three major collecting trips have been undertaken, all in early or late spring, between 1992 and 1996. Much of the earlier collecting concentrated on the area west of Safawi and north of the main road because this was where the transition between the two main local phytogeographical regions was most marked. To the west of Safawi the vegetation is largely of Mediterranean type, while that to the east is Irano-turanian. In practice, it was found that the cultivated areas towards the Syrian frontier north and west of Safawi were dominated by Mediterranean weeds, while the desert proper was the province of Irano-turanian species. The Mediterranean region is circumscribed by hot dry summers and mild moist winters, and has a largely annual weedy flora whose species pass the unfavourable growing season (summer) as seeds (but there are many exceptions). In contrast, the Irano-turanian region has similar hot dry summers, but markedly colder winters; the majority of species are perennial and pass the unfavourable season (winter) as underground rootstocks or bulbs (but again there are many exceptions). Some species occur throughout both regions, demanding only that winter be wet.

The checklist currently stands at 322 species in 46 vascular plant families. Many of the determinations of these species are provisional, awaiting examination by the appropriate specialists. The number appended to each species refers to the specimen number in the Badia Herbarium which currently contains 522 specimens. Those without numbers are either currently without appropriate specimens (0) or there are specimens awaiting processing and incorporation (A); the latter were all collected by Dr. Kevin Brown in 1996.

In addition to those listed below there are a further 49 species whose presence in the area under study has not yet been confirmed.

CHECKLIST OF THE BADIA FLORA

322 plant species in 46 families, ordered alphabetically

AIZOACEAE
Aizoon hispanicum (138)

ALLIACEAE
Allium ampeloprasum (508)
A. artemisietorum (60)
A. desertorum (305)
A. hierochuntinum (371)
A. sindjarense (401)
A. stamineum (486)

ANACARDIACEAE
Pistacia atlantica (259)

ASPHODELACEAE
Asphodelus tenuifolius (403)

BERBERIDACEAE
Bongardia chrysogonum (383)
Leontice leontopetalum (A)

BORAGINACEAE
Anchusa aegyptiaca (509)
A. milleri (348, 386)
Arnebia linearifolia (308, 394, 402)
Caccinia macranthera (385)
Gastrocotyle hispida (241, 362, 412)
Heliotropium ramosissimum (346)
Lappula spinocarpos (137, 208, 254)
Onosma aleppica (465)
Paracaryum rugulosum (80, 273, 335)
Rochelia disperma (472, 482)

CAPPARACEAE
Capparis ovata var. palaestina (210)
Maerua crassifolia (A)

CARYOPHYLLACEAE
Dianthus judaicus (439)
D. sp. (337)
Gypsophila capillaris (186, 264, 276, 306, 334)
Holosteum glutinosum (433, 488)
Minuartia picta (29, 63, 307)
Petrorhagia cretica (155, 185)
Polycarpon succulentum (A)
Robbairea delileana (257, 354)
Silene colorata subsp. oliveriana (62, 329, 495)
S. coniflora (435)
S. conoidea (314)
S. linearis (181, 424)
S. longipetala (467)
Spergularia diandra (28, 225, 252)

CHENOPODIACEAE
Agathophora alopecuroides (85)
Anabasis setifera (319)
A. syriaca (A)
Atriplex leucoclada (105)
A. tatarica (224, 361)
Bassia eriophora (315)
Chenopodium murale (407)
Halothamnus lancifolius (152, 284, 516)
Haloxylon salicornicum (347)
Hammada eigii (A)
Noaea mucronata (387)
Salsola villosa (73, 104)
Spinacia tetrandra (483, 484, 497)
Suaeda aegyptiaca (223)
Traganum nudatum (A)

CISTACEAE
Helianthemum lippii (269, 413)

COLCHICACEAE
Colchicum deserti-syriaci (449)
C. ritchii (341)

COMPOSITAE
Aaronsohnia factorovskyi (21, 65, 136, 154)
Acantholepis orientalis (506)
Achillea fragrantissima (31, 64, 111, 122, 250, 503, 507)
A. santolina (205, 379, 437)
Anthemis melampodina (169, 230)
A. pseudocotula (16, 295)

Artemisia herba-alba (34, 61, 115, 170)
Asteriscus hierochunticus (38, 72, 357)
Atractylis cancellata (267, 282)
A. mutica (33, 277, 285)
A. phaeolepis (A)
Calendula tripterocarpa (15, 114, 127)
Carduus pycnocephalus (41, 248)
Centaurea ammocyanus (23, 229, 290)
C. damascena (278)
C. iberica (378)
C. pallescens (12, 101, 139)
C. sinaica (190)
Cichorium pumilum (397)
Crepis aspera (97, 200)
C. kotschyana (206, 311)
Echinops glaberrimus (A)
E. polyceras (271)
E. spinosus (393)
Filago desertorum (221)
Gundelia tournefortii (0)
Gymnarrhena micrantha (67)
Koelpinia linearis (194, 367)
Lactuca serriola (414)
L. undulata (415, 519)
Lasiopogon muscoides (145)
Launaea nudicaulis (151, 249, 265, 296)
Leontodon laciniatus (215)
Leysera leyseroides (47, 166, 191)
Matricaria aurea (174)
Nauplius graveolens (125, 201)
Notobasis syriaca (247, 373)
Onopordum heteracanthum (356, 421)
Phagnalon rupestre (78, 107, 193)
Picris longirostris (14, 44, 266)
Pulicaria undulata (239, 514)
Reichardia tingitana (192, 261, 279)
Rhagadiolus angulosus (227)
Scariola orientalis (336)
Scorzonera judaica (310)
S. papposa (327, 340)
S. schweinfurthii (177, 343, 441, 476)
Senecio glaucus (131, 157, 316, 332, 360)
Tragopogon collinus (368)
Urospermum picroides (498)
Zoegea purpurea (96, 280)

CONVOLVULACEAE
Convolvulus fatmensis (120)
C. pilosellifolius (207, 355)
C. stachyidifolius (450)
Cuscuta gennesaretana (281, 291)

CRASSULACEAE
Umbilicus intermedius (77)

CRUCIFERAE
Alyssum marginatum (51, 268)
A. meniocoides (444, 479)
Biscutella didyma (391)
Brassica nigra (246)
Camelina hispida (431, 481)
Cardaria draba (374)
Carrichtera annua (406)
Chorispora purpurascens (253)
Clypeola jonthlaspi (490)
Diplotaxis harra (40, 117)
Eruca sativa (299, 398)
Erucaria boveana (410)
E. hispanica (156, 298, 330)
Erysimum crassipes (432)
Euclidium syriacum (372)
Farsetia aegyptia (204, 344, 453)
Isatis lusitanica (459)
Lepidium aucheri (113, 163, 217)
L. latifolium (513)
Malcolmia africana (416)
M. crenulata (232)
Matthiola longipetala (20, 159, 197, 302)
Notoceras bicorne (419, 510)
Sameraria armena (377)
Sinapis alba (A)
Sisymbrium irio (133, 233)
S. orientale (425)
S. runcinatum (375)
Texeira glastifolia (370)
Torularia torulosa (237, 301)
Zilla spinosa (83, 108)

CUCURBITACEAE
Citrullus colocynthis (69)

DIPSACACEAE
Cephelaria syriaca (448)
Pterocephalus brevis (118, 195, 422, 434, 493)
Scabiosa olivieri (161, 304, 423)
S. porphyroneura (90, 153)

EPHEDRACEAE
Ephedra transitoria (408, 491)

EUPHORBIACEAE
Andrachne telephioides (501)
Euphorbia chamaepeplus (475)
E. cuspidata (116, 272, 505)
E. falcata (48)
E. hierosolymitana (392, 452)
E. retusa (234)

GERANIACEAE
Erodium bryoniifolium (175, 187)
E. cicutarium (218)
E. deserti (168, 240)
E. glaucophyllum (A)
E. laciniatum
subsp. laciniatum (81)
subsp. pulverulentum (182, 409, 487)
Geranium molle (471)
G. tuberosum (454)

GRAMINEAE
Aegilops crassa (478)
A. cylindrica (321)
A. kotschyi (522)
A. searsii (442)
Aeluropus littoralis (349, 511)
Alopecurus utriculatus (469)
Arundo donax (0)
Avena barbata subsp. wiestii (91, 132)
Boissiera squarrosa (75, 389)
Brachypodium distachyum (42)
Bromus danthoniae (58, 110)
B. fasciculatus (148)
B. intermedius (188)
B. lanceolatus (447)
B. madritensis (2, 59, 214, 390)
B. rubens (128)
B. scoparius (7, 213, 243, 470)
B. tectorum (88, 144)
Crithopsis delileana (6, 56, 141, 165)
Cynodon dactylon (274, 512)
Dichanthium annulatum (351)
Echinaria capitata (451)
Eremopyrum bonaepartis (57, 135)
E. distans (143)
Heteranthelium piliferum (380)
Hordeum bulbosum (440)
H. marinum subsp. gussoneanum (518)
H. murinum subsp. glaucum (5, 55, 147)
H. spontaneum (457)
Loliolum subulatum (52, 134, 167)
Lolium temulentum (322)
Phalaris minor (8, 92, 164)
P. paradoxa (220)
Phragmites australis (0)
Poa bulbosa (219)
Polypogon monspeliensis (350)
Rhizocephalus orientalis (245)
Rostaria cristata (3, 142, 228)
Schismus arabicus (4, 54)
Stipa capensis (1, 53, 93)
S. hohenackeriana (426)
S. parviflora (76, 198)
Stipagrostis plumosa (345)

HYACINTHACEAE
Bellevalia sp. (A)
Ornithogalum neurostegium (489)

ILLECEBRACEAE
Herniaria hirsuta (9, 222)
Paronychia arabica (11, 358)
P. sinaica (39)
Pteranthus dichotomus (183)

IRIDACEAE
Gynandriris sisyrinchium (196, 211)
Iris atrofusca (0)

IXIOLIRIACEAE
Ixiolirion tataricum (381)

LABIATAE
Ballota undulata (388)

Eremostachys transjordanica (323)
Phlomis brachyodon (270, 288)
Salvia lanigera (179, 262, 309)
S. spinosa (121, 293, 328)
Teucrium polium (286, 338)
Thymus bovei (103, 339)
Ziziphora tenuior (477, 485)

LEGUMINOSAE
Alhagi maurorum (287)
Astragalus acinaciferus (292)
A. bombycinus (106, 173, 235, 294, 318, 445)
A. corrugatus (404)
A. deinacanthus (446, 480)
A. guttatus (430, 438)
A. hamosus (212)
A. spinosus (25, 84)
A. trachoniticus (160, 363)
A. tribuloides
  var. el-arishensis (37)
  var. leiocarpus (300, 312, 395)
  var. minutus (396)
  var. tribuloides (226)
Coronilla sp. (A)
Lathyrus inconspicuus (342, 520)
Lens orientalis (461)
Medicago laciniata (275, 333)
M. radiata (89)
Melilotus indica (418)
Onobrychis crista-galli (0)
O. ptolemaica (171, 180, 420)
Ononis sicula (184, 260)
Prosopis farcta (289)
Retama raetam (400)
Trigonella hamosa subsp. uncata (162, 236)
T. stellata (10, 66, 172)
Vicia cuspidata (A)
V. monantha (A)
V. peregrina (384)

LILIACEAE
Gagea sp. (A)

LINACEAE
Linum album (325)
L. mucronatum (463)

MALVACEAE
Althaea acaulis (443)
A. chrysantha (119, 500)
A. ludwigii (32, 124, 504)
Malva aegyptia (178)
M. nicaeensis (417)
M. parviflora (36, 123, 313, 502)
Malvella sherardiana (464)

PAPAVERACEAE
Fumaria parviflora (492)
Glaucium arabicum (455)
Hypecoum pendulum (35, 43, 209)
Papaver polytrichum (22, 202)
Roemeria hybrida (17, 46, 129)

PLANTAGINACEAE
Plantago afra (71, 263)
P. amplexicaulis (283)
P. coronopus (95, 199)
P. notata (231)
P. ovata (70, 150, 203)

POLYGONACEAE
Calligonum comosum (A)
Polygonum bellardii (255)
P. equisetiforme (A)
Rheum paleastinum (A)
Rumex cyprius (494)
R. vesicarius (27)

PRIMULACEAE
Anagallis arvensis (0)
Androsace maxima (429)

RANUNCULACEAE
Adonis aestivalis (366, 376)
A. dentata (50)
Ceratocephalus falcatus (244)

RESEDACEAE
Caylusea hexagyna (102, 303, 352)
Reseda alba subsp. decursiva (13, 79, 399)

RUBIACEAE
Callipeltis aperta (30, 45, 100, 405)

Crucianella ciliata (474)
Galium setaceum (49, 94, 130, 411)
G. tricornutum (382)

RUTACEAE
Haplophyllum blanchei (158, 326)
H. tuberculatum (238, 317)

SCROPHULARIACEAE
Kickxia azraquensis (26, 251)
Linaria confertiflora (462)
L. micrantha (364)
L. tenuis (A)
Orobanche cernua (189)
Scrophularia deserti (320)
S. hypericifolia (A)
S. peyronii (A)
Verbascum fruticulosum (82)
V. transjordanicum (109. 297)

SOLANACEAE
Hyoscyamus desertorum (324)
H. pusillus (86, 140)
H. reticulatus (458)
Solanum sinaicum (496)

TAMARICACEAE
Tamarix nilotica (427)
T. macrocarpa (A)
T. tetragyna (428)

UMBELLIFERAE
Anisosciadium isosciadium (18, 149, 359)
Bupleurum lancifolium (242)
Chaetosciadium trichospermum (19, 68, 112, 146)
Ducrosia flabellifolia (331)
Eryngium glomeratum (99)
Ferula blanchei (176)
F. communis (515)
Heptaptera crenata (456)
Lagoecia cuminoides (460)
Malabaila secacul (521)
Pimpinella eriocarpa (517)
Scandix stellata (466)
Turgenia latifolia (365, 369)

URTICACEAE
Parietaria alsinifolia (74)

VALERIANACEAE
Valerianella coronata (468)
V. oxyrrhyncha (24, 126, 216, 436)
V. pumila (499)

VERBENACEAE
Vitex agnus-castus (258, 473)

ZYGOPHYLLACEAE
Fagonia bruguieri (87, 98, 256)
Peganum harmala (353)

CHAPTER 15

Results of a Preliminary Survey of Livestock Owners

DARIUS CAMPBELL
ALAN ROE
(University of Durham)

1. INTRODUCTION

Owing to environmental factors, notably the erratic rainfall and wide ranging temperatures, pastoralism has been the primary economic activity of the Badia population for generations, the practice of which only began to change after the First World War. The development of the modern state and the extension of its infrastructure and authority into the Badia region, along with the arrival of technological innovations and availability of new services, is now reshaping many aspects of social and economic life.

The Badia Programme area falls within a climatic zone generally receiving less than 200mm of precipitation annually. In the villages to the north west of the Programme Area some rain-fed agriculture occurs, although rainfall diminishes moving south and east across the Programme Area, decreasing to less than 50mm. Much of the Programme Area is covered by a rolling plain of black basalt rocks, known locally as *al-harra*, which extends from the foothills of the Jabal Al-Arab in the north-west to the flat sandy plain of *al-hammad* in the east. In the far south and west of the Programme Area lie plains of black flint and gravel, *as-suwan*, onto which encroach a few areas of stabilized sand dunes in the district of 'ash-Shama'.

In addition to the village of Safawi, where the Jordan Badia Research and Development Programme's Field Centre is located, there are about thirty other villages and hamlets, grouped in the north-western corner of the Programme Area. A dispersed population of mobile pastoralists makes exact assessments of population within the area difficult, but the figure is judged to be about 15,000.

The Badia Programme area contains an important livestock resource in the country, consisting of 107,000 sheep and goats. Although many households have incomes drawn from mixed sources and the majority of people in the area are now sedentary, pastoralism continues to play a highly significant role in the economy. Whereas the population of the Badia cannot be described simply as pastoralist, most of the people in the Badia Programme Area depend, at least in part, on livestock for their livelihood; in 22 per cent of households livestock form the main source income, while 35 per cent of people list livestock production as their major activity.

2. AIMS OF SURVEY

The survey was designed and implemented as the combined first phase of the dual research proposals of Darius Campbell and Alan Roe, and as such has distinct objectives;

- Primarily to provide a preliminary contact with pastoralists in the Programme Area and enable the researchers to select a suitable sample group of pastoralists for long-term research flock monitoring, and residential work The questionnaire responses of' those pastoralists selected for further work would provide the necessary baseline data on individual flock management and structure.
- To gather initial data from a statistically significant number of respondents; providing the developing livestock-related research team at Safawi with an introductory overview of pastoral methods and practices within the Programme Area, and allowing for some initial data analysis to identify general patterns of livestock management and likely areas of constraints to the production system.
- To enable the researchers to divide their chosen sample group of forty flocks into two treatment groups with similar characteristics.

3. METHODOLOGY

A sample group of 105 livestock owners were interviewed during the course of data collection, representing 4.2 per cent of all Programme Area households, and 19.3 per cent of households whose main source of income is drawn from livestock production.

Despite an early intention to divide the Programme Area into geographical units and take a sample selection from each unit, it soon became apparent that such a system was impracticable and would serve no useful function; many of those sampled were mobile and ranged widely around the Programme Area.

The selection of pastoralists for interview was conducted with a bias towards those households for which livestock represented a major source of income, although not necessarily the sole source. A wide distribution of settlements and grazing areas was systematically visited (and in some cases revisited); the researchers would drive out to a selected area or village and call in wherever livestock, or signs of livestock ownership, could be seen. Structured questionnaires were used, although these were backed up by supplementary questions, discussion and an inspection of livestock. Throughout the fieldwork the researchers attempted to balance the number of days spent visiting village areas with those spent among more mobile tented communities.

The questionnaire itself was kept deliberately brief, as its objectives were limited, and the authors did not want to fatigue respondents already visited several times by other survey teams. The questionnaire comprised of 13 questions. These established background information on the respondent or owner, how the flock was being managed and structured, and its importance within the household economy, attitudes to milk production, lamb off-take from the flock, co-operation and organization with other livestock owners and use of supplementary feed concentrates. Finally, after further unstructured discussion, the respondent would be introduced to the idea of participation within the livestock monitoring programme, and asked if he would be prepared to discuss the matter further.

4. RESULTS

The owners of livestock were interviewed over a period of 3.5 months. Responses to the questionnaire indicated the 105 owners had a total of 40,749 animals; 37,780 sheep and 2,969 goats. Average numbers of animals owned were 388 (std dev. 415). Owners divided these into flocks of an average of 325 animals. Variation in the numbers of animals kept was large ranging from a few animals to over 2,500. In our survey of both villages and rangelands, there was a larger number of smaller flock sizes than would be expected by a normal distribution curve.

Estimates of concentrate feeding of animals accounted for a total of 1,274 tonnes per month for all 105 owners. This averaged out at 1.22kg concentrate/head of livestock per day (std dev. 0.80). Although an initial plot of feed/head/day against total animals owned appeared to indicate that smaller flocks received more feed/head, subsequent regression analysis did not indicate there was a strong correlation (R square = 0.11). No significant effects of mobility or other sources of income were found by analysis of variance.

The feeds quoted as used were whole barley grain and wheat bran. These were used in an average ratio of 3.1:1, but with a large variation (std dev. 1.9). Other hand-fed feeds observed in use were tomatoes, tomato waste, cabbages and Egyptian clover (*Beerseem*). Grazing was assessed to be important in certain seasons. Some, but by no means all owners, claimed that no concentrate feeds were needed in the best part of the wet season.

Of the 105 owners interviewed, 61 per cent moved with their flocks for at least one period (a few weeks) a year. These are referred to as mobile. The majority of these spent most of their time away from their home base, permanently inhabiting their tents. The other 39 per cent of owners were sedentary, only moving their animals a day's walk from their village. Sedentary owners averaged 169 animals, whilst mobile owners averaged 528 animals ($P<0.01$). No significant differences were observed in amount of bought feeds used by the two groups, although it appeared that shepherded flocks were fed higher ratios of barley to wheat bran (3.7) than family managed flocks (2.6) (ANOVA $P<0.01$). In seasons of poor grazing, i.e. the end of the dry season, owners said that the animals' ranging was necessary for the well-being of the sheep, even if very little actual grazing was being carried out. Goats were considered less in need of travelling each day. In seasons with some prospect of grazing, animals were often moved over considerable distances in response to reports of good grazing. Such movements were initiated even very early in the rainy season, when it appeared early grazing was still limited.

Hired shepherds, in addition to family members, were used for management of flocks; 2 Iraqi, 6 Jordanian and 27 Syrian shepherds were employed by owners in the sample, the remaining 68 per cent of the flocks were managed by family members. Naturally, shepherds were associated with owners with more animals, averaging 651 animals for owners with shepherds and 212 animals for family managed flocks ($P<0.01$). Such larger, shepherded flocks also more often tended to be divided into smaller groups; average numbers of flocks per owner were 1.3 for shepherded flocks and 1.1 for family managed flocks ($P<0.05$).

Sheep to goat ratios in flocks averaged 13:1. Mobile, and therefore larger, flocks were associated with larger sheep to goat ratios. Sheep to goat ratios were 25:1 in mobile flocks and 7:1 in sedentary flocks, however variation was large (ANOVA $P>0.\ 10$).

Some 80 per cent of livestock owners reported that they depended entirely on their animals for their income. Owners with other sources of income tended to have smaller flocks. Flocks averaged 181 animals where there was additional income and 440 animals where livestock were the sole source of income (however, ANOVA $P>0.10$). Only 6 per cent of mobile flock owners had other sources of income, compared to 39 per cent of sedentary owners (Chi^2 $P<0.01$).

Diseases observed by examination of animals in the field included PPR (peste des petits ruminants), blue tongue, foot and mouth disease, mastitis, lymphadinitis and pneumonia. The former three diseases were confirmed by JCO blood tests carried out during the survey. Additional faecal and other samples showed populations of intestinal parasites, blood parasites and ectoparasites. Owners did not necessarily recognize certain signs of ill health such as diarrhoea, but did report such diseases as sheep pox, brucellosis and clostridial disease as being important. Veterinary care was generally limited to advice from pharmacists and veterinarians in large towns who had not seen the sick animal. In a few cases, valuable animals were transported to towns for examination. Syrian shepherds were particularly respected for their knowledge of traditional cures for certain conditions. Several owners asked about intra vaginal sponges to allow synchronized oestrus.

Livestock owners all indicated that there was not a labour limitation to milking. That is to say, milking could be increased in response to increased production from ewes.

In all cases, in response to a postulated increase in numbers of lambs in a good year, livestock keepers said they would keep female lambs and sell males. It was always stated that under good conditions the flocks would be increased. Only in a few, generally smaller, flocks was it indicated that female lambs might be sold as well, in order to pay for the purchase of feeds.

5. DISCUSSION

The numbers of animals reported as owned by the livestock keepers in this survey represent a sizeable proportion of the total livestock in the region. According to the Ministry of Agriculture (1991) figures, the survey has covered 38 per cent of the livestock of the Programme Area. Using figures for total households and percentage of households involved in livestock rearing, the current survey would project the total numbers of sheep and goats as either 215,000 head or 326,000 head depending on whether the survey was considered as monitoring of those owners which call livestock their 'main income' or their 'main activity' (22 per cent and 35 per cent respectively of the Badia Programme human population according to the Programme census of February 1993). These figures are not put forward as accurate estimates of the livestock population of the Badia. We hope to first validate them in our more intensive studies of 40 flocks. This would give an indication of how accurately owners respond to surveys such as ours. The results also indicate how difficult it is to project total livestock ownership from sample surveys. In our more intensive studies we will hope to indicate net changes in livestock populations in the Badia through our monitored flocks. Such an index we feel may be of more importance in planning livestock policy.

The survey indicates how important concentrate feeds are in the livestock systems of the Badia. The average ratio of barley to wheat bran in the diet gives a ration with approximately 11MJ and 120g crude protein per kg fresh weight (from text-book estimates of feeding values of these concentrates). Animals receive on average 13.4MJ ME and 146g CP per day[1]. For a 50kg ewe at maintenance, walking 15km/day, such a ration would provide adequate energy and protein without need for additional nitrogen or UDP. Pregnant ewes and fattening lambs to 200g/day would also be adequately supplied with energy and protein. However it is likely that such feed sources may be inadequate for lactating ewes. Problems with low levels of calcium may, however, be encountered in all classes of sheep with such high levels of grain in their diets. The role of grazing and other feed sources in the diet needs more intensive study to clarify their relative importance.

Goats do play a role in the animal production system of the Programme Area, albeit of secondary importance to sheep. Prices for animals and milk are lower. The milk is recognized to have a lower fat content. However, goats are valued for their ease of shepherding, their role in leading the sheep, and are also valued by those owners who prefer not to range far with their animals. The presence of goats was further explained in that they provided milk for household consumption; owners pointed out that the advantage of goats was their extended milking season; they could provide a family with fresh milk for more than half the year. Sheep, however, produced milk of higher quality over a shorter period, more appropriate for sale.

If goat ownership is justified mainly for household milk production, as against commercial milk production from sheep, this may explain the lack of correlation between goat and sheep numbers within flocks; only sufficient goats to meet the household milk requirements are necessary, while the largest possible sheep numbers are desired.

Disease is still a very significant factor in animal production if our observations are typical of the flocks in the area. Many animals showed signs of scouring and nasal discharge indicating at least some loss of production potential. More significantly we observed many young lambs with severe signs of scouring, and also met several owners experiencing high rates of mortality in their animals. It was apparent that the current system of veterinary care was not meeting the needs of the livestock producers.

Answers to our question on response to increases in lambs born/surviving in a flock, although not surprising, are cause for the greatest concern. All the ideas for improvement of livestock systems in the Badia involve some degree of stasis in animal numbers. Whether it is by disease control, improved diets, range management, reproductive manipulation, increased commercialization of fattening lambs, it is inherent that some degree of control of animal numbers in flocks is achieved. This is not only to increase off-take and productivity from the Badia, but also to protect the already degraded environment. This is a fundamental issue which has to be addressed in any attempt to improve the livestock system of the Badia region.

Just over a third of owners interviewed employed shepherds to attend their flocks. The majority of these were Syrian, a nationality considered particularly well versed in sheep management; Bedouin themselves, but somehow dispossessed of their own flocks. It was clear through observation and discussion that many hired shepherds enjoyed some degree of managerial autonomy over their flocks. Often, shepherds are employed to maintain flocks at remote Badia locations, with only the necessary additional feed and watering visits by the owner; on a day-to-day basis it is the shepherd who effectively manages the flock.

---

[1] MJ = mega jules; ME = metabolizable energy; CP = crude protein.

One livestock owner claimed that his shepherd handled nearly all management decisions for him (including financial decisions). He seemed to have complete faith in his shepherd's knowledge and business sense. On another occasion we returned to the tent of one of our wealthier interviewees, only to find that he was absent and that his Syrian shepherd and family had been installed in his home, left to manage not only the flocks but also much of his estate. We also heard that shepherds are commonly trusted to treat sick animals with traditional remedies and simple surgery. However, while many shepherds seem to make substantial inputs into the management of flocks, there is equally evidence that in some households they provide little more than unskilled labour.

Professionally shepherded flocks tended to be somewhat larger than family managed flocks, and while the latter would sometimes have multiple shepherds (on occasion assisted by dogs) hired shepherds were observed to supervise flocks individually. Where shepherds managed smaller herds, it was usually in villages, where family members may have some alternative employment, and the children attend school.

Given the significant role that we found some professional shepherds played in livestock production, it is clear that future studies of livestock management should not focus upon owners alone.

Responses to questions concerning livestock and owner mobility, suggested that seasonal movements remain an important feature of pastoral production; over half respondents could be defined as mobile in as much as they spend at least one season at alternative grazing grounds. More surprising than this was the high number of pastoralists who were resident the whole year in tents, and were classified as 'travelling widely'. Many of these pastoralists owned houses in villages, but only referred back to them once or twice a year for brief visits, or to collect things. The Ahl Al Jabal tribe (Musa'id, Adhamat, Sharafaat, Zubaydi and Ghiath sections) which predominates within the Programme Area, traditionally migrates onto the eastern *hammad* plains during the winter and spring seasons to take advantage of ephemeral pastures which grow following rain[2]. Our discussions revealed that a majority of pastoralists still adhered to the same basic pattern of migration. Many of those pastoralists who lived permanently in tents grouped themselves on the southern *suwan* and *hammad* plains over the summer season, in the vicinity of the Grammar, Hams and Haze wells. The majority of these also described their migration as out to the eastern *hammad.* A small number of pastoralists living in the village areas claimed that they migrated down to the southern *suwan* plain for their migration, while others declared that they simply 'followed the rain' and could not be area-specific about their migration. In two or three cases, respondents explained that their annual migrations also included a westward movement to the agricultural areas surrounding Ajloun and Irbid to take advantage of crop residues.

Despite debate about the continuing relevance of pastoral migration, the livestock owners themselves all perceived it as serving a specific livestock related function, and necessary to profitable production. While spring grazing in *hammad* pastures is of unpredictable quantity and quality, all those owners who were able to travel (or send a shepherd) did so. However, justifications differed from individual to individual. Amongst semi-sedentary villagers, one of the most common reasons given for a migration was that the foothills of the Jabal Al-Arab became too cold for sheep in the winter, and many new-born lambs were lost.

---

2 During the course of fieldwork, members of the Sirdieh Sirhan and Bani Sakhar tribes were also encountered. Additionally, Saudi Arabian Bedouin, notably of the Shararat, can be found in the southern part of the Programme Area.

Other livestock owners explained that while their animals could subsist on feed concentrates, fresh pastures enhanced the quality of the diet; this was especially important for milk production. It was claimed young lambs would grow better and the milk would command a higher price if ewes were grazed on spring pastures. There was no clear indication from pastoralists as to how much of their flock's dietary requirements could be met by grazing during the spring months; some suggested that for three or four months it would not be necessary to feed the flocks any supplementary feeds at all, while others claimed that because the ewes were in milk, they required even more supplementary feeds than usual. However, the majority agreed that they were able to reduce the quantities of barley and wheat bran being fed to the animals, although according to responses given, there is no significant difference between the amount of supplementary feeds fed to mobile and sedentary flocks.

The first rains fell onto the northern Badia in mid October, and within only a matter of days many of the semi-sedentary flocks of the village areas had departed. While the family, tent, and household chattels are moved by vehicle, the flocks and some of the men, together with shepherd if they employ one, make the journey on foot. Owing to the need to graze the sheep *en-route*, progress is slow and the journey between the villages and *hammad* grazing grounds may take six days.

In most cases families travelled alone, although in the course of our fieldwork we came across several small groups of tents which travelled together on a permanent basis. These small groups usually, although not always, comprised kinsfolk or neighbours from a village. 'Linked' households of this type claimed that there was little, if any, co-operation between them with regard to livestock management. Each family would herd its own flocks separately, would collect supplementary feeds separately, and arrange the provision of water independent of the other households.

While some of the households we visited were divided all or part of the year, the majority migrated as a single unit. Female labour is especially significant during the milking season on the *hammad* pastures, although a few of the respondents acknowledged that, if necessary, the men could assist with this activity. Thus, in the majority of cases, labour did not seem to be a constraint upon milk production. However, we encountered a small number of livestock owners who claimed that they did not produce milk commercially at all. At least one of these had a very large flock, but seemed to feel the rewards from the sale of milk would not justify the effort of milking.

Answers to our question on milk production indicated there was potential for increasing milk production, i.e. the milking process did not seem to be a limiting factor at this stage of development of the livestock production system.

Selection of households for interview was biased towards those whose primary source of income was livestock production, although a significant number of owners (20) had additional sources. As might be expected, most of these households with additional sources of income were sedentary village dwellers; we encountered only two mobile respondents with additional incomes. The most common additional sources of income amongst the respondents to our questionnaire seemed to be remittances and military pensions.

6. FURTHER RESEARCH

Issues for further investigation, highlighted by this survey are:

Flock structures: the roles of sheep and goats in flocks.

Livestock feeds: how concentrates are used seasonally, and the role other feedstuffs play in the diet.

Mobility and grazing: patterns of mobility and grazing, both temporal and spatial.

Animal health: as an important component of any likely improvements to the system, attention has to be paid to how owners will respond to increased survival rates in flocks.

Stocking rates: any major changes in the system should involve new attitudes to increasing off-take rates from more efficient flocks, and limiting the growth of livestock numbers in the Badia region.

Innovations: Further discussion with owners who are already interested in innovations in animal production as to likely avenues for improvements to production systems.

These and other issues are to be investigated in our planned monthly flock monitoring experiments.

ACKNOWLEDGEMENTS

Our thanks to the Badia Programme for creating the opportunity to undertake this research and for providing logistic support in Safawi, and to DFID (ODA) and Smiths Industries who have funded, respectively, the work of Darius Campbell and Alan Roe.

CHAPTER 16

Population Fertility Trends

ALLAN M. FINDLAY
(University of Dundee, UK)

MOHAMMED MAANI
(University of Jordan)

1. INTRODUCTION

Population growth is widely seen as one of the most important factors influencing both economic development and environmental change (Harrison 1993; Preston 1994). Yet the relationships between these factors are highly complex and vary depending on the scale of analysis. This chapter is based on a study of population dynamics at the local scale of people surveyed by the authors in the early 1990s in two parts of the Badia. The Badia, with an average rainfall of less than 150mm, covers 80 per cent of the land area of Jordan, but is home to less than 2.5 per cent of Jordanians (Ministry of Planning 1988).

At the scale of analysis of this particular study, it can be assumed that population and the environment are interlinked on the one hand by the economic processes of production, and on the other hand by the demographic process of reproduction, which when balanced against mortality and migration regimes determines the pace of population growth or decline. Economic production systems are necessary in any particular place to make population possible through providing, either directly or indirectly, food, shelter and clothing. Amongst other factors, the production process requires both labour and land. As a result, analysis of production processes is linked to the study both of human reproduction as a means of providing future supplies of labour, and of aspects of the environment as the means and locus of production. But as has been pointed out elsewhere (Findlay 1991), the nature of linkages between production and reproduction are highly complex, being mediated by scale and the different ways in which production may be organized. This chapter commences by attempting to evaluate the longer term implications of current demographic regimes in the Badia of Jordan, before turning to consider the changing role of household labour as an issue influencing both production and reproduction in this arid environment.

In the 1980s, population growth in the arid areas of eastern and southern Jordan continued to be high and would have been even higher had it not been for significant out-migration. Fertility levels, although declining slowly, remained very high. Unlike the urban areas of Jordan, inter-generational wealth flows still appear to be from children to parents (Maani 1990) making it logical in Caldwell's (1982) terms for couples to maintain high fertility and to seek large families. From the mid-1980s onwards, opportunities for emigration declined substantially, and net return migration from the oil economies was recorded, producing new population pressures in this arid environment. This chapter will investigate these ideas from two surveys carried out by the authors in association with other Jordanian and British colleagues from the interdisciplinary Badia research team. The first survey described in this paper was undertaken in the eastern Badia early in 1993, and provides the opportunity for a very up-to-date analysis of the demographic regime of this most arid of environments. A second survey reported in the chapter was undertaken by the authors in an arid part of southern Jordan in 1990/1. This study specifically addressed Caldwell's ideas, by investigating family structure and wealth flow directions in relation to fertility attitudes.

2. FERTILITY DECLINE IN THE ARAB WORLD: SOME POINTERS IN THE LITERATURE

Fargues (1989) has explored the way in which four features - the near universality of marriage, the traditional ease of divorce, the institution of purdah and the effect of polygamy on age differentials between spouses - account for the culturally specific nature of Arab fertility behaviour. While in general these forces account for the persistence of very high fertility across the Arab world, Fargues notes the presence of clear-cut regional differences within the Arab world in terms of fertility trends. Fertility remains much higher in the countries of the Arabian peninsula, including Jordan, than it is in the Maghreb or Egypt. From an examination of the transitions in the age at marriage and in female education, he concludes that all Arab societies are in the process of moving to lower fertility levels, but that regional variations remain strong with regard to the transition in familial roles. One surrogate of this is the level of female participation in the labour force, which Fargues (1989, 168) notes is almost inverse to that of fertility rates. This should not be interpreted in the narrow way that Fargues does, leading to the erroneous conclusion that barriers to women's entry to the labour market remain the fundamental cause of spatial inequalities in fertility patterns. Instead it should be taken as an indicator of the need for wider research on the changing roles of both women and children within the family, and indeed of family structure itself, as an important key to understanding trends in fertility.

The now classic statement of Caldwell (1982) on the causes of fertility decline provide one line of approach to the topic. Caldwell has suggested that in so-called 'primitive' and 'traditional' societies there is a net wealth flow from children to parents, making it illogical to restrict fertility on economic grounds. This consequently favours high fertility levels. Inversely the circumstances of nuclear families make it economically rational to restrict fertility because of the net wealth flows from parents to children. This switch in wealth flow direction is produced by a complex matrix of forces relating to the costs and returns of having and rearing children. In simplistic terms, it can be represented as the balance between on the one hand the changing costs of supporting children as their historical involvement in formal education and training has been extended over time, and on the other hand the psychic income and measurable wealth produced by children in and for the household. For some rural households the link between these demographic and economic factors and the condition of the physical environment is a very direct one, since household labour is used in the production of crops or the tending of animals. It would be a mistake of course to imagine that all rural households depend on agricultural or pastoral production for their livelihood, since many are employed primarily in service activities. One might anticipate nevertheless that the value of child labour to parents would vary from one type of rural production system to another. This seems to be partly borne out by Randall (1994) who notes that pastoral populations often have lower fertility than peoples for whom cultivation is the chief source of food.

Caldwell's approach is very interesting in shifting analysis of demographic change away from diffusionist theories of 'modernization' towards concepts dealing with the household and labour relations, within the household, and between the household and its wider context (including the physical and social environment). It is not the purpose of this paper to review all the research which has been undertaken in relation to Caldwell's (1982) theory. Suffice it to note that his ideas have been tested, questioned and extended by a number of researchers (Smith 1981; Maani 1990) and found to have general validity. There are of course problems with his continual reference to concepts such as rationality in decision making and with his failure to specify clearly the ways in which household and societal modes of production are interrelated, but his work provides a useful starting point for research on the causes of fertility decline in the Arab world. The work of Pryor (1981) is also useful in the context of exploring fertility change in Jordan, schematizing the possible parallels between mobility trends (Zelinsky 1971) and inter-generational wealth flows. Pryor suggests that just as rural-urban migrants send back remittances to their areas of origin, so also wealth flows may be transferred at the same time between one generation and other as a result of the migration process. Remittances from or to migrants may, therefore, not only transfer finance between geographical regions, but also between generations, thus playing some part in bringing about changes in power relations within the family. Before these and other questions can be further addressed it is necessary to review the demographic circumstances in Jordan as a whole and in the Badia in particular.

3. THE DEMOGRAPHY OF JORDAN AND OF THE BADIA

As recently as 1990 Jordan had the second highest rate of natural increase in the world (Population Reference Bureau 1990). If sustained the demographic regime, with its extraordinarily high fertility rates and moderately low mortality rates, would have seen the population double in a period as short as 17 years. Added to this, Jordan has also experienced a very distinctive history in terms of its involvement in systems of international migration. Although estimates of migrant numbers are notoriously unreliable, it has been claimed that Jordan is the world's third most dependent country in terms of its reliance on migrant remittances (Findlay and Brown 1994). Throughout the 1980s Jordan's earnings from remittances, when expressed as a ratio of the value of exports, were exceeded only by the performance of the Yemen Arab Republic and PDR Yemen. Regardless of the details behind these 'league table' positions, it is evident that Jordan is an exceptional state in both demographic and economic terms and consequently is of particular interest to population geographers and demographers.

The population and urban geography of Jordan has been described elsewhere by one of the authors and is not repeated in detail here (Findlay 1986; Lavergne 1992). The population of Jordan has risen from a mere 680,000 in 1952 to 900,800 in 1961 and 2.13 million at the time of the last census in 1979. Since then population growth has continued at a very rapid rate, with the population estimated to have reached 3.45 million persons in

1990 according to National Population Commission of Jordan. Accepting these figures implies an average annual increase of population of 4.3 per cent.

Such rapid population growth has inevitably implied a change in the relationship between population and the environment. Until the second half of the twentieth century, nearly the entire population of the territory which was to become the state of Jordan was rural in its orientation, implying strong linkages between demographic systems and the capacity of the physical environment to provide a livelihood. In the north and west of Jordan this took the form of settled populations of cultivators and pastoralists, while in the east and south nomadism was an environmentally sensitive response to scarcer grazing resources. Lancaster (1981) has provided what is often considered to be the authoritative account of the way in which the Rwala Bedouin moved their animals across the Badia in relation to seasonal climatic variations. Since this livelihood form critically depends on grazing resources, any human population increase required either an increase in the number of animals in the pastoral system (and hence increased pressure on grazing) or the introduction of alternative livelihood forms. In practice, the latter appears to have been much more significant with by the mid-twentieth century an increasing proportion of Jordan's pastoralist families having one or more members working for the Jordanian army. Alternative livelihood sources, combined with the social and political forces favouring sedentarization, resulted in the settlement of increasing numbers of people in the Badia from the 1940s onwards. Sedentarization also meant less direct population-environment linkages for many of the people of the Badia, while introducing an increased spatial interdependence between the Badia population and the wider Jordanian economy.

Current numbers of nomadic pastoralists in the Badia are unknown, but they are generally perceived to be a shrinking minority. More significant than the numbers of pastoralists (nomadic, semi-nomadic and sedentary) is the change in environment-pastoral linkages which has taken place over recent decades as a result of the possibility of importing feedstuffs and of transporting water within the region to support the animals being grazed there. This development has broken the effective linkage of grazing resources and numbers of animals being grazed, and has removed the imperative for moving animals from one area to another to find grazing. At a local scale therefore, the absolute dependence on the physical environment of the ever-shrinking proportion of the population who look to pastoralism as their chief source of livelihood has disappeared. This is not to say, however, that serious overgrazing of natural grazing resources is no longer a threat or that population-environment linkages are no longer of importance. On the contrary, the increased scale of twentieth-century living simply means that population-environment links also are manifested at a different scale, as is amply illustrated by the remarkable fact that water from the Badia catchments is transferred to supply urban demands in the west of Jordan.

It is difficult to compare Jordan's national population growth with almost any other country because of the peculiar role of political events in affecting its recent demographic history. The conflicts with Israel in 1948 led to the massive displacement of Palestinian refugees into both the West and East Banks of Jordan, and the annexation of the West Bank by Israel in 1967 led to another influx of refugees into Jordan in the late 1960s. Refugee flows have therefore added to other forces favouring rapid population growth and have combined with natural trends towards rural-urban migration to encourage particularly strong population concentration in the largest urban centre, Amman. A sample survey by Samha (1979) found that the majority of migrants to

Amman had come either from Palestine or the West Bank, with only 22 per cent of migrants declaring their place of origin as being somewhere on the East Bank.

The 1983 Fertility and Family Health Survey suggested that Jordan had a total fertility rate (TFR) of 6.6, much higher than most other countries with an equivalent level of economic development. The overall TFR figure disguises however two important features of Jordanian fertility trends. First, substantial fertility differentials exist between urban and rural areas. In 1983 the TFR in urban areas was estimated to be 5.4, while in rural areas it was still 7.6. Second, the rate of fertility decline appears to have been much greater in recent years in rural than in urban areas, even though the latter have a lower overall fertility rate. Between 1976 and 1983 there was no appreciable reduction in TFRs in Amman (TFR = 5.1 in both years), while in rural Jordan over the same time period TFRs fell from 8.2 to 7.6. By the time of the 1990 Jordan Population and Family Health Survey (JPFHS), the TFR for Jordan had dropped to 5.6, but with a substantial fertility differential remaining in terms of the location and the educational status of women. In rural areas of Jordan the TFR was still as high as 6.8, while in the Amman conurbation it was down to 4.75. The general downward shift in age-specific fertility rates is shown in Figure 1 and discussed later with reference to the population surveyed by the authors in the Badia region in 1993.

Discussion of the causes of fertility decline has traditionally been approached at two levels - first, through the proximate or intermediate determinants of fertility (Bongaarts 1978; 1985; Hill 1985) and second, through underlying causal variables. The latter have already been discussed briefly above regarding the work of Fargues (1989) and others, but the former merit brief attention. Decomposition of the effect of the proximate determinants on fertility reduction based on the 1983 Jordan Fertility and Family Health Survey indicates that while for urban areas the index of contraception (Cc) accounts for the largest part of the recorded fertility decline (38 per cent), in rural areas the single most significant factor was the index of marriage (Cm), which accounted for 44 per cent of the total reduction in fertility. In other words, the increased age at marriage was the chief process responsible for fertility decline in rural Jordan, rather than planned fertility control within marriage (Maani 1990). This position is substantiated by behavioural data gleaned about desired family size, which points to the wanted total fertility rate remaining very high in rural areas in the early 1980s (a wanted TFR of 5.7).

The 1990 JPFHS substantiates some, but not all, of the patterns and relationships of the 1980s. In large urban areas use of modern contraceptive measures was reported as twice that of women in rural areas (34 and 17 per cent respectively). Interestingly the median age at first marriage in the JPFHS showed little variation between rural and urban areas (19.8 and 19.5 respectively), suggesting that by the 1990s national convergence on this important influence on fertility had taken place. This in part substantiates Fargues (1989) argument that age at marriage may no longer be a primary factor in explaining spatial differentials in fertility in the Arab world. According to the 1990 survey there was also surprisingly little rural-urban variation in the mean ideal number of children, with the average for ever-married women in rural areas being 4.8 while in large urban areas it was 4.3. However, it should be noted that a third of women included in the survey did not give a numerical response to the question, suggesting either that they did not comprehend the concept of ideal family size or that they did not perceive this as being something that lay within their control. It is necessary therefore to conclude that there may be only the weakest of relationships between stated ideal family size and action to plan the control of fertility to achieve a reduction in future childbearing. This is particularly true of rural areas where educational levels amongst women were much poorer than in urban areas and where the well known inverse relationship between educational attainment and fertility levels continues to militate in favour of a high fertility regime. In the comparisons made in the remainder of this chapter between the researchers' survey in the Badia and Jordan as a whole, the figures for the state of Jordan are drawn from the 1990 JPHS (Department of Statistics/Ministry of Health; 1992) unless otherwise stated.

4. SURVEY LOCATION, DEFINITIONS, METHODS AND LIMITATIONS

The Badia region of eastern Jordan covered by the joint research and development project of the Higher Council for Science and Technology and the Royal Geographical Society, London is an area of arid and semi-arid territory centred on the settlement of Safawi, some 156km to the north-east of Amman city. As well as Safawi there are 17 other recognized villages in the area as well as a dispersed population of pastoralists. The study area as defined by the project covers approximately 14 per cent of the land area of Jordan and largely comprises a rolling desert plain, covered with dark basalt rocks. Precipitation in the project area varies along a north-west to south-east axis, with annual rainfall highest on the edges of Jabal al-Arab by the Syrian frontier (200mm) but with most of the area receiving less than 100mm and with a significant proportion of the southern part of the area getting less than 50mm.

The land use of the area has been described by the Geomorphology and Physical Resources Team (1993, 10) of the project as follows:

> 'Extensive tracts of land have no current use or recognized value, although parts of the Badia, particularly in the north towards the Syrian border, are utilized on an ad hoc basis as natural pasture. Uncontrolled grazing frequently causes deterioration to the quality of the land and its ultimate abandonment because of the absence of management practices or strategies for maintaining soil quality.'

Whether 'bad management' is indeed the cause of problems in the physical environment of the Badia, or whether there are other factors relating more directly to rising population pressures is an issue which in the present authors' views remains open to question. Further research on what the implications of rapid population growth have been for the peoples of this area, in terms of their changing ability to make a livelihood from traditional systems such as pastoral nomadism certainly seems necessary. Caution is also required before imposing external value systems on the interpretation of the region's physical resources. One might well ask who it is that says that vast tracts of land have no recognized value?

The main tribe of the area is known as Al-Jabal, that is they are the people of the mountain (i.e. Jabal al-Arab). Within the tribe there are three main fractions or clans known as Masayid, Sharafaat and Adhamat. Although as noted earlier pastoralism was the traditional livelihood form in this area, with people moving seasonally in relation to the changing availability of grazing, the majority of people in the area are now sedentary, although pastoralism continues to play a significant role in the economy.

5. CHARACTERISTICS OF THE BADIA POPULATION: AGE, GENDER AND MAJOR ECONOMIC ACTIVITIES

Before these or any other in-depth research questions could be investigated, it was decided that a baseline survey of the demographic and socio-economic characteristics of the population should be attempted. The baseline demographic survey was piloted in late 1992 (Findlay 1992) and conducted in its full form in January and February 1993. The survey collected information at two levels: the household and the individual. Forms were completed for 2,398 households and data collected for 15,655 individuals. This meant that on average there were 6.5 persons per household, slightly fewer than the mean for Jordan as a whole (6.9) according to the JPFHS. By far the majority of households appear to be nuclear households. Some 96.2 per cent of individuals surveyed were either heads of household, or spouses or children of heads of household. Nearly everyone covered by the survey reported their nationality as Jordanian. Given the supposed strength of historic linkages with Syria and the proximity of the region to the Syrian frontier, it was surprising that only 36 persons of Syrian nationality were recorded by the survey.

The survey revealed the youthful age-sex structure of the population. Some 48.5 per cent were less than 15 years of age compared with 44.0 per cent in Jordan as a whole. Extremely high fertility rates combined with other aspects of the Badia's population structure meant that one in six of the population was under 5 years of age. Conversely mortality rates were high meaning that few of the population survived to old age, with only 2.9 per cent being 65 years of age or older. The sex ratio was 107.7 indicating a slight numerical dominance of men over to women, the figure for Jordan as a whole being 109.

Interpretation of the significance of the age and gender data described briefly above is dependent on the confidence which one has in the accuracy of the data. Experience indicates that accuracy of age reporting is highly variable from region to region and country to country. In particular there is a strong tendency towards age heaping on ages ending in the digits 0 and 5, the extent of which can be measured in a number of ways including the Whipple Index that calculates the proportion of respondents indicating an age ending with these digits compared with the population as a whole belonging to the age cohorts 23-62 years. This crude index is generally interpreted as indicating accurate age reporting when a value of less than 105 is achieved, with accuracy decreasing as the index rises (Newell 1988). The United Nations has published a scale to help with the comparison of values on the Whipple Index, which indicates that values of over 175 suggest that only very rough age estimates have been provided. In the case of the survey data, an index of 178 is achieved indicating extreme age heaping and highlighting the need for extreme care in the interpretation of demographic indicators based on age cohorts. This is most unfortunate since most demographic calculations, such as age specific mortality rates, life tables, life expectancy and population projections, use age related data as a basic input to their calculations. The implications of having a Whipple Index of 178 are therefore that all other demographic calculations based on the Badia survey have to be treated with the greatest care, since clearly the quality of reporting in the survey was very poor. This was so for a variety of reasons which have been partially discussed elsewhere (Findlay 1992). Offsetting this, it should be noted that nearly all other demographic surveys carried out in Jordan have faced difficulties with age heaping (Maani 1990) and, as in the Badia survey, the problem is most severe in cohorts of 30 years of age and over. Bearing these reservations in mind it is nevertheless interesting to note that the population pyramid for the Badia, like that for Jordan as a whole, indicates a very youthful population, reflecting a demographic regime with extremely high fertility rates, but with moderately low mortality rates in the adult cohorts.

Prior to considering the fertility and mortality regimes of the Badia population in more detail it is useful to briefly note the responses given in the household questionnaire concerning the major economic activities of the population (Table 1). Respondents had the possibility of listing several different major and minor sources of income. Over 75 per cent of households gave only one source, although it is probable that most households had multiple livelihood strategies.

Table 1. Major source of household income in the Badia

Income Source	Households	
	No.	%
Livestock production	534	22.3
Other agriculture	189	7.9
Self-employed (non-agricultural)	259	10.8
Waged employment	920	38.4
Remittance income	113	4.7
Other	359	15.0
No response/don't know	24	1.0

Many questions could be raised about the precision of this data and the interpretations which could be placed on it, especially in view of the large numbers of families in the area who have one or more family member serving in the army and the problems of disclosure which this implies. Despite this, a minimum of 30 per cent of households declared that either agriculture or pastoralism was a major source of their livelihood, raising the difficult, but important, question of whether current rates of population growth can be sustained by the primary sector or whether further population growth necessitates the expansion of other sectors of the Badia economy. Inversely, about 70 per cent of the population do not rely directly on the ecologically fragile resources of the physical environment for their livelihood; they live in a highly arid rural environment, but their population growth is not directly constrained by the immediate capacity of the physical environment to either support agriculture or pastoralism.

6. FERTILITY AND MORTALITY REGIMES IN THE BADIA, JANUARY 1993

Technical details of the relation between the baseline survey and the demographic indicators discussed below have been reported elsewhere (Findlay *et al.* 1994) and are not repeated here. In the year prior to the Badia survey it can be estimated that the population had a crude birth rate (CBR) of 48.3 per thousand compared with 36.1 in Jordan as a whole and 39.0 for rural areas (Department of Statistics 1992). The differential between the CBR of the Badia and that of other rural areas is suggestive of the presence of more than simply an age structure effect in explaining the extremely high fertility levels. Contributory factors to the high fertility rates are discussed in more detail below.

Turning to mortality, the crude death rate (CDR) was 11.6 per thousand, but as one would expect this statistic hides a vast inequality in the risk of death by age cohort. The study only collected information about deaths in the year preceding the survey (clearly a longer time span would have been beneficial in collecting a more robust data set), but the data remains very interesting in revealing the presence of high levels of infant and child mortality. For example the mortality rate for under fives was 34.1 per thousand, but it seems probable that some under-reporting of child deaths took place in the survey. The JPFHS study reported a figure of 38.8 for deaths reported in the five years before 1990. The overall mortality rate for under fives from rural areas in the JPFHS study was even higher at 47.3. Taken at face value this would suggest that child health in the Badia was better than elsewhere in rural Jordan, but a more likely interpretation is that the shortfall between the Badia survey figure and that of the JPFHS is a measure of the scale of under-reporting. In any case, the Badia 1993 survey indicated that children had a less than one in ten chance of surviving their first year of life, thus underscoring the very considerable health hazards facing the youngest members of the Badia population, as elsewhere in rural Jordan.

Taking together the CBR and CDR for the Badia, one arrives at a rate of natural increase for the region of 3.7 per cent per annum. This very high rate of natural growth would, other things being equal, result in the study area's population rising from around 15,700 in 1993 to 20,200 by the year 2000. On the same basis the total population would double in a little less than 20 years (i.e. by the year 2012). Such rapid rates of population increase would clearly have major implications for the development of the Badia region; not only the need for a very rapid expansion of basic infrastructure, services and employment opportunities, but also the demand for already scarce essential resources such as adequate supplies of water.

Like many other surveys, the Badia study derived its information on fertility from a birth history section of a survey questionnaire. This approach, while commonly adopted, has the disadvantage that it covers only the fertility behaviour of surviving mothers, thus adding the potential for another source of error to the more widespread problems of misreporting of child births and deaths. A further problem with the data which readers should bear in mind is the moderately small size of the population involved in the survey by comparison with the sample sizes in Jordan's national surveys of fertility. The effect of this is to increase the risk of error in estimates of age-specific fertility rates in cohorts such as the 45-49 age group which are based on populations with relatively few person-months exposure. As a result, data for this cohort have been excluded from Table 2 and the TFR value calculated instead for the 15-44 age cohorts. This provides a better basis for comparison with other surveys. Finally, it should be noted that the limitations of the baseline survey meant that data could only be collected for the year prior to the survey and not, as is often the case, for a longer time span. Despite these limitations, the data collected in the survey provides interesting insights of the demographic regime of the area in the early 1990s.

Table 2 and Figure 1 display the age-specific fertility patterns of the Badia survey alongside equivalent data from the 1976 Jordanian Fertility Survey and the 1990 JPFH survey. Considering the data for the Badia first, and independently from the other two data sets, it is evident that the region has a very high fertility regime. The TFR (15-44) for the population was 8.0, which can be thought of as the average number of children born to women across the fertile age cohorts. The age-specific fertility rate is particularly high in the age bracket 20-34 years with peak fertility in the 25-29 cohort.

Table 2. Age-specific fertility rates for the Badia, 1993

Cohort	Badia 1993	JPFHS 1990[1]	JFS 1976
15-19	33	49	71
20-24	303	219	300
25-29	400	296	367
30-34	350	264	332
35-39	308	188	240
40-44	196	79	112
TFR 15-44	8.0	5.5	7.1

(1) Based on the three years preceding the survey

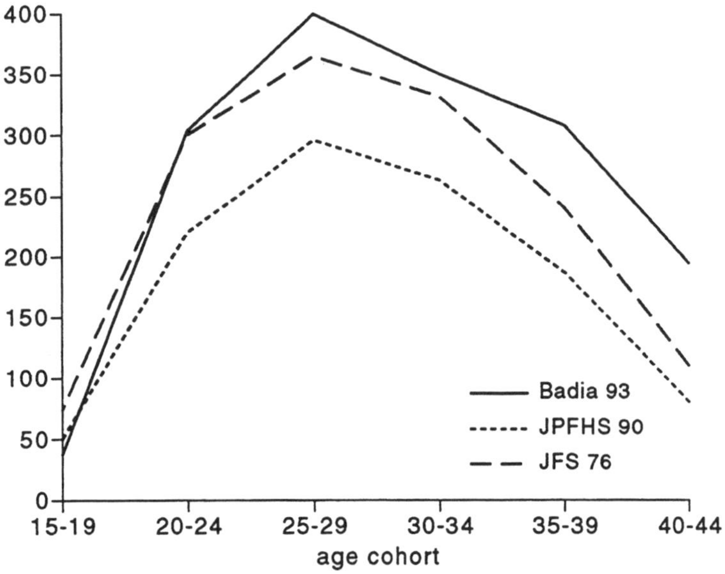

Fig. 1. Age-specific fertility curve for fertility surveys in Jordan in 1976 and 1990, compared with the 1993 Badia survey.

Fertility levels are clearly higher in the Badia than in either of the national surveys. In 1976 the TFR (15-44) for Jordan as a whole was believed to be 7.1, while the JPFHS reported a level of only 5.5 by 1990. It is clear that fertility across Jordan was much higher for all age cohorts in 1976 than it was in 1990. At a national level, it would seem possible to interpret the more or less even decline of fertility across all cohorts as some evidence that the decline in TFR is now resulting from a move towards increased control of fertility by women of all ages, rather than because of a postponement of childbearing, resulting from a later age at marriage, as was suggested earlier in this paper to be the case in rural Jordan during the late 1970s and early 1980s.

Comparison of the differential pattern of age-specific fertility evident in the Badia with that of Jordan as a whole is particularly interesting. The ASFR curve for the Badia in some respects resembles that of Jordan in 1976. This is particularly true for the age cohorts 20-24 years. Older women from the Badia however appeared from the 1993 survey to still exceed the national fertility picture from two decades earlier. Inversely younger women in the 15-19 age cohort, conforming to the nation-wide trends towards later marriage (only 28 per cent were or had ever been married), had as a result fewer children than might have been anticipated if the Badia population were viewed as simply a case of a fertility regime lagging 20 years behind the national picture.

Table 3. Marital status by age of women in the Badia population (% of cohort)

Cohort	Single	Married	Divorced	Widowed
15-19	71.6	27.4	1.0	0.0
20-24	30.9	67.9	1.2	0.0
25-29	10.8	88.1	0.8	0.3
30-34	3.9	91.4	0.4	4.3
35-39	0.9	92.5	0.9	5.7
40-44	1.0	90.5	1.5	7.0
45-49	0.7	82.1	1.3	15.9
50-54	0.0	87.4	0.0	12.6

Although the Badia survey is limited in several important respects, it contains one other useful piece of information relevant to the analysis of fertility trends: age at marriage. Since virtually all births in Jordan occur to women within marriage, age at marriage is a key intermediate or proximate determinant of fertility (Hill 1985). Data of the marital status of the female population (Table 3) contrasts with a benchmark figure for the age at first marriage for those in the 45-49 age cohort of almost exactly 16 years. However, in the population as a whole the median age at first marriage was 20, and as shown in Table 3 only 69 per cent of women in the 20-24 age cohort were or had ever been married, with the percentage dropping to 28 per cent for those aged 15-19. The clear implications are that a sustained trend towards later ages at first marriage is taking place and that as a result women are at risk of pregnancy over a shorter proportion of their time in the reproductive cohorts, which alone will lead towards lower TFRs, even in the absence of an increased adoption of contraceptive usage. In place of diffusionist models of fertility decline, the authors therefore interpret the Badia population of 1993 as belonging to a quite different demographic regime from that of the Jordanian population of 20 years ago.

7. POPULATION GROWTH AND POPULATION PRESSURE IN THE BADIA

Given the reservations expressed earlier about the quality of the mortality data gathered in the baseline survey, it is difficult to develop with any great confidence population projections for the region, based on the traditional demographic technique of using a components approach to estimating future demographic conditions. However, using the age-specific mortality rates reported earlier, survivorship ratios for each cohort were produced, as well as an estimate of life expectancy at birth. On this basis life expectancy at birth in 1993 was just over 62 years, a figure which reflects in particular the high infant mortality and child mortality rates of the area. Using ten year survivorship rates, the increase in the adult population aged 15-59 was also calculated. This is the age group for which under-recording of deaths would seem to be least likely and at the same time it represents the active age cohorts of the population in both production and reproduction. Projecting population growth of these age cohorts can also be done with much greater confidence than for younger cohorts, since over a ten-year time span the size of this segment of the population cannot be affected by downward shifts in the fertility rate. It is therefore interesting to note that comparison of the projected population aged 15-59 in the year 2003 with that in the same age range in 1993 suggests a 49 per cent increase, mainly because of the promotion into the active age cohorts of large numbers of persons from the younger part of the age pyramid.

An increase on this scale (and the assumptions surrounding such a projection can relatively easily be defended by contrast with the heroic assumptions necessary to make a projection for the whole population) seems highly probable and will place a tremendous burden on the resources of the region. There is no way that even the most ambitious rangeland development scheme can hope to provide a 49 per cent increase in employment opportunities in livestock management over the next ten years. Furthermore, even if a modest proportion of the growing population looks to pastoralism as its source of livelihood, it will still take a clear and well thought through initiative to sustain this trend without risk of further land degradation and loss of pasture land through overgrazing.

Given the conclusion that the livestock sector along with agricultural development cannot hope to absorb the region's growing population (and this is not a reason for not exploring improved livestock management methods), the Badia region will need to look to secondary and tertiary sector activities to provide the bulk of the employment growth to satisfy the 49 per cent growth in the active age population. Failing this, the net result will be the continued out-migration of population of the Badia into the towns and cities of western Jordan and in particular to the Amman-Zarqa conurbation.

A second dimension of population growth which must be mentioned, albeit briefly, is the implication for service provision. This might be illustrated with regard to any one of the many services which the population requires, such as access to health and education facilities. The pressure on service provision may however be even more starkly considered in relation to water provision. At the beginning of the 1990s Jordan and Israel had already been singled out by commentators as the two countries of the region facing the most severe water shortages (Shuval 1992). For example, it was estimated that in Jordan there were only 300m^3 of water available per person per annum compared with 1,200 in Egypt, 1,300 in Syria and 3,000 in Lebanon.

The Badia household questionnaire suggested that 82 per cent of households got their water piped into their homes by the water authority. In the country as a whole the proportion was 91 per cent and in Amman city 98 per cent. Over the next ten years it seems probable that the population of the Badia will seek to have similar access to piped water as other regions, thus increasing the pressure to provide water in this most arid part of the country. Again the 49 per cent growth rate in the active age cohorts, combined with the swelling cohorts of under 15 years of age which could result from sustained high fertility, will exacerbate this pressure of demand on one of Jordan's scarcest resources. In these circumstances the temptation to over-pump aquifers and to cause further environmental degradation through unwise exploitation of water resources can only be expected to increase.

8. POSSIBLE FUTURE TRENDS IN FERTILITY

Having noted above that the Badia has an exceptionally high fertility regime and that this has combined with low age specific death rates for the adult population to produce a very high population growth rate, it is important to balance this statement with an informed evaluation about future trends. It would clearly be more satisfactory to produce population projections for the Badia or any regional population of Jordan from an informed understanding of the likely trends which could emerge in fertility and mortality. It is neither reasonable nor acceptable to suggest that the future and in particular the longer term lying beyond the next ten years will be simply an extension of present conditions.

In an attempt to focus attention on possible reasons for change in fertility levels, consideration is now given to a second survey carried out by the authors in another, not too dissimilar, part of Jordan. The goal of this survey was to examine how changing roles within the family have emerged in the rural environment, and to consider what effects these changes might have on fertility behaviour.

Maani (1990) in a survey conducted in 1988 in several different parts of Jordan found, in the course of investigating Caldwell's theory of inter-generational wealth flows, that only a small minority of the rural families which he interviewed depended on agricultural production for their income 'while most of them depend partially on income from other sources including remittances. Money flows from children to parents are not therefore confined in Jordan to the immediate location of the household, nor even to the local labour market, but involve migration to urban areas, especially Amman' (Maani 1990, 190).

In order to investigate the links between inter-generational wealth flows, and fertility in more detail it was decided by the authors to undertake a new survey in rural Jordan in July 1990. This survey was needed to provide more detail of the migration histories of rural households. The data provided below relates to the village of Basta, some 25km west of Ma'an, and chosen as an example of a settlement from the rural hinterland of southern Jordan. Its population is made up of pastoralists from Bedouin society who have sedentarized during the twentieth century. All 96 households in the village were surveyed for information about the past mobility of the heads of household. A surprising 8.6 per cent of households had lived at their present address for a year or less, and 18.3 per cent had migrated during the last three years. Only 7.5 per cent of households had been at their current address for over 30 years. The pastoralist heritage of the some of the residents of Basta was evident from the fact that 6.5 per cent of households had lived as nomads immediately prior to settling at their current address, while 39 per cent of heads of households reported that their fathers had been pastoralists. A further 47 per cent had parents who had been agriculturalists.

Turning to the younger generation, this survey revealed that amongst the households studied there were 58 children of working-age, of whom 42 per cent were employed in urban areas of Jordan outside Ma'an governorate, mostly living and working in Amman city. Some 41 of the 96 households reported wealth flows of one kind or another between themselves and their non-resident working children. In the interviews some 67 per cent of households reported that at least some of their children had migrated to work in other parts of Jordan. A surprisingly small proportion (7 per cent) had children working abroad. This reflects both the difficulty of entering the local labour market, by comparison with employment opportunities elsewhere, as well as the tendency for many children to continue to live with their parents even if their employment is in the same vicinity as their family home. The proportions are in marked contrast with the results achieved by Maani (1990, 203) in his 1988 survey of Amman, where 82 per cent had working-age children employed in Amman itself, and a further 13 per cent had sons working abroad.

Before the survey data can be discussed in greater detail it is necessary to return once again to consider the concept of wealth flows and how they may be measured. Wealth flow direction within a household was determined using a scoring system devised in relation to the responses received to a series of questions included in the survey on aspects of wealth transfer. These questions inquired into the level of general economic assistance provided to parents by their working-age children, the nature of the family budget and the level of earnings which fathers expected to receive from their children in the future. The responses to these questions were organized so that where wealth flow could be inferred to be occurring in an upward direction from children to parents high scores were recorded, and where working children were making no contribution or only a small contribution to the household budget low scores were registered. The summation of the scores on all the questions produced a range of values from 10 to 26, where 10 was taken to mean a strong downward flow of wealth and where 26 was the extreme case of an upward flow of wealth. Having examined the statistical distribution of scores resulting from this method it was decided for the purposes of this analysis to define all scores between 16 and 26 as reflecting a net upward transfer of wealth between the generations. There are clearly difficulties in using this kind of scoring system, but faced with the need to investigate abstract and complex concepts such as wealth flow direction the approach seems to have some merits.

As Table 4 shows, just over 90 per cent of urban households in Amman had net downward wealth flows. In terms of Caldwell's (1982) theory, these households were experiencing a situation in which the investment in children exceeded the returns, and in which it would be no longer economically rational to maintain high fertility levels. Discussion of the significance of this specific situation is treated elsewhere by Maani (1990). By contrast, in the rural households included in the Basta survey slightly under 30 per cent of all households had downward wealth flows while the remainder sustained the traditional pattern of net wealth transfer from children to adults. The cross-tabulations of Table 4 show that for those households with migrant children of working age, the process of population movement may be a mechanism affecting the direction of wealth flows. The table shows that in the majority of cases young migrants leaving Basta (and another village, Ketm) continue to provide significant support to their parents, but less than for the sample population as a whole. This is to say that migrant children were in a transitional position returning less to their parents than children domiciled with their parents, but nevertheless sustaining a net upward wealth flow.

It is worth asking how and why migration should have an effect on wealth flow directions and what the possible consequences may be on family structure and on vital rates in rural areas. To do this, it is helpful to inquire into the context of rural-urban migration in Jordan.

Table 4. Wealth flow direction by area of interview

Area	Net upward wealth flow (%)	Net downward wealth flow (%)	No. surveyed (households)
Amman city	9.8	90.2	418
Basta	70.2	29.8	96
Rural to urban migrants from Basta and Ketm	62.0	38.0	61 (114 migrants

The commercialization of rural Jordan, as in other countries (Stark 1981), had a profound effect on lifestyle, economy and demography. Investment in Jordanian agriculture by urban families seeking to orientate production to urban needs has led to the introduction of new technologies such as the use of chemical fertilizers and plastic greenhouses which have made production more capital intensive, and also tended to marginalize those rural producers without access to capital. Inevitably poorer farmers and pastoralists looked to other sources of income both for security and as a means of becoming competitive. Encouraging their children to migrate either to find work in Amman or perhaps in the Jordanian army was one mechanism by which income could be diversified away from pastoralism and agriculture, at the same time as providing a means of earning larger capital sums from waged employment. When questioned on their aspirations for their children virtually all rural respondents indicated that they wanted them to enter paid employment not involving pastoralism or agricultural cultivation. Equally they saw the means to achieve this as being through their children remaining longer in the education system, thus raising the costs of child rearing to the parents and delaying the date when their children could begin to make a contribution to the household economy. In turn the emphasis on having well qualified children with the potential for secure and perhaps professional urban employment meant that effectively a switch in aspirations had occurred from the traditional pattern of having a very large family to a pattern of desiring relatively few 'high-quality' children. The receipt of remittances from working-age children in Amman and elsewhere was perceived to be not only of financial assistance, but also enhanced the parents' status and reputation, thus introducing a rather different type of wealth flow which children entering employment in the local agricultural community could not hope to offer.

In summary, the opportunities for waged employment in the secondary and tertiary sectors has opened up for rural communities much more than an outlet for surplus labour. The transition in employment patterns may also prove in the longer term to be a catalyst for societal change in the study areas in the Badia. It may favour, in Caldwell's terms, the reversal of the compass needle from a situation favouring high fertility in order to sustain traditional rural production, to one of lower fertility in order to be able to rear and support a small number of children throughout the full education system in order to maximize their opportunities on entry to the urban labour market. Changing modes of production have affected mobility circumstances which in turn are directly influencing people's perception of desired family size. Patterned regularities in mobility and temporal trends in fertility are clearly linked, but only further research into the components of wealth flows resulting from rural to urban migration and the consequences of that migration for family structures in peripheral areas will elucidate the nature of these important linkages.

9. CONCLUSION

The research reported in this chapter has sought to advance demographic understanding in two ways. First, it has tested the validity of some aspects of Caldwell's (1982) theory of inter-generational wealth flows, as an influence on fertility trends. It has been shown that Amman city has both the lowest total fertility rates in Jordan and according to the survey work which has been reported is also a location where most households experience net downward wealth flows. Conversely rural areas of the Badia have been shown to have much higher total fertility rates than Amman, and according to the survey results still experience net upward wealth transfers. It should be evident to all that fertility levels in Amman, while lower than in rural areas, remain at a high overall level by comparison with Western nations. The powerful effect of Arab culture continues to militate against a complete fertility transition taking place, but as the paper has shown the barriers to this are much more complex than the simple invocation of female workforce participation rates as proposed by Fargues (1989).

Second, the chapter has argued that there is a need to merge concepts of the fertility transition with ideas about population mobility. Children of working age who had become rural to urban migrants were shown to represent an intermediate position between the net downward wealth flows of urban households and the net upward wealth flows of rural households. Furthermore the migration act, as well as other factors operating over time in Jordan, appeared to be responsible for producing changes in the composition of inter-generational wealth flows. The research presented certainly would lend credibility to the view that future work on the role of inter-generational wealth flows should explicitly consider the role of mobility and migrant remittances.

ACKNOWLEDGEMENTS

The authors would like to acknowledge financial assistance from the John Robertson Bequest (University of Glasgow) which facilitated part of this research. They are also grateful to colleagues at URBAMA, University of Tours for permission to reproduce Table 4 from a paper given by the authors to an URBAMA conference in 1990.

REFERENCES

Bongaarts, J. (1978) A framework for analyzing the proximate determinants of fertility. *Population and Development Review,* 4, 105-32.

Bongaarts, J. (1985) The fertility inhibiting effects of the intermediate fertility variables. In Shorter, F. and Zurayk, H. eds. *Population Factors in Development Planning in the Middle East*. Population Council, Cairo, pp. 152-69.

Caldwell, J. (1982) *Theory of Fertility Decline*. Academic Press, London.

Department of Statistics, (1992) *Jordan: Population and Family Health.* Department of Statistics/Ministry of Health: Amman.

Fargues, P. (1989) The decline of Arab fertility. *Population* (English Selection), 4, 147-75.

Findlay, A. (1986) Amman: urbanization in a charity state. *Bulletin de la Société Languedocienne de Géographie,* 109, 211-22.

Findlay, A. (1991) Population and environment: reproduction and production. In P. Sarre (ed.) *Environment, Population and Development.* Hodder, London, pp. 3-38.

Findlay, A. (1992) Report on the Badia baseline pilot survey, Jordan. Applied Population Research Unit Report. University of Glasgow.

Findlay, A. and Brown, J. (1994) International migration and remittances. *Atlas of World Development.* Wiley, Chichester, 224-6.

Findlay, A., Maani, M. and Briggs, J. (1994) Fertility trends in the Badia of Jordan. Jordan Badia Research and Development Programme Demographic Team Report, University of Durham: CORD.

Geomorphology and Physical Resources Research Team. (Jordan Badia Research and Development Programme) (1993) Report No 1, RGS/HCST.

Harrison, P. (1993) *The Third Revolution: Population, Environment and a Sustainable World.* Penguin, Harmondsworth.

Hill, A. (1985) A practical guide to estimating the Bongaarts indices of the proximate determinants of fertility. In F. Shorter and H. Zurayk (eds) *Population Factors in Development Planning in the Middle East.* Population Council, Cairo, pp.170-94.

Lancaster, W. (1981) *The Rwala Bedouin Today.* Cambridge University Press, Cambridge.

Lavergne, M. (1992) Aménagement du territoire et croissance urbaine en Jordanie. *Maghreb-Machrek,* 140, 35-58.

Maani, M. (1990) Recent Changes in Family Structure and Fertility in Jordan. Unpublished Ph.D. thesis, University of Glasgow.

Ministry of Planning, (1988) International Bank Report on the Water Sector in Jordan. Ministry of Planning: Amman.

Newell, C. (1988) *Methods and Models in Demography.* Belhaven, London.

Preston, S. (1994) Population and Environment: From Rio to Cairo. IUSSP offprint of its distinguished lecture series on 'Population and Development'. IUSSP, Liège.

Pryor, R. (1981) Population redistribution: the demographic and mobility transitions. In J.I. Clarke and L.A. Kosinski (eds) *Redistribution of Population in Africa.* Heinemann, London, pp. 25-30.

Randall, S. (1994) Are pastoralists demographically different from sedentary agriculturalists? In B. Zaba and J.I. Clarke (eds) *Environment and Population Change,* Ordina, Liège, pp. 325-49.

Samha, M. (1979) Migration to Amman. Unpublished Ph.D. thesis, University of Durham.

Shuval, H. (1992) Le problème du partage de l'eau entre Israel et les Palestiniens. *Maghreb-Machrek* 138, 27-37.

Smith, R. (1981) Fertility, economy and household formation in England over three centuries. *Population and Development Review,* 7, 595-618.

Stark, O. (1981) The asset demand for children during agricultural modernization. *Population and Development Review,* 7, 671-76.

Zelinsky, W. (1971) The hypothesis of the mobility transition. *Geographical Review,* 61, 219-49.

CHAPTER 17

Demographic Change and Population Projections, 1976-2013

MOHAMMED MAANI, HARB HUNAITI
(University of Jordan)

ALLAN M. FINDLAY
(University of Dundee, UK)

1. INTRODUCTION

Population trends in Jordan are of great interest to demographers given the country's distinctive demographic regime. Until very recently the rate of natural increase was amongst the highest in the world (Findlay and Maani 1995). In the mid-1990s the population doubling time for Jordan remains only 18 years. As well as having a very high rate of natural increase Jordan has also been a major destination for refugee movements. Palestinians fleeing from conflicts in neighbouring territories to the west, have formed the majority of refugees, but other influxes of population have also occurred, including the recent arrival of Palestinian labour migrants and their families from Kuwait in the aftermath of the Gulf War of 1991. The size of Jordan's refugee population is a matter of some dispute, but few would contest that in world terms Jordan has one of the highest proportions of refugees in relation to total population. In addition to these exceptional circumstances, in the 1970s and 1980s Jordan experienced an intense involvement in the system of international labour migration which focused on the Arab oil economies. A very significant proportion of skilled Jordanians emigrated during these years resulting in Jordan's national earnings from migrant remittances rising rapidly to account by the 1980s for the equivalent of a quarter of the value of the Gross Domestic Product. In short, Jordan is a state which has experienced remarkable demographic and economic circumstances over the last few decades. It is as a result a site of particular interest to demographers and population geographers.

This paper is concerned with the demographic regime of the peoples of the arid eastern part of Jordan. The study area represents just one part of the much wider territory of eastern and southern Jordan known as the Badia. The Badia, having one of the highest fertility rates of any region in Jordan, provides an extreme example of the demographic pressures experienced in the rest of the country. In the Badia a demographic regime with an exceptionally high rate of natural increase combines with a physical environment which is one of the most arid in Jordan. This combination has led to fears in some quarters of population growth triggering environmental degradation.

Historically the economy of the Badia has been based on pastoralism. Employment opportunities have not diversified to the degree that they have in the major conurbations of

Jordan, such as Amman and Zarqa. The rapid rate of population growth presents a difficult challenge, requiring new opportunities for sustainable development to be initiated in the area to combat the continued heavy out-migration from the Badia. The detailed demographic picture, presented here, of the situation in the Badia seeks not only to provide an essential data base for planners concerned with this region, but also to raise questions of wider significance to planners concerned with the future welfare of populations in arid environments in general (Findlay 1997).

2. AIMS AND OBJECTIVES

Previous research, reported by the demographic team of the RGS/HCST Badia Research and Development Project, has summarized the main results of a survey of all households in the study area conducted in 1993 (Findlay and Maani 1995). This paper seeks to provide a more in-depth and authoritative account of the demographic regime of the Badia. This will be achieved in three ways. First, the 1993 survey has been analysed in more detail, testing for inaccuracies in survey reporting and applying standard demographic techniques to correct for errors. Second, a longer time period is covered with regard to the demographic history of the Badia study area. Analysis is attempted of demographic data from surveys undertaken in 1976 and 1987. In addition, the 1979 and 1994 census records for the region have been studied and re-evaluated. Therefore this paper offers for the first time an evaluation of the demographic regime of the Badia based on detailed information for five different years spread across a 20-year period. Thirdly, geographical variations in population trends within the region at a range of different scales have been examined.

The information collected through the five surveys and censuses varies in type and quality. Comparable head count statistics for all five studies were available however, and provide the basis for monitoring the most basic aspects of demographic change at a detailed geographic level. A common areal definition of the villages and their surrounding lands has been used, with settled and nomadic peoples being recorded relative to the nearest village place name. The 1976 survey provides a valuable early source of demographic information, and comparison of the 1976 and 1993 surveys reveals some interesting demographic changes.

3. DATA SOURCES AND DATA QUALITY

Standard demographic techniques have been applied in this paper to evaluate and clean the data sets. These include sex ratio techniques, male and female ratio scores, accuracy of data indices and smoothing procedures such as the Carrier and Farrag (1959), King/Newton (UN 1982), Brass (1971;1975) and Trussell (1975) methods, which have been widely evaluated elsewhere in the demographic literature (Shryock and Siegel 1976; UN 1982), so no further discussion is attempted here.

The data sets available to the authors included a survey undertaken in 1976 by the University of Jordan initiated at the request of Crown Prince Hassan. The survey covered all of the population of the Badia and not just the study area of the RGS/HCST project. Populations were surveyed in the north, east and southern Badia. The survey was based on a representative sample of the population and involved around 15,000 persons. There were three questionnaires in the survey covering: (a) housing characteristics, (b) demographic information, and (c) economic characteristics.

The second data source used in this report is the December 1979 Census of Housing and Population conducted by the Department of Statistics for the entire population (Hashemite Kingdom of Jordan 1979). Although the census has been widely discussed in other publications, this report provides the first detailed disaggregation of headcount information for the Badia study area.

A third snapshot of the population of the Badia is possible for the year 1987 when a population estimation was prepared by the Ministry of Planning, based on a comprehensive survey of socio-economic data designed by the Ministry of Planning for regional planning purposes.

The fourth survey was undertaken in January 1993 under the auspices of the RGS/HCST Research and Development Project in the Badia, and was designed as a baseline survey covering all households in the Badia and study area. Full details of the methods of data collection have previously been described by the authors (Findlay and Maani 1995).

Finally the researchers have been fortunate to access the headcount statistics from the December 1994 Jordanian National Population Census.

Details of the relative quality of the 1976 and 1993 data sets are presented later in this report. As in all developing countries there is evidence of reporting inaccuracies which require some correction and smoothing. In general the information reported suggests that the quality of the demographic data in the 1993 survey was higher than that collected in 1976. A very high correlation exists between the headcount data classed by settlement for the 29 enumeration areas used in January 1993 and the comparable units from the December 1994 census ($r = 0.99$). This indicates a strong degree of continuity between these two independent data sources and increases confidence in the 1993 survey. Detailed comparison of the distribution of population at the two dates leads to the conclusion that the most serious under counting in 1993 took place at Umm al-Quttayn.

4. POPULATION DISTRIBUTION AND POPULATION GROWTH, 1976-94

The Badia has experienced dramatic changes in the pattern of population over the past 20 years, as shown in Table 1. Population trebled during the intercensal period 1979-94 from 5,497 to 16,267. The other survey years of 1987 and 1993 fit within this broad trend confirming annual growth rates of 3.1 per cent during the 1980s and early 1990s. The statistics in Table 1 are disaggregated at two levels: as three geographical areas (defined by the Ministry of Planning as development units), and as 29 more detailed units mostly focused on the villages of the Badia (Fig. 1). The three larger areas comprise:

(i) a southern region stretching parallel to the main route from Irbid to Ruwayshid, which includes the major settlement of Safawi;
(ii) a cluster of villages centred on Umm al-Quttayn, situated to the south-west of Jabal al-Arab; and
(iii) a line of villages in the north-east of the study area, on the eastern flank of Jabal al-Arab, located quite close to the Syrian border.

The distribution of population from the 1994 census data, reveals that much of the highly arid southern part of the Badia, which receives on average less than 50mm of

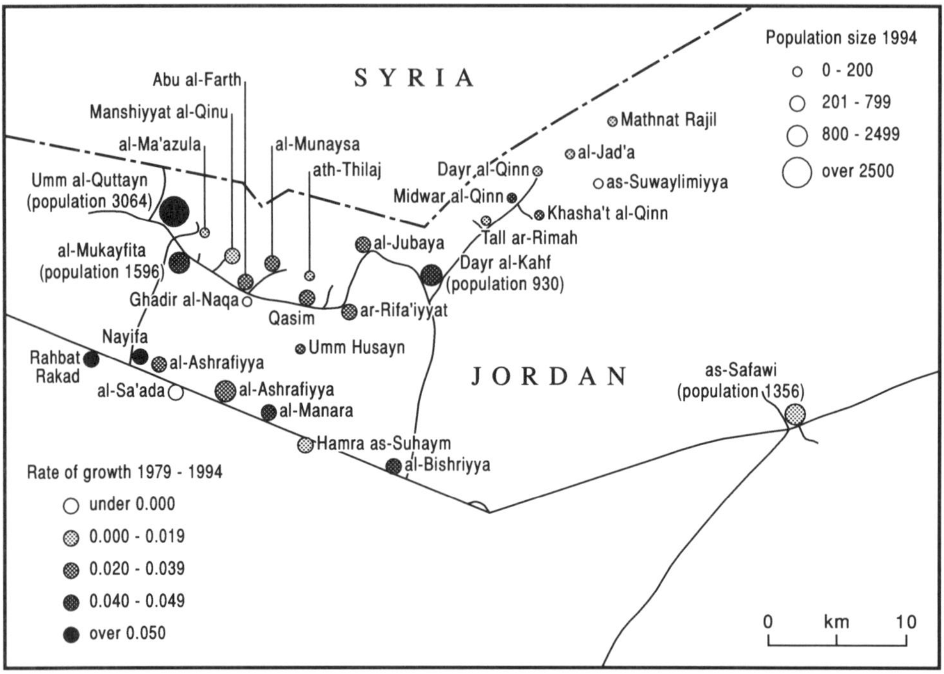

Fig. 1. Village population growth in BRDP study area, 1976 to 1993.

precipitation, has very little by way of settled population. The majority of the population of the southern part of the study area is concentrated in the small settlements which are strung along the main highway to Iraq. By contrast, the northern fringes of the study area receive a higher rainfall reaching 200mm on Jabal al-Arab. This area has significant levels of rural settlement across the lower slopes of the upland area and with some quite populous areas such as around the main centre of Umm al-Quttayn. Interviews with the inhabitants of these villages indicate that many of their ancestors were nomadic pastoralists who settled on a more permanent basis in the villages of Jabal al-Arab in the 1960s. Although nomadic pastoralism was the traditional livelihood, the majority of the population is now sedentary. However, seasonal movement of herds still takes place, such moves taking place in relation to the villages where the pastoralists maintain a semi-permanent presence.

The uneven pattern of growth evident from Table 1 and Figure 1 is very striking. In 1976 the settlements in the southern part of the study area, stretching from Rahbat Rakad to Safawi accounted for more than half (57 per cent) of the total population. By contrast the villages to the north west around Umm al-Quttayn were at this time very small settlements with a total population of 1,315, or approximately a quarter of the Badia population. The 1994 census records a dramatic change in the demography of the area. All regions experienced an increase in population. However the population in the north-west exploded with five and half times as many persons in 1994 as in 1976. The 1994 total of 7,268 persons accounted for 45 per cent of the Badia study area total, while the southern region had shrunk in relative terms to only 39 per cent of the total. The north-east region remained the least populous with less than 1,000 persons in 1976 (18 per cent of the total) and 2,612 in 1994 (16 per cent of the total). The overall picture of significant population growth throughout the Badia during the last 20 years has therefore to be qualified by recognition of the fact that growth was accompanied by substantial population redistribution.

Table 1. Population distribution and growth in the Badia, 1976-94

Village	1976	1979	1987	1993	1994	rate 1979-94
Umm al-Quttayn	201	208	1,727	2,504	3,064	0.077881
al-Mukayfita	250	304	1,029	1,596	1,596	0.048010
Abu al-Farth	176	181	211	533	612	0.035271
Qasim	125	143	167	433	499	0.036184
al-Munaysa	49	56	107	196	222	0.039877
ath-Thilaj	30	35	50	69	69	0.019652
Ar-Rifa'iyyat	189	231	235	511	669	0.030787
al-Ma'azula	90	108	153	93	147	0.008926
Umm Husayn	32	37	43	131	129	0.036159
Manshiyyat al-Qinn	158	164	204	202	244	0.011503
Ghadir al-Naqa	15	20	25	16	17	-0.00470
Subtotal	1,315	1,487	3,951	6,284	7,268	0.045940
Dayr al-Kahf	149	167	558	817	930	0.049717
Jubaya	150	179	209	544	669	0.038171
Tall al-Rimah	150	156	185	274	269	0.015775
Mathnat Rajil	100	113	133	143	119	0.001497
Dayr al-Qinn	150	167	194	128	188	0.003429
Midwar al-Qinn	25	30	41	137	156	0.047733
al-Jad'a	100	116	134	93	116	0
al-Suwaylimiyya	29	31	36	17	6	-0.04754
Khasha't al-Qinn	69	72	84	125	159	0.022937
Subtotal	922	1,031	1,574	2,278	2,612	0.026914
Rahbat Rakad	50	56	65	385	456	0.060718
Nayifa	35	38	44	500	672	0.083172
al-Ashrafiyya	301	308	383	582	634	0.020902
al-Manara	305	309	360	718	920	0.031588
Hulaywat al-Masariha	136	140	166	596	629	0.043501
Hamra al-Suhaym	308	320	613	428	469	0.011068
al-Bishriyya	240	134	273	708	703	0.047990
Safawi	968	1,067	1,325	1,103	1,356	0.006939
al-Sa'ada	581	607	755	432	548	-0.00296
Subtotal	2,924	2,979	3,984	5,452	6,387	0.022081
Grand Total	5,161	5,497	9,509	14,014	16,267	0.031412

The pattern which emerges from Table 1 is of a major net population shift from the south and east to the north-west of the study area (i.e. away from the most arid area towards the area with the highest rainfall). This is not to say that the differential pattern of growth was caused by a migration from the one area to the other, since clearly the differential pattern of growth resulted from the net effects of migration from the region to other parts of

Jordan and to other countries. It would appear that heavy out-migration from the east and southern part of the Badia to other areas of Jordan and abroad has reduced the rate of population growth in these areas more than from the north-west.

At the more detailed level of analysis of the villages of the Badia region, it can be seen from Figure 1 that the highest levels of absolute population growth recorded between the two censuses were in the west of the area. Figure 1 maps settlement size according to the 1994 census in terms of four categories of proportional circles and also shows the pace of annual population growth during 1979-94 for each settlement in terms of the shading within each circle. Thus an annual growth rate of 0.05 or more as mapped for the settlement of Rahbat Rakad can be thought of in terms of a 5 per cent growth per annum.

By far the most rapid rate of growth was recorded for Umm al-Quttayn, the largest settlement in the area. Umm al-Quttayn grew from 208 persons in 1979 to 3,064 persons in 1994 (an annual growth rate of 0.064), and neighbouring al-Mukayfita sprouted from 304 persons to 1,596 over the same period. These settlements accounted for 72 per cent of all population growth in the north-west region between the two censuses. This suggests that they were the focus of net in-migration. Of the 10,770 extra persons recorded in the study area in 1994 compared with 1979, 4148 or 38 per cent were living in these two villages. The other two settlements picked out on Figure 1 as having an average annual growth rate of over 5 per cent per annum are the two neighbouring settlements of Rahbat Rakad and Nayifa, both located on the main Irbid highway and beside the road junction for the Umm al-Quttayn feeder road. By contrast, the average population growth was only 2.6 per cent, and in the third largest settlement of the study area (Safawi) growth was only 0.6 per cent per annum. Figure 1 also shows very vividly that the settlements of the north-east of the study area are very small, but also that their growth rates are very low or negative, implying that they have experienced considerable net out-migration.

5. AGE-SEX STRUCTURE OF THE POPULATION

Not only has the spatial distribution of population in the Badia changed dramatically during the last 20 years, but so also has the age-sex structure. Identification of these changes is very important to planners, since they have profound implications for the demand for services such as education and health. Since the Badia data, like that of many developing countries, encounters problems of inaccurate reporting, the robustness of the data needs to be evaluated and appropriate methods for correcting and smoothing the data selected, before analysis can be attempted of the significance of structural changes.

5.1. Data Evaluation, 1976

There are four indices commonly used to measure the accuracy of age-sex data. These are (1) the sex ratio score, (2) male age ratio score, (3) female age ratio score, (4) accuracy index. The higher the value of these indices the lower the accuracy of the data. The lowest attainable value of zero indicates perfectly accurate data. The simplest way to evaluate the data is to plot the age structure on a graph to show whether heaping occurs at certain ages. Accordingly the age structures for 1976 and 1993 have been plotted, and Figures 2 and 3 show a considerable degree of heaping at ages ending in the digits 0 and 5. This is confirmed by the four age-sex accuracy indices shown in Table 2.

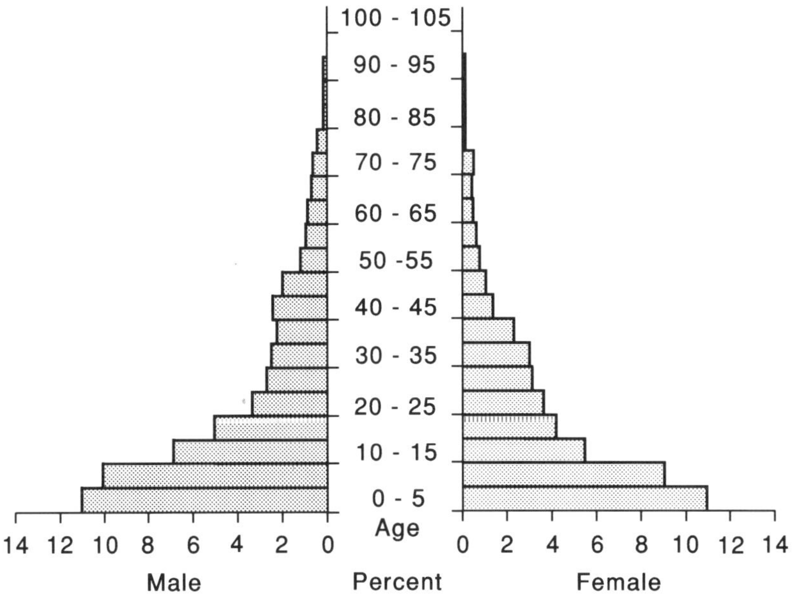

Fig. 2. Age-sex pyramid for the Badia population, 1976.

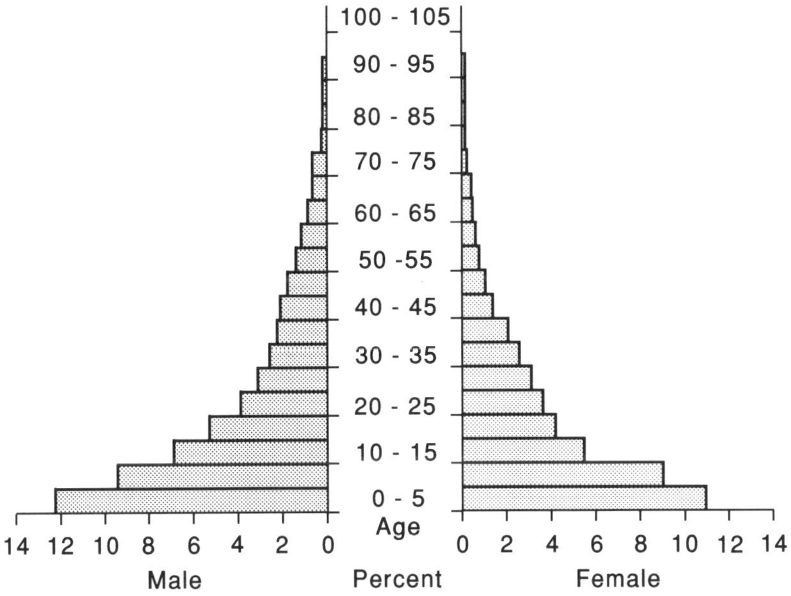

Fig. 3. Age-sex pyramid for the Badia population, 1993.

Table 2. Age-sex accuracy scores for the Badia, 1976 and 1993

Index	1976	1993
Sex ratio score	40.67	12.65
Male age ratio score	21.44	10.68
Female age ratio score	38.55	10.55
Accuracy index	181.99	59.19

5.1.1. Sex ratios, 1976

Table 2 reveals a high level of misreporting of gender data. The sex ratios for 1976 are particularly imbalanced among young cohorts aged between 15 and 34 years, as shown in Table 3. Imbalances in these age cohorts tend to be interpreted as reflecting high out-migration. The high imbalances in the sex ratio for the groups above 40 years of age can be explained by under-reporting of females and an age exaggeration by men relative to women. As Table 3 shows, similar imbalances are reported in the national data sets for Jordan, especially in the 1976 Jordanian Fertility Survey.

Table 3. Sex ratios for Jordan and the Badia, 1976

Age	Badia	Jordan
0-4	100.7	
5-9	110.6	159.0
10-14	133.6	196.0
15-19	98.4	99.2
20-24	89.4	79.4
25-29	84.0	78.1
30-34	74.3	85.6
35-39	100.8	89.4
40-44	148.8	98.0
45-49	179.6	115.1
50-54	128.6	108.8
55-59	206.3	114.5
60-64	126.2	115.0
65-69	200.0	122.6
70-74	121.4	119.9
75+		118.7

5.1.2. Age ratio scores and accuracy index, 1976

In 1976 the age ratio scores for males and females were 21.44 and 38.55 respectively. These figures indicate that there was a greater age reporting bias for females than for males, a not unusual fact in developing countries. The 1976 accuracy index of 181 implies that the raw data are highly unreliable and need some correction, before further analysis is possible.

5.1.3. Data correction, 1976

Six different data smoothing procedures were applied to the data set (Table 4) including the Carrier/Farrag, King/Newton, Arriaga, United Nations and Strong methods. The results of these procedures were evaluated in terms of the four indices listed in Table 5. Strong's smoothing procedure was selected as it produced the lowest sex ratio score (9.49), male ratio score (3.57), female ratio score (3.72) and accuracy index (35.76). This procedure was the most helpful in generating an age-sex distribution of the kind which one might have expected to find in the Badia study area in 1976 if data had been reported accurately. The output of this series of data manipulations can be seen in Tables 5 to 8.

Table 4. Reported population by age and sex in the Badia, 1976

Age	Male	Female
All ages	2,705	2,457
0-4	560	556
5-9	510	461
10-14	428	320
15-19	179	182
20-24	161	180
25-29	142	169
30-34	110	148
35-39	129	128
40-44	119	80
45-49	97	54
50-54	72	56
55-59	33	16
60-64	53	42
65-69	28	14
70-74	34	28
75-79	18	5
80+	32	18

Table 5. Summary of indices measuring the accuracy of age-sex data in the Badia, 1976

Index	Reported	Smoothed				
		Carrier/ Farrag	King/ Newton	Arriaga	United Nations	Strong
Sex ratio score	40.67	15.33	17.60	15.40	14.43	9.49
Male age ratio score	21.44	8.57	9.48	8.39	9.41	3.57
Female age ratio score	38.55	7.08	8.01	7.06	7.70	3.72
Accuracy index	181.99	61.65	70.28	61.66	60.40	35.76

Note: The accuracy index is the sum of the male and female age ratio scores plus three times the sex ratio score, all calculated using data for ages 10-14 through 65-69

Table 6. Sex ratios of the reported and smoothed populations of the Badia, 1976

Index	Reported	Smoothed				
		Carrier/ Farrag	King/ Newton	Arriaga	United Nations	Strong
0-4	100.7			102.0		104.0
10-14	133.8	123.4	120.0	122.1	125.4	111.0
15-19	98.4	117.6	122.1	119.4	109.5	111.1
20-24	89.4	90.2	92.5	90.7	86.1	101.2
25-29	84.0	82.9	80.1	82.4	81.6	98.7
30-34	74.3	80.7	82.5	81.2	79.1	95.6
35-39	100.8	94.1	91.6	93.3	99.9	100.9
40-44	148.8	152.4	145.9	151.4	144.4	122.2
45-49	179.6	173.6	183.6	174.9	167.2	136.1
50-54	128.6	147.6	149.1	148.1	148.7	149.8
55-59	206.3	143.6	141.6	143.0	145.7	160.6
60-64	126.2	143.4	143.9	143.6	146.5	145.6
65-69	200.0	146.2	145.5	145.9	144.6	154.5
70-74	121.4			151.7		155.6
75-79	360.0			166.3		160.0
80+	177.8					

Table 7a. Reported and smoothed male populations of the Badia by age, 1976

Index	Reported	Smoothed				
		Carrier/ Farrag	King/ Newton	Arriaga	United Nations	Strong
Total: 0-79	2,673			2,673		2,673
Total: 10-69	1,551	1,551	1,551	1,551	1,546	1,551
0-4	560			600		602
5-9	510			471		468
10-14	428	351	351	348	395	343
15-19	179	256	256	259	218	261
20-24	161	169	175	170	147	189
25-29	142	134	129	134	137	150
30-34	110	125	125	124	119	125
35-39	129	114	114	115	123	108
40-44	119	119	116	118	120	98
45-49	97	97	100	98	98	83
50-54	72	59	61	59	67	65
55-59	33	46	44	46	44	53
60-64	53	44	44	44	42	41
65-69	28	37	37	37	36	34
70-74	34			30		28
75-79	18			22		24
80+	32					

5.1.4. Age-sex structure indices, 1976

In order to examine the main structural characteristics of the population of the Badia in 1976, two standard demographic descriptors were calculated: the dependency ratio and the mean age of the population. The percentages of the population of 15 years of age or less, 15-64, and 65+ were 53, 44 and 3 respectively. This implied that the population was extremely youthful, with all the implications which this had for demands on the education and health services. The general poor level of provision of basic infrastructure especially in the southern part of the study area in the 1960s and 1970s may have been one of the factors encouraging strong out-migration. The dependency ratio in 1976 was 129, which indicates a large number of dependants in the population relative to the number of persons of working age, capable of generating a livelihood for the households of the Badia. The mean age of the population was 31 years, confirming the youthful nature of the population.

Table 7b. Reported and smoothed female populations of the Badia by age, 1976

Total: 0-79	2,439			2,439		2,439
Total: 10-69	1,389	1,389	1,389	1,389	1,397	1,389
0-4	556			588		579
5-9	461			429		438
10-14	320	284	293	285	315	309
15-19	182	218	209	217	199	235
20-24	180	188	189	187	171	187
25-29	169	161	160	162	168	152
30-34	148	154	151	153	151	131
35-39	128	122	125	123	123	107
40-44	80	78	80	78	83	80
45-49	54	56	54	56	59	61
50-54	56	40	41	40	45	44
55-59	16	32	31	32	30	33
60-64	42	31	30	31	29	28
65-69	14	25	26	25	25	22
70-74	28			20		18
75-79	5			13		15
80+	18					

Comparing these figures with the national equivalents for Jordan in 1976, differentials are found which reflect the distinctive demographic position of the Badia. Table 9 shows that the Badia community is more youthful than Jordanian society as a whole, although the sex ratio is much the same. The sex and dependency ratio in the Badia is 129 compared with 120 in Jordan. In the future, if Jordan proceeds through the demographic transition, fertility should decline and accordingly the percentage contribution of younger cohorts will be reduced, while that of the working age population group should increase. If the age structure of the Badia area follows the same pattern as Jordan, the percentage of the working age group will increase at the expense of younger cohorts.

Table 8. Age ratios of the reported and smoothed populations of Badia, 1976

Index	Reported	Smoothed				
		Carrier/ Farrag	King/ Newton	Arriaga	United Nations	Strong
MALE						
5-9	103.2			99.3		99.0
10-14	124.2			95.5		94.2
15-19	60.8	98.5	97.2	100.0	80.6	98.0
20-24	100.3	86.7	90.9	86.4	82.8	92.1
25-29	104.8	91.2	85.8	90.9	103.1	95.5
30-34	81.2	100.3	103.0	99.9	91.5	96.9
35-39	112.7	94.0	94.5	94.9	103.1	96.7
40-44	105.3	112.7	108.9	111.1	108.1	102.9
45-49	101.6	108.9	112.4	110.3	105.5	101.4
50-54	110.8	82.4	84.8	82.3	93.8	96.3
55-59	52.8	89.5	84.1	89.0	81.2	100.0
60-64	173.8	106.1	107.8	105.9	104.2	93.2
65-69	64.4			100.6		99.2
70-74	147.8			100.7		96.6
FEMALE						
5-9	105.3			98.3		98.6
10-14	99.5			88.3		91.9
15-19	72.8	92.2	86.9	91.8	82.1	94.7
20-24	102.6	98.9	102.1	98.7	93.0	96.6
25-29	103.0	94.4	94.3	95.4	104.7	95.7
30-34	99.7	109.1	106.3	107.2	103.3	100.9
35-39	112.3	104.5	107.8	106.6	105.5	101.3
40-44	87.9	88.1	89.2	87.2	91.0	95.8
45-49	79.4	94.6	89.9	94.8	92.1	98.2
50-54	160.0	90.8	95.8	90.7	100.8	93.0
55-59	32.7	90.9	87.3	91.0	82.5	92.3
60-64	280.0	107.1	107.4	106.4	103.3	101.2
65-69	40.0			101.2		96.0
70-74	294.7			101.5		97.3

Table 9. Indices of main characteristics of the age-sex structure, 1976

Area	Main age groups			Sex Ratio	Dependency Ratio
	-15	15-64	65+		
Badia	53.5	43.7	2.8	110	129
Jordan	52.0	45.4	2.6	109	120

5.2. Data Evaluation, 1993

5.2.1. Accuracy scores for the 1993 Baseline Survey

Table 2 shows the value of indices measuring the accuracy or otherwise of the Baseline Survey. Since these scores are significant in the evaluation process of the data each score will be discussed individually. The sex-ratio score 12.65, implies a misreporting to some extent of gender related data. Table 10 shows that the sex ratios for the age groupings from 20 to 49 are less than 100, once again implying an outflow of the male population to other parts of the country, given that there is no reason to anticipate higher male mortality in these cohorts. Interviews conducted with households in the Badia found that a considerable proportion of the young population work as soldiers outside the area. The sex-ratios for groups aged 50 and over are much greater than 100. This suggests a systematic reporting bias in favour of males over females. There is also evidence of age heaping due to age preference for years ending in 0 and 5 (shown in Figure 4). The age ratio scores for men and women in 1993 were similar (Table 2), suggesting in particular increased equality between men and women in educational opportunities. The accuracy index which summarizes the male and female age scores represents the sum of these scores plus three times the sex ratio score. It can be seen that the value of the accuracy index is much lower in 1993 than in 1976, indicating a great improvement in the quality of the data set.

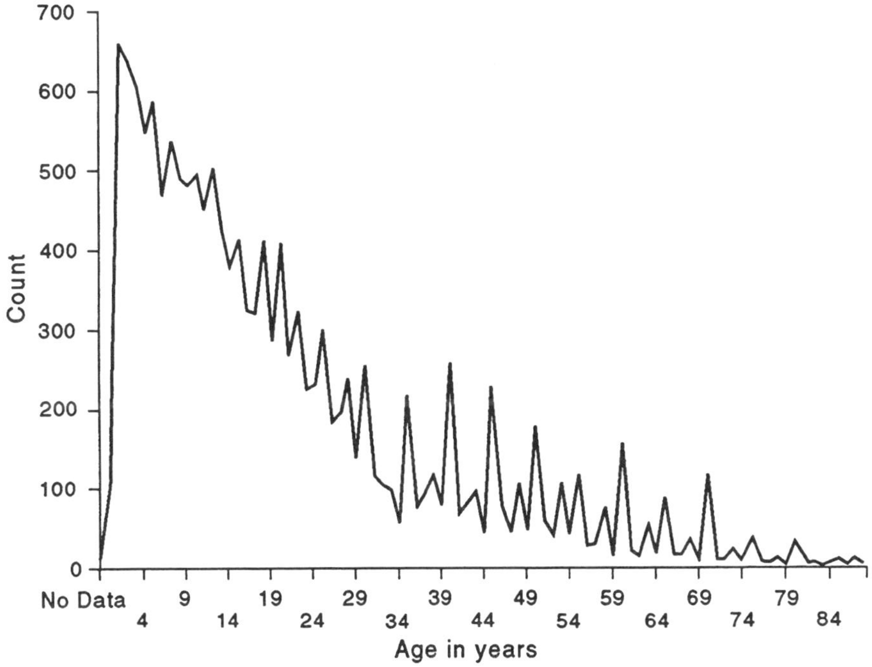

Fig. 4. Age heaping.

In summary the accuracy scores reported in Table 2 once again point to the need for some cleaning of the data set and the application of standard demographic techniques to transform the reported data into a more smoothed format. Having said this, it is also important to consider the relative accuracy of the 1993 survey by comparison with that undertaken in 1976. There can be little doubt from studying Table 2 that the data collected in the 1993 survey shows less sign of misreporting than that for 1976. The sex-ratio scores dropped from 40.6 in 1976 to 12.6 in 1993. By comparing the sex-ratios in Tables 3 and 10 one can see that the sex ratio in 1993 is more balanced than in 1976. This probably occurred because of educational improvements and increased awareness of date of birth.

Table 10. Sex ratio by age, 1993

Age	Sex Ratio
0-4	104.8
5-9	116.7
10-14	111.9
15-19	106.2
20-24	97.0
25-29	99.4
30-34	79.3
35-39	86.3
40-44	96.8
45-49	94.6
50-54	118.7
55-59	128.3
60-64	131.0
65-69	176.7
70-74	135.8
75-79	152.4
80+	196.2

5.2.2. *Data correction, 1993*

The same age smoothing procedures as applied in the previous section were used to correct the 1993 data (Table 11). The results of these procedures were evaluated in terms of the four indices used above and listed in Table 12. The best smoothing procedure was once again Strong's method. Data manipulated by this procedure is included in Tables 13 to 15.

5.2.3. *Age Structure of the Population, 1993*

The main characteristics of the age sex structure of the Badia population in 1993 can be summarized by the sex ratio, percentage of population in the main age groups, and dependency ratio. Table 16 shows that the percentage of population in each of the main age bands as 47.3, 50.1 and 2.6 respectively, implying that around half of the population in 1993 were of school age. The dependency ratio of approximately 100, indicates that each person in the working age group had to support one dependent. The sex ratio for the Badia, of 106, was less than the figure for Jordan as a whole at 109. Comparison of the 1993 indices for the Badia with the national figures reveals that the percentage of the age group of 15 years or less in the Badia remained higher than that for Jordan. The percentage of people of working age, in the Badia, varies little from the national level.

However, proportionately, there are fewer elderly people in the Badia. In summary, the Badia study area had a more youthful age structure than Jordanian society as a whole in 1993.

Table 11. Reported population by age and sex in the Badia, 1993

Age	Male	Female
All ages	7,890	7,407
0-4	1,308	1,248
5-9	1,382	1,184
10-14	1,192	1,065
15-19	905	852
20-24	715	737
25-29	523	526
30-34	280	353
35-39	270	313
40-44	269	278
45-49	244	258
50-54	229	193
55-59	145	113
60-64	148	113
65-69	106	60
70-74	91	67
75-79	32	21
80+	51	26

Table 12. Summary of indices measuring the accuracy of age-sex data in the Badia, 1993

Index	Reported	Smoothed				
		Carrier/ Farrag	King/ Newton	Arriaga	United Nations	Strong
Sex ratio score	12.65	8.72	9.40	8.85	8.98	6.16
Male age ratio score	10.68	5.81	6.57	6.35	5.52	2.80
Female age ratio score	10.55	5.03	5.65	5.62	4.67	2.22
Accuracy index	59.19	37.02	40.43	38.53	37.13	23.50

Note: The accuracy index is the sum of the male and female age ratio scores plus three times the sex ratio score, all calculated using data for ages 10-14 through 65-69

Table 13a. Reported and smoothed male population by age, Badia, 1993

Index	Reported	Smoothed				
		Carrier/ Farrag	King/ Newton	Arriaga	United Nations	Strong
Total: 0-79	7,839			7,839		7,839
Total: 0-79	7,839			7,839		7,839
Total: 10-69	5,026	5,026	5,026	5,026	5,041	5,026
0-4	1,308			1,408		1,432
5-9	1,382			1,282		1,258
10-14	1,192	1,150	1,139	1,145	1,190	1,083
15-19	905	947	958	952	923	902
20-24	715	722	716	712	712	703
25-29	523	516	522	526	502	549
30-34	280	305	320	307	312	390
35-39	270	245	230	243	258	307
40-44	269	269	268	270	265	257
45-49	244	244	246	243	251	220
50-54	229	203	203	203	214	200
55-59	145	171	171	171	163	170
60-64	148	145	143	143	135	139
65-69	106	109	111	111	115	107
70-74	91			78		77
75-79	32			45		46
80+	51					

Change in the age-sex structure from 1976 to 1993 indicates a reduction in the weight of young cohorts over time, and a slight increase in the proportion of working age groups. This implies that either fertility or migration rates, or both, declined in the intervening years. Other research (Findlay and Maani 1995) reveals that fertility did not decline over this time period, although mortality was reduced a little as a result of improved health services and hygiene. Accordingly there was no reduction in rates of natural increase. A high return migration rate amongst those employed in the military sector during the time period helped to increase the numbers in the working age population. The reduction in the sex ratio from 110 to 106 and the change in the dependency ratio are compatible with this interpretation of events. The mean age of the population in the area increased from 31 in 1976 to 35 in 1993, confirming the trends described above and indicating a slight reduction in the youthfulness of the population. The upward trend in average age apples to both sexes (Table 17).

6. FERTILITY AND NUPTIALITY

6.1. Fertility

Fertility is an important component of population change in the Badia. The information on fertility collected in the 1993 survey, appears to be more accurate than the data for 1976. However, it was still necessary to apply the Trussell P/F ratio method to produce a more reliable set of age specific fertility rates for 1993. As a result of applying this standard demographic technique an adjusted total fertility rate was produced for the 1993 population. This provides a better basis for making demographic projections than previous calculations of the area's total fertility rate reported by Findlay and Maani (1995). As a consequence of the adjustments, the figure for total fertility in 1976 is increased from 6.4 to 7.0, while the same figure for 1993 is reduced from 8.5 to 7.5 (Table 18). Even after the adjustments have been carried out, the time trend on the adjusted data still indicates an increase in the total fertility rate during the 1970s and 1980s.

Table 13b. Reported and smoothed female population by age, Badia, 1993

Total: 0-79	7,381			7,381		7,381
Total: 10-69	4,861	4,861	4,861	4,861	4,866	4,861
0-4	1,248			1,275		1,290
5-9	1,184			1,157		1,142
10-14	1,065	1,037	1,032	1,034	1,051	995
15-19	852	880	885	883	876	847
20-24	737	714	710	709	717	690
25-29	526	549	553	555	528	560
30-34	353	369	378	369	367	425
35-39	313	297	288	297	304	341
40-44	278	294	291	293	282	276
45-49	258	242	246	243	252	224
50-54	193	174	176	174	189	181
55-59	113	132	130	132	127	142
60-64	113	100	100	99	98	104
65-69	60	73	73	74	74	77
70-74	67			53		54
75-79	21			35		34
80+	26					

Birth parity in 1993 increased with age across the cohorts 15-49; in 1976 parity increased with age to 39 years and then declined for the 40-49 age cohort (Table 18), but this may reflect poor recall in the 1976 survey concerning cumulative fertility behaviour. The disappearance of this feature by the time of the 1993 survey presumably reflects improved educational levels amongst the female population.

The most striking feature of the Badia's fertility regime, is the increase in total fertility between 1976 and 1993. The most likely explanation for this is that in the Badia, as in many transitional societies, improvements in health and nutrition have led to a decrease in the proportion of conceptions ending in miscarriage, leading to a rise in the fertility rate. An alternative partial explanation, which would need empirical verification, is that in 1976 the high rate of male emigration from the area resulted in fertility being slightly depressed, compared with the situation in the early 1990s when many former emigrants had returned home and when new opportunities for emigration were very scarce.

Table 14. Sex ratios, reported and smoothed populations for the Badia, 1993

Age	Reported	Smoothed				
		Carrier/ Farrag	King/ Newton	Arriaga	United Nations	Strong
0-4	104.8			110.5		111.0
5-9	116.7			110.8		110.2
10-14	111.9	110.9	110.4	110.7	113.3	108.9
15-19	106.2	107.6	108.2	107.9	105.4	106.5
20-24	97.0	101.0	100.8	100.5	99.4	101.9
25-29	99.4	94.1	94.4	94.8	95.0	98.0
30-34	79.3	82.8	84.6	83.2	85.0	91.7
35-39	86.3	82.3	79.9	81.8	84.8	90.0
40-44	96.8	91.5	92.1	92.2	93.8	93.0
45-49	94.6	100.9	100.0	100.0	99.5	98.2
50-54	118.7	116.6	115.7	116.8	113.4	110.6
55-59	128.3	129.7	131.1	129.4	128.1	119.7
60-64	131.0	144.7	142.5	144.2	138.5	133.3
65-69	176.7	149.7	152.7	150.4	155.1	139.0
70-74	135.8			148.9		142.6
75-79	152.4			126.3		135.3
80+	196.2					

The comparison of age specific fertility rates (ASFRs) for 1976 and 1993 (Table 18) shows that the peak fertility level increased from the 20-24 age cohort to the 25-29 age group. This change occurred in Jordan as a whole, but at the national level the change was accompanied by fertility decline. The difference between the Badia study area and the rest of the country cannot be explained by a lower age at first marriage (almost the same in the 1993 Badia survey as in the national survey for Jordan in 1990). The differences in ASFRs and total fertility rates (TFRs) must therefore lie in other determinants of fertility levels, such as contraceptive use, which reflect differences in attitudes towards child-bearing and the role of women in society.

The general trend towards higher fertility levels in the Badia over time obscures the fact that ASFRs dropped amongst women in the younger cohorts. Not until the 25-29 cohort is there a rise in adjusted ASFRs between 1976 and 1993. Adjusted ASFRs are higher for all cohorts between the ages of 25 and 44 and decline for the oldest female cohort, suggesting a greater concentration of childbearing in the middle years of life. The ASFRs for 1990 for Jordan as a whole indicate that the same pattern exists at a national level. There is little indication of planned fertility control taking place within marriage amongst the current population of the Badia. There is therefore scope for considerable fertility decline in the future, should the social and cultural norms of other parts of Jordan with regard to ideal family size begin to influence the Badia. Ideational variables (social and cultural value systems affecting aspirations and behaviour) seem likely in the near future to change the structure of power relations in the family, including changes to the role and status of women in society (Findlay and Maani, 1995). These changes should produce an accelerated fertility decline.

Table 15. Age ratios, reported and smoothed populations for the Badia, 1993

Sex and Age	Reported	Smoothed				
		Carrier/ Farrag	King// Newton	Arriaga	United Nations	Strong
MALE						
5-9	110.6			100.4		100.0
10-14	104.2			102.5		100.3
15-19	94.9	101.2	103.3	102.6	97.1	101.0
20-24	100.1	98.6	96.7	96.4	99.9	96.9
25-29	105.1	100.6	100.8	103.2	98.1	100.5
30-34	70.6	80.2	85.2	79.8	82.0	91.1
35-39	98.4	85.3	78.1	84.4	89.5	95.0
40-44	104.7	110.0	112.6	110.8	104.0	97.4
45-49	98.0	103.4	104.3	103.0	104.8	96.3
50-54	117.7	98.1	97.6	97.8	103.5	102.7
55-59	76.9	98.1	98.8	99.0	93.3	100.3
60-64	117.9	103.2	101.2	101.3	97.3	100.1
65-69	88.7			100.4		99.3
70-74	131.9			100.6		100.7
FEMALE						
5-9	102.4			100.3		100.0
10-14	104.6			101.4		100.0
15-19	94.6	100.5	101.7	101.3	99.2	100.6
20-24	107.0	100.0	98.7	98.6	102.0	98.0
25-29	96.5	101.3	101.7	103.0	97.5	100.5
30-34	84.1	87.1	90.0	86.6	88.1	94.3
35-39	99.2	89.8	86.0	89.9	93.8	97.3
40-44	97.4	109.0	109.0	108.2	101.4	97.7
45-49	109.6	103.3	105.3	104.4	107.1	98.0
50-54	104.0	93.4	93.5	92.4	99.5	99.0
55-59	73.9	95.9	94.5	97.0	88.8	99.6
60-64	130.6	97.6	98.6	96.1	97.0	95.0
65-69	66.7			97.4		97.5
70-74	165.4			96.3		97.3

Table 16. Indices of main characteristics of the age-sex structure, 1993

Area	Main Groups			Sex Ratio	Dependency Ratio
	0-15	15-64	65+		
Badia	47.3	50.1	2.6	106	99.6
National 1990 Jordan	44.0	51.6	4.6	109	94.2

Table 17. Changes in indices of age-sex structure, 1976-1993

Indices	1976	1993	Changes
0-15 age group	53.5%	47.3%	-6.2%
15-65 age group	43.7%	50.1%	+6.4%
65+ age group	2.8%	2.6%	-0.2%
Sex ratio	110	106	-4
Dependency ratio	129	99.6	-29.4
Mean age	31	35	4

Table 18. Reported and adjusted age specific fertility rate (per 1,000), total fertility rate and birth parity, Badia 1976-93

Age	Reported ASFR		Adjusted ASFR		Birth parity		
	1976	1993	1976	1993	1976	1993	Jordan 1990
15-19	43	3	62	40	0.7	0.01	49
20-24	264	301	300	285	1.5	0.9	219
25-29	246	376	267	334	2.5	2.5	296
30-34	222	334	242	294	2.9	4.5	264
35-39	228	335	244	290	5.5	6.7	188
40-44	132	185	95	138	2.1	6.7	79
45-49	151	140	192	125	4.0	7.8	19
TFR	6.4	8.5	7.0	7.5			5.6

Crude Birth Rate 1976 = 39.9; Crude Birth Rate 1993 = 48.9

6.2. Nuptiality

Marriage marks the beginning of child-bearing for women in the Badia and in the rest of Jordan. Factors such as age at first marriage, proportion of the population married, and frequency of marriage are therefore very important to the understanding of fertility. Table 19 compares data on marital status of women, by age for 1976 and 1993. The percentage of single women, aged 15-24, increased between 1976 and 1993. The proportion of single women by age in the Badia area in 1993 was very similar to the pattern for Jordan as a whole (Table 20).

The figures for unmarried women in the cohorts 25-49 for 1976 are suspiciously high and are not consistent with the other information reported. However, there can be little doubt that adult mortality in the 1970s was much higher than in the 1990s and this produced higher proportions of widowed persons in the population as shown in Table 19, and it may be that the marital status of some widows was incorrectly recorded as 'single'. Table 19 also indicates that the proportion of divorced women rose between 1976 and 1993, reflecting significant social changes in the society of the Badia. Compared to the national situation, however, divorce for younger women (15-29) in the Badia in 1993 remained lower than in the rest of Jordan (see Table 20). Amongst older women (30-49) the divorce rate was higher.

Table 19. Percentage distribution of women by current marital status in the Badia, by age, 1976 and 1993

Age	Single		Married		Widowed		Divorced	
	1976	1993	1976	1993	1976	1993	1976	1993
15-19	70.2	84.9	28.9	14.8	0.0	0.4	0.9	0.0
20-24	34.2	38.5	64.7	60.5	0.5	1.1	0.6	0.0
25-29	25.3	15.4	72.0	84.0	1.7	0.6	0.0	0.0
30-34	24.4	5.4	74.8	92.1	0.8	0.6	0.0	1.9
35-39	13.1	1.3	83.2	93.0	2.8	1.0	0.9	4.8
40-44	38.5	0.4	53.8	92.4	7.7	1.8	0.0	5.4
45-49	26.7	2.7	65.0	88.8	8.3	0.8	0.0	7.8

Table 20. Percentage distribution of Jordanian women by current marital status, according to age, 1990

Age	Single	Married	Widowed	Divorced
15-19	89.4	10.4	0.0	0.2
20-24	54.8	44.5	0.2	0.4
25-29	26.3	71.2	0.4	1.9
30-34	10.9	86.0	1.4	1.5
35-39	5.4	89.3	3.5	1.6
40-44	2.7	90.8	5.2	1.3
45-49	2.0	88.3	8.7	0.8

The mean age at first marriage in the Badia was 16.4 years in 1976 and 19.3 in 1993. This latter figure was very close to the national situation, which in 1990 was 19.6. Clearly a trend to later marriage is taking place in the Badia as in Jordan as a whole.

There is little alteration to the frequency of marriage from 1976 to 1993. Table 21 reveals that approximately 90 per cent of women in both years had been married only once.

Table 21. Percentage distribution of married men, according to times married, 1976 and 1993

Year	Number of times married				
	Once	Twice	3 times	4 times	5 + times
1976	89.8	8.8	0.6	0.7	0.1
1993	89.5	8.4	1.9	0.2	0.0

7. MORTALITY

7.1. Infant mortality

The infant mortality rate in 1976 was 174, a little higher for male than for female children (Table 22). By 1993, although infant mortality rates had fallen to half the level of 1976, there was still almost a 1 in 10 chance of death during the first year of life. This pattern is not surprising for an isolated region where health services were not widely available until recently and where travel to health centres involved a long journey on relatively poor roads. Disposable income in the area was very low in the 1970s and early 1980s and even today lags well behind the rest of Jordan.

After 1976, an infrastructure programme introducing public water services, electricity and health services was initiated in the Badia. Road communications were also improved, in particular between the study area and the city of Zarqa. This latter change made it easier to market livestock products from the area and helped to increase the average income of the population dependent on livestock production.

Table 22. Infant mortality rates (per 1,000) for the Badia 1976 and 1993

Year	Both sexes	Male	Female
1976	174	179	167
1993	86	102	69

Infant mortality rates for the Badia do not compare favourably with the national figure for Jordan, which in 1990 was only 34 per 1,000, less than half the rate of that in the study area. Therefore although infant mortality rates are decreasing in the Badia, there remains an urgent need to improve the health services and health education in the region.

7.2. Life expectancy

Increasing life expectancy is often seen as an indicator of health improvements in an area. In the Badia, significant increases occurred for both sexes, between 1976 and 1993, as shown in Table 23. In 1976 the life expectancy of 41.4 for males and 42.0 for females indicates just how poor life chances were in the region. Although the 1993 life expectancies were still low by international standards, the improvements within the Badia were substantial, representing a sustained annual increase of one year per annum for women over the 16 year period.

Table 23. Life expectancies in the Badia, 1976 and 1993

Year	Life expectancy		
	Both sexes	Male	Female
1976	41.1	40.3	42.1
1993	54.7	51.1	58.8

Table 24. Crude death rate in the Badia, 1976 and 1993

Year	Both sexes	Male	Female
1976	21.1	21.9	20.2
1993	10.4	12.5	8.1

7.3. Crude death rate

The crude death rate is one of the most cited demographic statistics, despite its shortcomings as a measure of mortality. Improvements in life prospects are evident from Table 24, which reveals that the crude death rate halved over the period 1976 to 1993 to approximately 10 per 1,000, with the rate for females declining more than that for males.

Table 25. Age specific death rates and survival ratios for the Badia, 1976

Age, x	Width, n	Nmx	nax	nqx	1x
0	1	0.18119	0.553	0.16761	100,000
1	4	0.04154	1.239	0.14907	83,239
5	5	0.00711	2.500	0.03495	70,830
10	5	0.00397	2.500	0.01965	68,355
15	5	0.00560	2.500	0.02762	67,011
20	5	0.00707	2.500	0.03473	65,161
25	5	0.00773	2.500	0.03762	62,897
30	5	0.00808	2.500	0.03961	60,512
35	5	0.00870	2.500	0.04259	58,115
40	5	0.00929	2.500	0.04538	55,640
45	5	0.01029	2.500	0.05017	53,115
50	5	0.01379	2.500	0.06665	50,450
55	5	0.01921	2.500	0.09163	47,088
60	5	0.03153	2.500	0.14613	42,773
65	5	0.05005	2.500	0.22240	36,523
70	5	0.08296	2.500	0.34355	28,400
75	5	0.13317	2.500	0.49954	18,643
80+		0.23159	4.318	1.00000	9,330

Nmx = Age-specific central death rate
nax = Average person-years lived by those who die between ages x a
nqx = Probability of dying between exact ages x and x+n (age-specific mortality rate)
1x = Number of survivors at age x

7.4. Life tables

Using the age specific death rates for 1976 and 1993 (Table 25 and 26) life tables were constructed and evaluated with reference to a range of model life tables. It was found that the Badia life tables approximated most closely to the Coale/Demeny South region model table. This was judged to be the most appropriate model life table for the region and was used subsequently in making demographic projections.

8. MIGRATION

Detailed migration data for the Badia was not available in either the 1976 or 1993 surveys. Thus it was impossible to calculate gross migration rates at the regional or settlement levels. Instead migration could only be estimated using the residuals method (the difference between the growth rate and the rate of natural increase). Application of the residuals method would seem to indicate a slight reduction of net migration over time. Between 1993 and 1994 the net rate was only 0.006. Interpretation of this change is very difficult because of the problems associated with the data and methods being used to make these estimates. The shift could represent a real decrease in out-migration resulting from a decline in migration opportunities for those seeking to live in other parts of Jordan or to work abroad, but it is equally possible that the change in the residuals reflect a decline in reporting errors over time.

Table 26. Age specific death rates and survival ratios for the Badia, 1993

Age, x	Width, n	nmx	nax	nqx	1x
0	1	0.07358	0.259	0.06978	100,000
1	4	0.01218	1.619	0.04734	93,022
5	5	0.00415	2.500	0.02052	88,619
10	5	0.00246	2.500	0.01220	86,800
15	5	0.00289	2.500	0.01433	85,741
20	5	0.00356	2.500	0.01764	84,512
25	5	0.00413	2.500	0.02046	83,022
30	5	0.00472	2.500	0.02332	81,323
35	5	0.00544	2.500	0.02685	79,426
40	5	0.00652	2.500	0.03207	77,294
45	5	0.00749	2.500	0.03674	74,815
50	5	0.01004	2.500	0.04895	72,066
55	5	0.01355	2.500	0.06552	68,538
60	5	0.02061	2.500	0.09802	64,047
65	5	0.03275	2.500	0.15136	57,769
70	5	0.05277	2.500	0.23310	49,025
75	5	0.08363	2.500	0.34583	37,597
80+		0.18994	5.265	1.00000	24,595

Nmx = Age-specific central death rate
nax = Average person-years lived by those who die between ages x a
nqx = Probability of dying between exact ages x and x+n (age-specific mortality rate)
1x = Number of survivors at age x

9. POPULATION PROJECTIONS

Three population projections for the population of the Badia study area have been prepared by the authors of this paper. These are based on three different assumptions relating to fertility trends, involving different explanations as to why the Badia currently has a total fertility rate similar to that in the rest of Jordan 20 years ago. The three assumptions were:

(1) a 'low' fertility projection for the area. This was defined as a fertility regime in which rates for the Badia would converge with those for the rest of Jordan within 20 years.
(2) a medium fertility projection. In this scenario convergence between the Badia and the rest of Jordan also occurs, but at a slower rate with the fertility level in the Badia by the year 2013 still lagging ten years behind the country as a whole.
(3) a high fertility projection. Under this scenario fertility rates decline but remain 20 years behind those of Jordan.

Although these three scenarios form the basis for the three different projections presented in Tables 27 to 29, all also anticipate an improvement in life expectancy. It is expected that further improvements in health will be sustained over the next 20 years with female life expectancy rising from 58.8 in 1993 to 61.1 years by 1998 and then rising annually by a further 0.4 years to reach 67.1 years by 2013. Male life expectancy was assumed to rise from 51.1 in 1993 to 53.6 by 1998 and to 60.9 by 2013. As specified earlier, the projections of mortality were based on model life tables following matching of the adjusted age specific mortality schedules for the area. Migration is by far the hardest element of the demographic regime to model, given the absence of any way to predict external economic and social conditions over the next 20 years. As Tables 27 to 29 show, standard net migration trends are held constant across all three projections, with a heavier net out-migration of males than females and with a gently declining net loss over time due to migration. These migration assumptions are based on analysis of the five data sets studied by the authors which indicate that throughout the last 20 years there has been a continuous net loss of population from the Badia through out-migration. The lowest level of loss appeared to be that recorded between 1993 and 1994 and this level was therefore used to provide an estimate of the minimum loss (as opposed to the most likely level) which can be expected to occur in the future. Applying this rate of net migration loss over the twenty year period provides estimates of absolute net out-migration. These figures were distributed by age group across the male and female populations in relation to the estimates of net migration achieved from sex ratios for the 1993 survey. The life expectancy assumptions and these migration estimates when combined with the fertility assumptions discussed above provide high, medium and low estimates of the future population of the Badia to the year 2013.

Table 27a. Projected indices of main characteristics of the age-sex structure in Badia, 1993-2013 (based on a high fertility rate)

	1993	1998	2003	2008	2013
POPULATION					
Total	15,222	17,894	21,009	24,409	28,366
Male	7,840	8,998	10,405	11,909	13,730
Female	7,382	8,897	10,604	12,500	14,635
0-4	2,722	3,721	4,174	4,334	4,869
5-14	4,478	4,978	6,164	7,604	8,250
15-64	7,627	8,719	10,118	11,830	14,508
65+	395	476	553	642	739
Females 15-49	3,363	3,961	4,659	5,532	6,866
PER CENT:					
Male	51.5	50.3	49.5	48.8	48.4
Female	48.5	49.7	50.5	51.2	51.6
0-4	17.9	20.8	19.9	17.8	17.2
5-14	29.4	27.8	29.3	31.2	29.1
15-64	50.1	48.7	48.2	48.5	51.1
65+	2.6	2.7	2.6	2.6	2.6
Females 15-49	22.1	22.1	22.2	22.7	24.2
Sex ratio	106.2	101.1	98.1	95.3	93.8
Dependency ratio	0.944	0.9	1.022	1.009	0.904
Median age	16.2	15.6	15.4	15.5	16.5
ANNUAL RATES OF CHANGE					
	1993-98	**1998-03**	**2003-08**	**2008-13**	
CBR/1000	51.8	48.8	42.8	40.8	
CDR/1000	13.0	11.6	9.7	8.3	
NIR per cent	3.9	3.7	3.3	3.3	
GR per cent	3.2	3.2	3.0	3.0	

Table 27b. Projected demographic indicators for Badia, 1993-2013 (based on a high fertility rate)

	1993	1998	2003	2008	2013
FERTILITY					
TFR	7.50	7.50	6.60	5.60	5.60
GRR	3.62	3.62	3.19	2.71	2.71
NRR	2.90	2.99	2.70	2.34	2.39
Mean age of child-bearing	29.6	29.6	29.6	29.4	29.4
C/W ratio	0.81	0.94	0.90	0.78	0.71
AGE DISTRIBUTION OF FERTILITY TABLE = UN ARAB					
Mortality					
Male LE	51.1	53.6	55.6	57.9	60.9
Female LE	58.8	61.1	63.1	65.1	67.1
Total LE	54.8	57.3	59.4	61.6	64.1
IMR	114.1	104.6	96.8	88.5	78.8
CMR	741,976.1	837,730.6	776,963.9	712,235.4	638,628.5
LIFE TABLE = COALE-DEMENY SOUTH					
Migration					
Male Migration	-88.0	-77.0	-73.0	-71.0	-64.0
Female Migration	-26.0	-26.0	-25.0	26.0	-26.0
Total Migration	-114.0	-103.0	-98.0	-45.0	-90.0
BIRTHS AND DEATHS					

	1993-98	1998-03	2003-08	2008-13
Births	858.1	948.6	972.1	1,077.0
Deaths	215.1	225.2	220.5	218.2

Table 28a. Projected indices of main characteristics of the age-sex structure in Badia, 1993-2013 (based on a medium fertility rate)

POPULATION	1993	1998	2003	2008	2013
Total	15,222	17,655	20,225	23,328	26,522
Male	7,840	8,877	10,008	11,363	12,796
Female	7,382	8,779	10,216	11,965	13,725
0-4	2,722	3,482	3,620	4,018	4,094
5-14	4,478	4,978	5,934	6,839	7,408
15-64	7,627	8,719	10,118	11,830	14,280
65+	395	476	553	642	739
Females 15-49	3,363	3,961	4,659	5,532	6,753

Table 28a (cont.). Projected indices of main characteristics of the age-sex structure in Badia, 1993-2013 (based on a medium fertility rate)

PER CENT	1993	1998	2003	2008	2013
Male	51.5	50.3	49.5	48.7	48.2
Female	48.5	49.7	50.5	51.3	51.8
0-4	17.9	19.7	17.9	17.2	15.4
5-9	29.4	28.2	29.3	29.3	27.9
15-64	50.1	49.4	50.0	50.7	53.8
65+	2.6	2.7	2.7	2.8	2.8
Females 15-49	22.1	22.4	23.0	23.7	25.5
Sex ratio	106.2	101.1	98.0	95.0	93.2
Dependency ratio	0.944	0.970	0.944	0.918	0.805
Median age	16.2	15.9	16.2	16.6	17.7
ANNUAL RATES OF CHANGE					
	1993-98	**1998-03**	**2003-08**	**2008-13**	
CBR/1000	48.8	43.4	41.4	36.3	
CDR/1000	12.6	11.0	9.6	8.0	
RNI per cent	3.6	3.2	3.2	2.8	
GR per cent	3.0	2.7	2.9	2.6	

Table 28b. Projected demographic indicators for the Badia, 1993-2013 (based on a medium fertility rate)

	1993	1998	2003	2008	2013
FERTILITY					
TFR	7.50	6.60	5.60	5.60	4.00
GRR	3.62	3.19	2.71	2.71	1.93
NRR	2.90	2.63	2.29	2.34	1.71
Mean age of child-bearing	29.6	29.6	29.4	29.4	29.3
C/W ratio	0.81	0.88	0.78	0.73	0.61
AGE DISTRIBUTION OF FERTILITY TABLE = UN ARAB					
Mortality					
Male LE	51.1	53.6	55.6	57.9	60.9
Female LE	58.8	61.1	63.1	65.1	67.1
Total LE	54.8	57.3	59.4	61.6	64.1
IMR	114.1	104.6	96.8	88.5	78.8
CMR	741,955.8	837,730.6	776,963.9	712,235.4	638,628.5

Table 28b (cont.). Projected demographic indicators for the Badia, 1993-2013 (based on a medium fertility rate)

LIFE TABLE = COALE-DEMENY SOUTH					
Migration					
Male Migration	-88.0	-77.0	-73.0	-71.0	-64.0
Female Migration	-26.0	-26.0	-25.0	26.0	-26.0
Total Migration	-114.0	-103.0	-98.0	-45.0	-90.0

BIRTHS AND DEATHS				
	1993-98	**1998-03**	**2003-08**	**2008-13**
Births	803.0	822.7	901.3	905.6
Deaths	207.8	208.3	209.1	199.4

Table 29a. Projected indices of main characteristics of the age-sex structure in Badia, 1993-2013 (based on a low fertility rate)

	1993	1998	2003	2008	2013
POPULATION					
Total	15,222	17,894	20,697	23,176	25,772
Male	7,840	8,998	10,247	11,283	12,413
Female	7,382	8,897	10,450	11,892	13,360
0-4	2,722	3,721	3,862	3,402	3,485
5-14	4,478	4,978	6,164	7,302	7,041
15-64	7,627	8,719	10,118	11,830	14,508
65+	395	476	553	642	739
Females 15-49	3,363	3,961	4,659	5,532	6,866
PER CENT					
Male	51.5	50.3	49.5	48.7	48.2
Female	48.5	49.7	50.5	51.3	51.8
0-4	17.9	20.8	18.7	14.7	13.5
5-9	29.4	27.8	29.8	31.5	27.3
15-64	50.1	48.7	48.9	51.0	56.3
65	2.6	2.7	2.7	2.8	2.9
Females 15-49	22.1	22.1	22.5	23.9	26.6
Sex ratio	106.2	101.1	98.1	94.9	92.9
Dependency ratio	0.944	0.998	0.991	0.905	0.726
Median age	16.2	15.6	15.7	16.7	18.3

Table 29a. (cont.). Rates of change

	1993-98	1998-03	2003-08	2008-13
ANNUAL RATES OF CHANGE				
CBR/1000	51.8	45.5	34.8	31.5
CDR/1000	13.0	11.2	8.9	7.5
NIR per cent	3.9	3.4	2.6	2.4
GR per cent	3.2	2.9	2.3	2.1

Table 29b. Projected demographic indicators for the Badia, 1993-2013 (based on a low fertility rate)

	1993	1998	2003	2008	2013
FERTILITY					
TFR	7.50	7.50	5.60	4.00	4.00
GRR	3.62	3.62	2.71	1.93	1.93
NRR	2.90	2.99	2.29	1.67	1.71
Mean age of child-bearing	29.6	29.6	29.4	29.3	29.3
C/W ratio	0.81	0.94	0.83	0.61	0.51
AGE DISTRIBUTION OF FERTILITY TABLE = UN ARAB					
Mortality					
Male LE	51.1	53.6	55.6	57.9	60.9
Female LE	58.8	61.1	63.1	65.1	67.1
Total LE	54.8	57.3	59.4	61.6	64.1
IMR	114.1	104.6	96.8	88.5	78.8
CMR	741955.8	837730.6	776963.9	712235.4	638628.5
LIFE TABLE = COALE-DEMENY SOUTH					
Migration					
Male Migration	-88.0	-77.0	-73.0	-71.0	-64.0
Female Migration	-26.0	-26.0	-25.0	26.0	-26.0
Total Migration	-114.0	-103.0	-98.0	-45.0	-90.0

BIRTHS AND DEATHS				
	1993-98	**1998-03**	**2003-08**	**2008-13**
Births	858.1	877.8	763.1	770.8
Deaths	215.1	216.7	196.0	183.9

The range of values indicated by the population projections suggest that the Badia by 2013 will have to provide accommodation, jobs and service functions for between 25,800 and 28,400 people, compared with the 15,200 residents recorded in the 1993 survey. Even under the lowest fertility assumptions, the population will remain very youthful with over 40 per cent of the population under the age of 15 in the year 2013. And under all scenarios, the reduction of the fertility rate fails to translate into an absolute decline in population numbers in the younger cohorts. Thus for example in the medium growth scenario (Table 30), the cohort of 15-19 years (those seeking to enter the local labour market) grows from 1,749 persons in 1993 to 3,304 persons by 2013, indicating the rising requirement to offer new opportunities to earn a livelihood for the young people of the Badia. The scale of economic expansion necessary to match these demographic pressures requires a development programme based, not only on new job opportunities in the primary sector, but significant employment growth in the local secondary and tertiary sectors. The policy implications of these population projections are therefore very great indeed. Attention needs to be paid not only to the policy implications which flow from the projected large increase in the aggregate numbers of people living in the Badia, but also to the changing structure and composition of the population which is anticipated.

Table 30a. Population projection for the Badia, 1993-2003, male and female (based on a medium fertility rate)

Age	1993			1998			2003		
	Total	M	F	Total	M	F	Total	M	F
Total	15,222.0	7,840.0	7,382.0	17,655.3	8,876.6	8,778.7	20,224.7	10,008.3	10,216.4
0-4	2,722.0	1,432.0	1,290.0	3,482.3	1,763.7	1,718.6	3,619.9	1,837.2	1,782.7
5-9	2,400.0	1258.0	1142.0	2,601.7	1,358.0	1,243.7	3,353.4	1,686.2	1,667.3
10-14	2078.0	1083.0	995.0	2,376.4	1,243.3	1,133.1	2,580.3	1,344.5	1,235.8
15-19	1749.0	902.0	847.0	2,022.6	1,055.6	967.0	2,321.2	1,216.1	1,105.1
20-24	1,393.0	703.0	690.0	1,629.8	822.2	807.7	1,903.8	976.3	927.5
25-29	1,109.0	549.0	560.0	1,277.4	616.7	660.7	1,509.7	731.6	778.1
30-34	815.0	390.0	425.0	1,019.0	484.9	534.1	1,197.4	565.5	631.8
35-39	648.0	307.0	341.0	721.5	313.7	407.9	939.0	422.8	516.2
40-44	533.0	257.0	276.0	548.4	224.0	324.3	625.4	234.7	390.7
45-49	444.0	220.0	224.0	472.0	212.5	259.5	496.7	186.9	309.8
50-54	381.0	200.0	181.0	407.9	196.0	212.0	439.3	192.3	247.0
55-59	312.0	170.0	142.0	345.7	175.3	170.5	375.8	172.6	203.3
60-64	243.0	139.0	104.0	274.2	146.7	127.5	309.8	152.5	157.3
65-69	184.0	107.0	77.0	204.7	112.1	92.5	234.2	119.8	114.4
70-74	131.0	77.0	54.0	144.3	81.3	63.0	163.1	86.4	76.7
75-79	80.0	46.0	34.0	86.5	48.7	37.8	97.3	52.4	44.9
80+	0.0	0.0	0.0	40.9	22.1	18.9	58.5	30.5	28.0

ACKNOWLEDGEMENT

The work could not have been completed without the support of a wide range of colleagues and institutions. Allan Findlay is very grateful to the John Robertson Bequest, University of Glasgow, which helped to fund the early phases of the project and to the University of Dundee which met the costs of preparing this report. All the authors are grateful to the corporate sponsors of the Badia Research and Development Project, without whom it would have been impossible to undertake the 1993 Demographic Survey and whose assistance made it possible to meet to analyse the results. At a personal level the authors are indebted to Rodney Stobbs, University of Durham, who assisted with place-name research necessary to the preparation of Figure 1 and to Jim Ford, Nicola Ross and Pat Michie, all of the University of Dundee, who contributed in various ways to the technical production of this research paper. Hopefully, the demographic projections presented in this paper will form a significant contribution to other researchers involved in evaluating the policy options for the development of the Badia region of Jordan.

Table 30b. Population projection for the Badia, 2008-13, male and female (based on a medium fertility rate)

Age	2008			2013		
	Total	Males	Females	Total	Males	Females
Total	23,328.1	11,363.5	11,964.6	26,521.7	12,796.6	13,725.4
0-4	4,017.9	2,043.1	1,974.8	4,094.0	2,088.7	2,005.3
5-9	3,508.0	1,769.1	1,738.9	3,918.8	1,983.5	1,935.4
10-14	3,330.7	1,672.0	1,658.7	3,489.0	1,757.3	1,731.7
15-19	2,547.0	1,318.0	1,229.0	3,303.8	1,652.5	1,651.3
20-24	2,238.1	1,138.9	1,099.2	2,473.2	1,249.2	1,223.9
25-29	1,801.0	882.3	918.6	2,135.1	1,044.6	1,090.6
30-34	1,449.3	679.7	769.6	1,740.0	829.8	910.2
35-39	1,129.1	505.2	623.9	1,382.3	620.9	761.4
40-44	854.7	346.4	508.3	1,048.0	432.4	615.7
45-49	588.2	205.0	383.1	816.0	316.4	499.6
50-54	470.2	168.5	301.7	560.6	186.6	374.0
55-59	409.9	172.3	237.6	444.5	153.3	291.2
60-64	342.4	151.0	191.4	376.8	151.9	224.9
65-69	268.1	125.9	142.2	300.4	125.9	174.5
70-74	189.4	93.5	95.9	218.1	99.7	118.3
75-79	112.3	56.7	55.6	133.6	62.7	70.8
80+	71.8	35.7	36.1	87.3	40.8	46.5

REFERENCES

Brass, W. (1971) On the scale of mortality. In W. Brass (ed.) *Biological Aspects of Demography*, Taylor and Francis, London, pp. 69-110.

Brass, W. (1975) *Methods for Estimating Fertility and Mortality from Limited and Defective Data*, Chapel Hill, University of North Carolina.

Carrier, N.H., and Farrag, A.M. (1959) The reduction of errors in census populations for statistically underdeveloped countries. *Population Studies,* 12(3) 240-85.

Findlay, A.M. and Maani, M. (1995) Fertility trends in an arid environment: The case of the population of the Jordanian Badia. *Discussion Paper, Centre for Overseas Research and Development*, University of Durham.

Findlay, A.M. (1997) Policy implications of population growth in arid environments. In J.I. Clarke, and D. Noin (eds) *Population and Environment in Arid Regions*, UNESCO & Parthenon.

Hashemite Kingdom of Jordan (1979) *Results of the Housing and Population Census 1979: East Bank, Vol.2, Population Characteristics*, Department of Statistics, Jordan.

Shryock, H.S. and Siegel, J.S. (1976) *The Methods and Materials of Demography*, Academic Press, London.

Trussell, T.J. (1975) A re-estimation of the multiplying factors for the Brass technique for determining childhood survivorship rates. *Population Studies,* 29(1), 97-107.

United Nations (1982) *Model Life Tables for Developing Countries.* Department of International Social and Economic Affairs, Population Studies, No.77.

CHAPTER 18

Development Anthropology: A Primary Survey

CYNTHIA BRANDENBURG
(School of Oriental and African Studies, University of London)

1. INTRODUCTION

This report's objective is to outline the scope of the second stage of Phase I of the development anthropology component in the Jordan Badia Research and Development Programme; the fieldwork, methodology employed, findings, and way forward. Field-work conducted was in co-operation with Jordanian counterparts as part of a continuing joint-programme between Jordanian and British researchers.

1.1. Objectives/Timeframe

Following the successful completion of the baseline survey piloted in November 1992 and conducted in January and February 1993 by a Jordanian and British team of demographers and social scientists, stage two required an expansion of the information gathered to gain an understanding of inhabitants' views of what development of this area means to them and the role, if any, they expect to play in this process.

A two-month period was decided on to undertake the initial overview of the villages with a view to identifying a village or villages to reside in as the first of a series of in-depth case studies on local life, examining issues key to the sustained development of the area: relationships between people and service providers; land tenure and access to resources; current settlement practices and migratory patterns; changing modes of production; reception toward technological innovation; division of labour - the role of women and children in the household economy, and the contributions women make to the decision-making process; and the impact of change on social relations at the household level. Field visits began on 21 July 1993 and were concluded on 10 September 1993.

1.2. Programme Area

For the purposes of the tribal survey of the development anthropology component it was necessary to delimit the area studied to that of the 34 villages within the total Programme Area. Along the Mafraq/Baghdad road this includes the villages of Safawi towards the eastern Programme Area border, and the village of Rahbat Rakad on the western border, as well as the villages bordering Syria reached from the Bishriyya road.

1.3. Tribal People of the Area

The primary inhabitants of the villages trace their lineage to a common descent from the Ahl Al-Jabal of southern Syria. They then identify themselves as belonging to one of three tribes: Masayeed (considered to be the largest group), Sharafaat, or Ahdhamat. The three tribes are headed by a 'Shaykh of Shaykhs' from the Masayeed. Each tribe has a tribal shaykh who represents the tribe in interactions between the tribe and the government and with other tribes. The villages themselves may also have a local governor (who may or may not also be the tribal shaykh) and some villages may also have a Mukhtar (whose responsibilities vary from one village to the next but may generally be said to be primarily administrative). Villages with a low population number may not have a local shaykh or governor and so must rely on the tribal shaykh residing in a distant village.

A notable exception to the tribal population is the minority presence of Druze within the region encountered in Umm al-Quttayn.

1.4. Development Anthropology Co-ordinates

During the report period from 21 July to 10 September 1993 the development anthropology component including myself was comprised of the following members: Nahid Trabaishat and Maysoon Mughrabi; both of Yarmuk University. Nahid Trabaishat directly contributed to the interviews conducted within the villages of the Programme Area from 21 July to 28 July and then from 10 to 12 August, and Maysoon Mughrabi from 21 July to 5 August.

1.5. Methodology

Inherent within the domain of development anthropology is the objective of direct interference. While obvious, one should not lose sight of this through the output of the anthropological method of study and analysis. The development anthropologist finds him or herself in the position of having accountability to both 'sponsor' and 'beneficiary'.

As a joint programme the methodology employed benefited from the participation of both the so-called 'native fieldworker' and the 'foreign fieldworker'. This gave rise to a two-stage approach to the field: one highly structured and survey-oriented and one more reliant on participant observation and discussion rather than on interview. The latter was relied upon most during the period of residence within the host community.

The need for dialogue becomes paramount if a realizable development programme is to be established. Information gathering is only one part of the process; provision of information on the part of the sponsor is equally important. With this in mind, the interviews conducted offered information about the programme objectives and suggested scenarios of a particular 'future' on which to comment as well as eliciting requisites for personal participation in such a 'future'.

1.6. Schedule

Following a visual inspection of all 34 villages on the first day of the report period, my Jordanian colleagues and I then began daily field visits from the centre which began on 22 July 1993 and were concluded on 12 August 1993. The villages selected were based on a random categorization by population number and one or more household/service providers were interviewed each day. During this period 20 villages were selected and 35 interviews were conducted. During the week of 13 August to 20 August, I entered into discussions with a household previously visited to confirm their invitation to reside with them. I remained within the village from 21 August until 7 September 1993 returning to the field centre only to deal with administrative matters.

2. FIELDWORK

2.1. The Disinterested Fieldworker

Though my own ideal would have been to be presented as a researcher unbiased towards one party or another, the people of the Badia clearly saw me and my co-researchers as individuals with vested interests; individuals with any number of interests other than those of the people themselves. This was certainly understandable considering that we introduced ourselves as Programme researchers requiring information for a detailed study of the area to be used to 'develop' the area. The motivations of researchers and the worth of studies were highly suspect in the minds of most of the inhabitants we spoke with. Many were aware of the Hammad Basin Study (ACSAD 1984) which was consistently quoted as being 'without benefit', so why should we be any different?

Our own rhetoric confirmed our position as representatives of the Programme at the same time declaring the altruistic intention of representing the inhabitants through their participation in dialogue with us. Visually our affiliation as representatives of an influencing authority was clear; we arrived and departed in logoed Programme vehicles and it was often a Programme driver who made the initial introductions seeking the participants' full co-operation.

If we were not viewed as instruments of this authoritative organization we were viewed as self-interested; eliciting information for academic benefit. Motivation was often the focus of counter-interrogation on the part of the inhabitants. One village leader not satisfied with my motivation for involvement in the programme pressed for an admission of using the fieldwork data for my doctoral thesis.

Mutual interrogation seemed to set the stage for the fieldwork. The inhabitants were suspicious of our motives generally and my personal affiliations as a non-Jordanian. Likewise, I suspected the all too ready and homogeneous responses given over tea during the survey-style interviews by the men, and by the women when in the presence of their male kin and affines. We each wanted the other to disclose their hidden truths. I do not mean to imply that this was counterproductive though admittedly it was at times an aggressive form of interaction. To the contrary, I believe that it was a vital part of the process of establishing credible relationships based on identity and positioning. Both participant and researcher were seeking to meet the 'inner other' as well as the 'projected other'. It would be inappropriate to couch the process in terms of mere dialogue, dismissing the interrogative element as unavoidable when seeking structured and less structured responses. It is only to be expected and in my view desired that the interrogator likewise becomes the subject.

2.2. Projections: Native and Foreign

When the foreign fieldworker meets her own reflection in the native fieldworker and the native fieldworker meets her own reflection in the foreign fieldworker both may recoil.

As a foreign fieldworker my approach to dress and conduct in the field differed from that of my Jordanian colleagues. Throughout the fieldwork I modified my approach to suit the particular circumstances, I saw no threat to my own self-identity in wearing modest clothes, covering my hair, and altering my body language to show respect. However, this in effect succeeded in offending the very individuals whose co-operation and mutual respect I equally sought, the native fieldworkers, my counterparts. Regrettably, offending someone seemed inevitable.

However, after listening to the reasons why my colleagues felt compromised by this projection I decided to draw on my identity as a foreign researcher and not wear a head cover on the interviews that my colleagues and I conducted together. Later, when I moved to the village on my own I asked the senior woman of the village for her advice and she told me that she would prefer me to cover in front of men to show respect. My deference, though by no means mandatory, was important to her and facilitated my introduction to this particular community.

This initial clash of projections made me more aware of the precariousness regarding notions of primitiveness and modernity. Does a Bedouin cease to be a Bedouin because he relies on a Mercedes lorry rather than a camel? Does a woman become more modern because she removes her *hijab* and wears jeans?

2.3. Advantages and Limitations of Being A Female Fieldworker

The obvious advantage in being a female researcher is the ability to enter into both domains of men and women. It is the level of entry which is the limiting factor. Whereas a male researcher would be largely excluded from the women's domain, I had unrestricted access. Though my dress was conformist, my identity as a foreign researcher also gave me further freedom to be in the presence of men that would have been otherwise inappropriate and to ask sensitive questions that would have been viewed as impertinent. But, there were definite areas of exclusion. I could not attend the tribal meetings debating the position of tribal members' claims to land and conflicts with other tribes over the same land. It was only by my bond with the senior woman that I even knew that such a meeting had taken place, and was then allowed to listen to the relation of the meeting's agenda, those present, and its outcome by this woman's eldest son. I could not accompany the men during 'business trips' and my presence was not welcome when visitors arrived wishing to conduct business. A male fieldworker might not also have been welcome in such circumstances but input by a male researcher on the political, social, and economic spheres in which men participate would clearly make a complementary contribution.

2.4. One Family? The Tribes of the Ahl Al-Jabal

Over the period of fieldwork I became increasingly intrigued by the concept and dynamics of the 'tribe'. What are the codes of membership? What function does it serve? What differentiates one from another? Initially, when an individual introduced him or herself as 'X' providing the relative noun rather than stating 'from X', I had no concept of what this label meant. As time went on, when an individual introduced him or herself as 'X' I had the beginnings of a constructed image about the way this person oriented him or herself; I found myself relating with this person as definitely X and not Y.

Then one day while accompanying a women 'from Ahl Al-Jabal' outside the Programme Area we met a woman from the Beni Hassan tribe. To my astonishment my companion introduced herself to this stranger as a member of the Beni Hassan tribe! I remember wondering how she could possibly say such a thing when she appeared, even to me, as so obviously from Ahl Al-Jabal. The woman from the Beni Hassan then, undaunted, proceeded to query the specific family she was supposed to belong to saying that there were many, reciting a long list, all well known amongst the Beni Hassan. My companion offered the name of one of the families, to which the stranger responded, 'But, you look like someone from Ahl al-Jabal'. Only at this point did my companion admit to being from Ahl Al-Jabal. Following our departure, the only comment made with regard to this incident was that the Beni Hassan woman was honourable, *muhtarima.*

So is a tribe here merely an identifier to differentiate between self and other where one relies on ever-increasing circles of identity when confronting the other? This is an area to be explored further. At the present time I can only talk about the tribes as they presented themselves to me.

The only tribes from Ahl Al-Jabal that I came across were the Masayeed, Al-Sharafaat, and Ahdhamat. Whether or not there are further tribal distinctions in the Programme Area I am not sure. I had little interaction with any other than Masayeed and Sharafaat though I did meet a leading figure amongst the Ahdhamat. A smaller village may be said to be primarily one tribe or another but the wives may well have been taken from another tribe.

3. GOVERNMENT ADMINISTRATION

3.1. The Interests of the Kingdoms - The Interest of the Tribe

One of the most intriguing areas was the inter-connectivity between Kingdom and tribe at the administrative level. Tribal dynamics seem to have become incorporated into local government administration with tribal shaykhs acting as mediators between tribes and the government. The government has clearly recognized the importance of supporting existing local systems of hegemony. The shaykh thus retains and perhaps increases his value to the community he represents while the government benefits from his allegiance and return-support. The ties between tribal leaders and the Hashemite Kingdom of Jordan were clearly testified to by the proud production of photographs of many shaykhs we interviewed standing with King Hussein.

As stated earlier, the tribal shaykh may serve as the local governor for a large village, in which case he may have responsibilities for smaller villages which do not have a shaykh or leader resident in the village. There is also a Head of the Municipality who, as in Safawi, acts as a local government administrator while also being the tribal shaykh. In some villages the local administrator may be a Mukhtar chosen by the people for his leadership qualities. A Mukhtar may also assist a Head of the Municipality in which case his responsibilities are principally administrative rather than inter-tribal. It may not be taken as a 'given' that the shaykh or administrator of one area will have the same responsibilities as that of another. Also, depending on the gravity of the circumstances requiring mediation, shaykhs and senior men of tribes outside the Programme Area may become involved; such an occurrence was related by one of the wives of a shaykh who told of a plea for tolerance made by representatives of the Beni Hassan, the Beni Sakhar and one of the large tribes within the Programme Area on behalf of a man who had caused a death within the shaykh's family and was offering payment of blood money to avenge the death.

The above illustrates the importance of understanding the areas and relationships outside the Programme Area which impact on the lives of the inhabitants of the Programme Area. One of the greatest deficits I had was due to the insular method in which the fieldwork was conducted: focusing on the area defined as the Programme Area independent from inhabitants, service providers and areas of relevance outside these borders. Time constraints did not allow much work outside the Programme boundary but it is suggested that future research should include the review of documentation and interviewing of individuals with direct relevance to the area.

My colleagues and I conducted interviews with local government administrators representing all of the above referenced roles. General responsibilities were detailed as follows:

3.1.1. Local Governor

1) Deals with inhabitants' concerns over provision of services.
2) Administration of land distribution/mediation of disputes over land distribution.
3) Road construction/maintenance.

3.1.2. Tribal Shaykh

1) Mediator/arbitrator in disputes between family members/members of tribe.
2) Represents tribe in disputes between tribes.
3) Provides moral obligations to members of tribe: When a man wishes to marry, the shaykh acts on the man's behalf as a Jah and accompanies him to the family of the bride to negotiate the engagement.
4) Protector/provider of safe asylum: We heard about an instance of a girl who was sent by the police to the shaykh to stay until the pronouncement of sentence after she had committed a 'crime' with her male paternal cousin. In another instance a girl took refuge in the house of the shaykh for a six-month period; her mother did not get along with the neighbours (in an area where her husband's family lived) and the father's family disapproved of the woman going out of the area. The shaykh is also called upon to provide asylum for couples who wish to marry against the bride's father's will. In this case, the shaykh mediates between the two families while the couple remain under his protection. When agreement is reached, he will then preside over the marriage. Some of the inhabitants told us that the shaykh may refuse to aid the couple if he believes that they have behaved disgracefully, in which case the bride's family may take the law into their own hands against the couple before they reach the police.

3.1.3. Village chief/Mukhtar

1) Administrative/recording of official documents.
2) Official signatory.

3.2. Police Intervention

Information about police intervention came from the inhabitants themselves. The tribe serves as self-policing in all but exceptional circumstances. The tribe fills the moral obligations of its people and the police seem to be viewed as undermining the authority of the tribe. When individuals fear the severe punishment (death) which may be meted out in accordance with tribal values, police intervention is sought. Moral infringements with fear of tribal reprisals are the primary reasons for police involvement in what were formerly considered purely tribal concerns.

In addition to the case of elopement, we heard of other circumstances where police intervention might be required due to disagreements over marriage. If a girl promised to her paternal cousin refuses this marriage and becomes engaged to another man, the uncle's family may threaten to kill the prospective bride and/or men of her family. While this threat may merely be a demonstration of the indignation of the cousin's family, the bride's family clearly takes the threat seriously. My colleagues and I witnessed such an event on the day the groom's family was to take the bride to her new home. Five men of the bride's family were threatened with death and the bride was also to be killed if her father allowed her to be removed from his home. The men of the household were heavily armed with Kaleshnikovs and I understand the police had been alerted to remain on stand-by. The tribal system of negotiation triumphed on this particular day with the bride's father agreeing to compensate financially his brother for the loss of the bride to his family.

Even when the act committed is a criminal act rather than a moral infringement, the police may be the last to be summoned. I heard that some sheep had been stolen and that the owner of these sheep knew who the perpetrator was; later gathering that the suspect was a long-standing enemy. When I asked him what he was going to do about it, he said that he and some other men of the tribe would take care of it; the police did not need to be involved. He did not provide me with further details.

3.3. Land Distribution and Tribal Concepts of Ownership

Land ownership is currently the focus of much attention by both the government and the tribes. According to the answers given by the local inhabitants where land distribution has been implemented, this programme seems to have begun some four to five years ago. Administration of this scheme is carried out by the local council for the areas concerned. Prices given for 500 square metres of land ranged from JD200 to JD250. Once distributed, the owner is given a deed detailing all the particulars of the property. I also saw a document which I was told precedes the deed; this had a very detailed map of the property, boundaries, and access to available resources such as wells.

When I first arrived in the Badia one of the first things I noticed and queried was the dotting of rock piles in the middle of what looked to be unsettled land. Later, the inhabitants of villages around the area confirmed that these were markers indicating possession of land. There was no grid pattern conforming to expectations of indicating property boundaries, and the placement seemed to be at random intervals without any distinguishable pattern. The only linear pattern I saw was placement parallel to a major road. These markers alert nomadic pastoralists to the perceived rights of ownership by someone else, but does not prevent them from living on this land and grazing it. The practice seemed to be accepted by the 'owners' who were aware of their presence and who visited their temporary neighbours. I do not know whether this is always the case or whether this was a special relationship between these parties. Also, when walking across one such marked property, I was told by one of the children accompanying me that the rock pile we were now coming across marked land belonging to one of her uncles. The division was clearly known to her but I am not sure whether someone outside the family would know that they had passed from one owner's land to another. These rock piles seem to be the first step in securing land for oneself by publicly communicating one's perceived rights to the particular land. A further step is to build houses on at least part of this same land in which the owner's wife and children may live while the husband tends livestock elsewhere. Further steps are for the husband to remain on the land as well and to bring relatives to settle in the area. Settlement equals rights of ownership in the eyes of the inhabitants; the greater the number of settlement indicators, the stronger the rights are to the settlement. Distribution then becomes an approval process by the government for land settlement and a government recognition of rights the inhabitants already perceive are theirs. Not perhaps surprising then, most people I spoke with who possessed deeds to their land were more than satisfied with the governmental distribution; getting just what they wanted. Yet, while seemingly straightforward for most, there are problems and disputes over who gets what.

More often than not, such disputes over land are between members of a family; between brothers. The local shaykh has the responsibility for arbitrating in such matters. However, I learned of a heated dispute between two tribes vying for land which to the eye would seem of little importance to anyone as it is one of the driest, most desolate locations in the Programme Area with little growth of any kind. Grazing this area would at the current time be pointless. So what makes this land of potential interest to these two competing tribes? This land lies east of the village of Safawi on the recently constructed road heading north towards the Syrian border from the main Mafraq/Baghdad road. This road, when completed, will lead to a large range preservation whose borders have already been marked out and where grazing is prohibited. It is hoped that this land will then have the opportunity to regenerate itself for future, controlled grazing. Perhaps a more crucial factor to the dispute over land south of this region is the well recently drilled in the area known as Shubayka, which is seen by key Programme administrators as the catalyst for generating revitalization of the area. Plans were circulating to offer water to anyone within the programme region wanting to fill a tanker for private use, though implementation was not explained or a schedule given. No concerns were expressed as to the sustainability of this water source over time with respect to the tremendous demand such changes will bring. This well is seen as the proverbial carrot before the horse enticing people back into this now uninhabited, barren region. But who is supposed to return to this land?

Remains of a village may be seen where people once inhabited the area, but I received no absolute answers from anyone as to who used to live in the village or what happened to them. Members of the Sharafaat tribe explained to me that this land has 'always' been considered their land. The characteristic rock piles seem to have been placed on the disputed land to support this 'fact'. However, as I learned while listening to a discussion between the senior woman of the village I resided in during the latter half of my fieldwork and her eldest son, a tribal gathering had been held where both tribes were represented by the head shaykh of each and with at least one influential member of the third tribe. The Sharafaat tribe was represented by Shaykh Swaylim, the head of the municipality and shaykh of Safawi. The Masayeed do not recognize the historical right of the Sharafaat to this land, and are appealing to the higher authority of the government to recognize their own claims to the land. I do not know the actual outcome of the meeting held of the Ahdhamat which was also seeking a claim to the same land. Later a member of the Masayeed tribe suggested that the upcoming election for the Minister of Public Works will be influenced by this very land dispute, and that Shaykh Swaylim had promised his tribe, the Sharafaat, this land as an election incentive.

Tribal concepts of what is 'law' and the dictates of the government become intermingled, whereby government interests and methodologies are being incorporated and used to further the interest of the tribe. On the one hand, the Sharafaat appealed to the government system to confirm what they believe is historically theirs; on the other, the Masayeed appealed to the same governmental system to confirm that a 'historical right' is not necessarily a 'right by law', in order to benefit their own tribal perspective. Of even greater interest, why are people being enticed into this area? Even if people do return to the area to settle, how long will they stay?

Near Azraq I learned from a Haj that he had once benefited from a similar enticement programme in the south of Al-Jafar. He settled in this area with his family because he knew that the government was going to distribute land and provide incentives for cultivating this land by provision of water for irrigation. The area was to be developed and the Haj wanted to be a part of that development. After settling he ran into problems with his alfalfa crop so he simply left the area and returned to relying on the mobile lifestyle that tending livestock afforded. Though he now has land in the south he gave no indication of returning there to live again and had begun to make a new claim to this new area with marked boundaries. According to him, the development of this area was not enough to keep him and others in the area. Clearly, drawing people into an area for the short-term is a costly endeavour and may not achieve the desired results. It seems that the current Programme would benefit by reviewing the success of other areas within Jordan where settlement has been encouraged by enhanced infrastructure and distribution of land, to assist a better understanding of what may be adopted and what may not.

4. SERVICE PROVISION: THE INHABITANTS' PERSPECTIVE

This section is from the inhabitants' perspective, because of little opportunity to deal directly with service providers during this initial stage. Service provision must be understood both from the providers' position and in terms of the reception and adoption of these services. In future, it will be necessary systematically to interview service providers as well as reviewing policy and procedure with regard to service delivery.

4.1. Utilities

4.1.1. Electricity

Few villages in the Programme Area have not been supplied with electricity. Those who have not yet been supplied have for the most part been contacted by the electricity board and I saw much planning activity, such as surveying land for placement of lines etc., in those smaller villages where a main line was present but no house-to-house connection. A project team was also assessing the potential for solar powered electricity for villages where traditional electrical installation will not be made available for some time. The downside of this pilot project is that while installation will be free of charge for the residents of the chosen village, any future installation will cost some JD17,000 per house. Future residents would then most likely have to have electricity supplied by the standard means.

One of the real frustrations of inhabitants waiting for the provision of electricity is that they are asked for some JD150 or more per building, to be supplied well before the planning stage, and they often complained of being given a date of installation which was not adhered to and then being subsequently given new dates which they no longer believe would be fulfilled. However, when a representative of the electricity board was asked for a date of installation for a particular household, no date could be provided.

4.1.2. Water

Ground and surface waters are of vital importance to the region. Evaporation, however, is one of the key problems to be examined with regard to water harvesting. Syrian dams pose another problem with regard to the availability of surface water to be used for animal and human consumption. A number of wells formerly used by the inhabitants of the region have been capped posing further hardship for the people, who also has a sense of 'ownership' over these resources and felt a conviction of their 'rights' to these resources being one day restored.

The pumping station at the junction of the main Mafraq/Safawi road with the Bishriyya road supplies water to the surrounding villages. Lorry loads may be purchased for a nominal charge which are used to supply both village households and livestock needs. The main line leading north on the Bishriyya road in the area known as Al-Asfar had been damaged by gunfire caused presumably by passing livestock herders with an immediate demand for water. The only repair that had been made at the time of the fieldwork was the placement of stones over the top to slow the loss of water from this point.

While the pipeline supplies water for the settled inhabitants of the region, all but a few complained of water not reaching them on a consistent basis. In extreme cases, water was only available a few days per month; in others, water was available two to three days out of every week or two weeks. In one particular village bordering a much larger village which had good access to drinking water, one woman bitterly complained about their own dependence on well-water which was not fit for human consumption and the ongoing ill health they experienced in comparison with their more fortunate neighbours.

When water is available from the supply line it is stored in large metal containers for human consumption and household use. Large concrete open or covered containers are also filled for livestock. Surprisingly, no rationing attempts were made with regard to agricultural usage for households combining livestock production with cultivation of the land. Water was led via a hose from the central tap on the supply line to orchards where the water was left to run virtually all day with no concern for the surplus flooding into the surrounding barren land; 'if available use it' seemed to be philosophy, with no connection being made between current use and future availability.

4.1.3. Waste disposal/sewage

One of the problem areas observed but which none of the inhabitants ever complained to me about was the available means of waste disposal. Refuse seems to be dealt with in much the same way in many of the villages as it has been in the past: openly dumped on the land at a distance from the dwelling, partially because of the lack of refuse collection in the smaller villages. The problem is that settlement brings about a higher concentration of 'modern refuse' largely comprised of man-made materials such as plastics which are not as biodegradable as former organically-based waste. This directly impacts the lives of individuals within the community, particular those who own livestock; animals eat the plastic, become ill and die. Even in the larger villages, I often saw a window being flung open and household refuse flying out onto the ground at the back of the house.

Public sewage systems require a consistent supply of running water. This presents a problem to most of the region's inhabitants. The only plumbed toilets I came across were at the Safawi Field Centre which is one of the few places where water flows copiously from the taps. Houses/buildings with no running water usually have the toilet outside, at a distance from the building. The inhabitants said that they preferred this because they considered it more hygienic than indoor toilets. When asked if plumbed toilets were available whether they would have the toilet indoors, most said that they would. In built-up villages the toilet usually consisted of a walled, open enclosure, i.e. with no ceiling or door, built on a foundation with a deep cesspit. In the small village I stayed in for the latter half of fieldwork, no walled toilet was necessary as there were so few houses. They told me that they would build walled enclosures as the village grows.

At the risk of sounding too 'politically correct', it seems that with the need to conserve water in the region, the feasibility of installing alternative sewage systems (i.e. dry systems) should be considered in future development plans, rather than conventional plumbed systems. They would also be in keeping with the current lifestyle of the inhabitants.

4.2. Health

One of the greatest difficulties encountered was in discussing an issue like health care, its provision and adoption without advocating and even interpreting the responses according to one's own personal bias. This conflict of interests became most evident when discussing birth practices and health care provision to children. The hospital is gaining popularity as the best place to give birth and some women that we interviewed supplemented breast milk with prepared milk saying that their own milk was not sufficient for the child's well-being. My counterparts supported trends for institutionalizing birth practices, seeing such trends as 'progressive'. My own bias favoured hospital birth on an as-needed basis, rather than being the optimum choice in all cases.

These messages seem to be received most strongly by women in their 20s who generally expressed fear of home birth as dangerous for both mother and child. In contrast, their mothers nearly always expressed the case for home birth, even demonstrating the semi-seated posture taken and the position of the midwife, often the woman's own mother or other female elder experienced in such matters, behind the woman. Most women acknowledged that labour in the hospital took longer than at home comparing either their own experience or that of neighbours who chose or required hospital delivery. Many advocated a moderate stance: hospital birth only in the case of necessity, in the event of a problem with the birth. Health care as in other areas of service provision in the Badia is a matter of successfully selling 'expert intervention' as being beneficial.

As may be expected, views on health care differ between family members (with women expressing different concerns and interests from those of the men, particularly with regard to family planning), households, villages, and even tribes. As with the effect of education regarding Islamic practice (e.g. facial tattoos being rare among young girls and women in 20s to 30s, and an increasing number of girls wearing *hijab*), education certainly plays a major role in the changing attitudes to health care. This is reflected in the attitudes of inhabitants within larger villages having access to more and better services than those in smaller villages or amongst tent-dwellers, who seem to have a *laissez-faire* attitude towards health care which may mistakenly be constructed as negligent where children are concerned. If inroads are to be made with regard to improved health care for women and children, it will be important to understand rather than condemn such attitudes. In practice, this can be very difficult when a child's teeth are eaten away by decay at the age of six and the mother points out the fact that new teeth will replace those rotten ones; or when parents refuse to see any urgency to take a seriously ill infant to a doctor and who simply comment that if the child dies it is meant to be.

Medical care, when sought, is generally for curative rather than preventative treatment. Though I did not have access to enough data during this brief period, the staff at the health care centre complained that children are not regularly brought in for vaccinations, particularly amongst the less-educated and more nomadic groups. Some post-natal services were available at this centre but little used. It is also interesting to note that women often went to health care centres outside the Programme Area in the larger market towns of Mafraq and Zarqa rather than to their own or neighbouring centres.

More service providers in general, and particularly within health care, should be interviewed to provide a more balanced view of the provision and adoption of services by of the local inhabitants. In the case of the health centre that was interviewed, we learned about specific issues of concern to women regarding polygamy, and about the concerns of elder women over changing family relations with regard to the increasing number of nuclear families and its impact on the parental household, as well as learning that homeopathic (traditional medicine) remedies are not seen as 'alternative medicine' in the Western sense of the word and may be employed side by side with modern medicine.

The health centre visited was in the village of Mukayfita, where the nurse's assistant interviewed had trained at the hospital in Mafraq for two years. Her responsibilities included administering vaccinations and talking to the people about general health care. The clinic is open daily until 2.00pm and a doctor is available every day except Thursday and public holidays. Patients must go to the centre in Umm al-Quttayn if the Mukayfita clinic is closed or the doctor is not available. Vaccinations are given for diphtheria, tetanus, whooping cough and, for adults, hepatitis B.

Primary childrens' complaints were said to be cold and diarrhoea. Women suffered from fatigue and from gynaecological problems. The nurse's assistant said that the fatigue was due to the harsh working conditions for women tending livestock, cultivating land and working within the home. I had also noticed that women appeared gaunt. The gynaecological problems were attributed to the predominance of polygamous marriage; with the man passing an illness from one wife to the next. In a few cases women also suffered from breast infection due to breast feeding. The elderly were reported as having few problems; few incidences of high blood pressure and no cases of diabetes.

When I asked about family planning means, I was told that the women never ask for contraceptives although they are available. The only method for spacing children is the practice of breast feeding which suppresses ovulation. Women from other villages spoke to me of their desire to limit the number of children that they have, but said that their husbands refuse to allow them to take birth control pills. While women seem prepared to act against their husbands' wishes in many other areas, taking birth control measures without their husbands' permission did not appear to be one of them.

Women who come to this centre following childbirth receive information about care of newly born babies including bathing instruction. Confirming what my colleagues and I had seen elsewhere, I was told that the mothers do not diaper their babies. I had seen a number of young babies who had been swaddled tightly from the waist down. The women told me that this was to prevent the bones in the legs from growing crooked. The period may be anything from between one and three months up to a year. Unfortunately, I neglected to ask about this practice at the centre.

4.2.1. Homeopathic medicine

At the health centre I was given some samples of herbs used for certain general aliments; these were:

- colic, nausea, abdominal pain, and headache: *Artemisia herbalba*
- cold, rhinitis, chesty cough: *Mafricania chamomilla* L.
- swollen glands (due to cold), headache: *Thymus serpylim* wild thyme
- dry cough: aloeswood
- diarrhoea, fatigue (children, adults), sleep trouble (used to send children to sleep), upset stomach: *Pimpinalla anisum* L. aniseed
- menstrual discomfort: *Trigonella foenum-graecum* and mint.

These are prepared as herbal teas.

Most of the women that I asked about the use of traditional medicine confirmed that they used various homeopathic remedies to treat common ailments. Many said that they preferred using traditional medicine to going to a doctor. Some of the same treatments, such as *Artemisia herbalba* 'tea', are used for sick animals but I personally saw little positive effect; the animals I saw treated this way all died. However, they appeared to be near death when the treatment was given so I don't know whether earlier administration would have altered the outcome.

Some plants are said to be available in specialist retailers in the market towns of either Mafraq or Zarqa, including (a) for childhood ailments: *Mafricaria chamomilla,* and *Artemisia santonicum*, and (b) for adults: *Artemisia abrotanum*. Plants readily available off the land were said to be *Thymus serpyllum*/wild thyme, sage and mint.

Stones may also be purchased to ease the pain of childbirth. I saw one woman wearing a green stone which she said is rubbed during childbirth to ease the pain. When I asked her why she still wears it round her neck she simply replied that she liked wearing it. She called this an amulet.

4.2.2. Disability

Disability presents certain problems which distance the sufferer from the rest of the community. A number of individuals in the region have a skin discoloration where the extremities, at least, appear patchy. I was told that such individuals are not allowed to marry anyone who does not have this same condition. Disabled persons seem to be perceived on some level as weakening the fitness of the whole. I saw physically disabled persons who worked and seemed valued in the community; they just have limited opportunity to marry and thereby reproduce the same disability.

4.2.3. Lifestyle and health

Most of the inhabitants interviewed appeared overall to be in good general health, and the interview with the Health Centre in Umm al-Quttayn seemed to support this. There was little evidence of stress-related diseases that are common in industrialized regions. Senility did not seem to be common among the aged members of the community and most of the aged were physically very fit, often putting me to shame in terms of physical strength. The elder members of the community are not ‘put out to pasture’, and are highly valued contributors to society. I often walked with the Hajja of the village I resided in across the rough terrain to visit neighbouring tent-dwellers. One of her first tasks when I went to live in the village was to teach me ‘how to walk’, springing from rock to rock while balancing a heavy load on my head. Other elderly women I met also talked about the great distances that they walked to visit neighbours. Tending livestock and cultivating land seems to have its positive effects on the emotional and physical well-being of the inhabitants.

4.3. Education

The opportunity to educate one’s children has played an important role in the inhabitants’ motivations for settling and in many instances has influenced the decision to settle in a particular area. My colleagues and I met a family who had moved into a village just a few months prior to the interviews we conducted in July. One of their primary stated reasons was so that the women and children could live in the village and the children could go to school; the men continued to live in tents away from the village tending their livestock. I also understood from a number of inhabitants that the government had promised to provide a school for a settlement which had a certain minimum number of ten to twelve children of school age residing within the settled area. The frustration was clearly evident in the voices of those who believe they are entitled to this provision but have not yet seen the arrival of a teacher. One village was willing to provide the building themselves; they can’t provide the teacher. Teachers are not necessarily ‘permanent’, and are not necessarily from the Badia itself. Complaints often centred around a desire to have permanent teachers. With more children having access to higher education the region will, it is hoped, be able to provide for its own teachers.

Currently the school system is divided into primary and secondary. The primary school is for grades one to ten and the secondary school for grades eleven and twelve. The former school system of some two years prior was comprised of a primary (grades one to six), a preparatory (grades seven to nine), and a secondary (grades ten to twelve) school. Of course, not all of the villages within the Programme Area have a secondary school, if they have a primary school. Primary school is perceived by the settled inhabitants as being a minimum for their children and send both their male and female children. If one drives along the main Mafraq/Baghdad road around 7.30 to 7.45 in the morning, groups of children will be seen standing along the roadside in their uniforms waiting for the public bus to take them to a neighbouring village to attend school. While both sexes usually attend primary school, it is particularly important for boys to do so, as the military provides a primary source of income for the inhabitants. The disparity between the sexes appears to arise at secondary school level.

4.3.1. For all? The disparity between the sexes

One of my dearest young female friends fervently wishes to become a teacher. This, at least for her, is not likely to happen; her father will not allow her to continue her education to secondary level. He said that this was because the nearest secondary school is some 40km away, and although the public buses go there he will not allow his daughters to travel this distance. However, I believe that the distance is not the full, or even primary, reason for not allowing her to continue her education. Other men in this same area had the opportunity to take their wives from near kin (the preferred choice for this particular group), women who had grown up in a large town outside the Programme Area. Rather than do this, they chose wives from a neighbouring tribe because they did not want women who had no experience with life and work in the region. Women whose heads are filled with educated ideas but whose hands are idle are not of use of men who rely primarily on the land for their living.

Interestingly, there seem to be certain unexpected areas of 'equal opportunity' in education within the home. Girls as well as boys are taught, in the less populated areas, gun maintenance and marksmanship. Women carry guns with them when they walk alone as well as having other weapons to hand in the house. Once I thought that a baby was lying on top of a bone and asked what this was doing underneath the baby. It turned out that the bone was, in fact, a small club fashioned from a piece of hard wood. I also know of girls from the age of 11 being taught to drive though I do not believe that this is so they can one day get a licence. I have never seen any women driving on the main roads in the Programme Area; though Jordanian women in urbanized areas obtain licences and drive. I do not know whether teaching girls to shoot and drive is something peculiar to this particular group or whether it is commonplace throughout the region where necessity might require it. In isolated areas perhaps there is greater perceived need to ensure that all members of the community are capable of defending themselves.

At the health centre I visited, I was told that one of the frustrations expressed by women is the unwillingness on the part of their husbands to allow their daughters to continue their formal education. The women generally seemed to have greater hopes for their daughters with regard to education than their husbands, and this was one of the causes for argument within the home. This does not represent a homogenous view of women in the region, as many women elsewhere said that they saw no need for continuing education for their daughters; at least not the kind of education that would be available at secondary school. If they expressed a desire for continuing education for their daughters it was usually related to education that would be perceived as benefiting the girl within the home. Many such women expressed a desire to have a centre for learning/improving craft techniques focusing on sewing, embroidery, and carpet-making.

4.3.2. Women and vocational education?

One of the areas that it would seem beneficial to examine as part of a future case study on education would be the role that vocational education might play in the lives of the women of the region. Many have heard of projects in other areas which featured the introduction of a craft learning centre and said that this was something they felt would benefit them directly. Some women said that they would be interested in selling their crafts to market towns, like Mafraq and Zarqa. Women who are not allowed to work outside the home said that this would not compromise their traditions and that they would be interested in boosting their household income in such a way. Yet, most of the young girls that I spoke with are not learning the traditional crafts that their mothers and grandmothers know; the very crafts which this future institutionalized centre would teach.

Rather sadly, this seems to be due partially to notions of modernity in which the inhabitants are beginning to believe. Woollen items made from the available livestock resource are seen as outdated; synthetic yarn bought from Mafraq or Zarqa is considered more valuable and is considerably more costly. Beautiful woollen carpets and cushion covers hand-dyed in vivid colours woven by the grandmothers are now becoming virtually obsolete; replaced by plastic floor coverings and synthetic facsimiles. 'Better' seems to equal 'new' regardless of the inherent qualities of the material to hand. The same notions of modernity that influence the handicraft industry also apply to the building industry. Limestone or concrete are the desired building materials in preference to the readily available and formerly used basalt.

The idea of providing an institutionalized centre with external administrators seems doubtful when the knowledge and capability for vocational education is clearly available within the region. What perhaps needs to be addressed are the reasons why the inhabitants view such as institution as providing a better education than the grandmothers and many of the mothers are capable of providing for their daughters themselves.

4.3.3. Television and public service information

Television has, without doubt, begun to play an educational role in the region. Those who do not as yet have access to electricity plug the television into the car battery to watch the news in the evening. In the family I lived with during the latter half of fieldwork, only feuding members of the family absented themselves from this communal gathering. Some of the tent-dwellers I visited did not have television but seemed to be informed about current events through communications with other family members residing in settled areas.

I was curious about the impact, if any, of public service information which covers such topics as health, agriculture, livestock, and water utilization. While I did not have an opportunity to watch these broadcasts during my stay to be able to comment on presentation, content, and slotting, I questioned the women I visited about whether they watched such announcements and if they applied the information to any areas of their lives. One teacher residing in the area said that she used in the classroom the information she gained from these announcements, particularly with regard to health care and safety matters. She remarked that she believed only the more formally educated inhabitants of the region would actually apply the information, if they listened to it at all.

There seemed to be a lot of ignorance about safety matters particularly with regard to children. It was common to see children playing with plastic bags which they put over their heads or faces, or chewing on ammunition castings. I also met a child who was severely scarred and quite lucky to be alive after an uncle was playfully tossing him up into the air when the boy's head met the rotating blade of an overhead fan. Greater emphasis on disseminating public service information may prove advantageous in preventing accidents within the home caused purely by a lack of knowledge about such matters.

4.3.4. Marketing the future: hands-on training

The tribal people of the Badia are independent, strong-willed, free-thinking individuals who will not simply accept changes to their lifestyle without being firmly convinced of the advantages inherent within such change. This may not be achieved merely through the medium of discourse and argument. The men, in particular, reiterated over and over again that any technological 'advancements' in either livestock or agricultural modes of production would need to be demonstrable before being adopted. Pilot projects involving the direct participation of the individuals targeted to benefit would, I believe, be well received. As has been previously stated, the inhabitants are frustrated with 'studies'; a programme of 'action' which would unite the community members in working for a common goal would also stimulate feelings of personal pride and self determination, and would, it seems, be well received.

These pilot projects could run the gamut from maintaining/refurbishing existing facilities (schools, water reservoirs, etc.) to range management, land cultivation, and improving modes of production combining livestock and agriculture.

4.4. Transportation

There were only a few complaints about the public bus service that runs on the main road between Safawi and Mafraq. Complaints usually came from the outlying regions where bus services were limited and in some cases non-existent. It was even suggested by inhabitants of a large village that the smaller village from which they had moved is experiencing an exodus of inhabitants due to the lack of transportation and communications services.

4.5. Communications

4.5.1. Telephone

Small villages do not usually have a public telephone installation and so the people must travel to a neighbouring village which may be at a great distance. The public phone may also be located in the same building as the post office which may limit the access to it (see below). We sometimes thought that the only phone in a village was the public phone, later finding that the shaykh might also have a private phone. For villages with few or no connective services, the absence of a public telephone was more strongly felt.

4.5.2. Postal Service

Post offices are available in most of the larger villages but the service even in these villages was often the subject of dissatisfaction and frustration. We learned that the postal administrator was not necessarily from the same village, resulting in limited service. Many do not open on a regular basis; when the village phone is in the same building this presents an even greater problem. Service was generally described as being poor.

4.6. Goods

The importance of the market towns of Mafraq and Zarqa should not be underestimated. While small shops with food and household items are available in a number of villages, there is a heavy reliance on shopping in one of these major market towns. The lack of availability of goods within the region was a common reason for travelling this greater distance. Women travel alone by bus to purchase material for clothing, yarn for handicrafts, household items such as basins, pulls, toys for the children, makeup, etc.

The objection to paying inflated prices for goods sold house-to-house by Syrian traders also played a part in women's decisions to purchase goods from these towns. Cursing the need to purchase anything at all from these Syrians usually followed their departure.

5. THE LOCAL ECONOMY

While the analysis of data available from the baseline survey will provide an empirical representation of the local economy in the region and reveal who is doing what work, some general comments and observations may be made to support this analysis. It became evident very quickly that the people do not generally depend on a single source of income but draw on as many opportunities as are available to them. So while someone may be viewed as tending livestock for a living they may also cultivate the land when rainfall and/or other water sources facilitate such activity. Families also look to the military as either a primary or secondary source of income, and pensions of various kinds supplement local incomes. Trade, in its accepted and euphemistic definitions, provides additional sources of income. This is one of the reasons that it will be useful to gain a better understanding of trade links between the border areas of Syria, Iraq and Saudi Arabia, the primary areas of exchange.

5.1. Who Does What Work? The Division of Labour

As alluded to previously, the inhabitants are influenced by the tribe in general and the family in particular with regard to the division of labour. One of the disparities found when looking at the data from the baseline survey and comparing it with that found in the field concerned the work of women and children and their contribution to the household economy. Even in the field, replies to questions on the work that these members of the community contribute revealed the confusion as to the definition of work. Many seemed to equate work as being directly related to an exchange of services or commodities for monetary return. With such a definition, women and children may not be viewed by either the inhabitants themselves or the interviewer as contributing to the local economy. So, in a purely empirical analysis, the only women who work are those who receive a salary, i.e. teachers, clerical workers, etc. Children would not even be considered a part of the equation. It was often only persistent questioning and through observation that revealed the importance and extent of the contribution that these individuals afford the community. While there was a separate activity matrix included in the baseline survey questionnaire, the answers provided regarding women and children were far from representative.

Men who primarily depend on livestock production and/or land cultivation usually herd the livestock at a distance from the home. But women and children tend to be responsible for this same activity near the home. Some adolescent boys also herd the livestock from one area to another. Women with the assistance of the children are responsible for feeding the livestock and milking. Women convert the milk into dairy products consumed by the household and/or milk is sold to dairy traders who dwell and work in a tent on the outskirts of the village. A dairy trader purchases the surplus milk from the village, milking the animals himself, and sells the products to the market towns.

Women also work for their husbands in land cultivation and harvesting. One of the most common reasons given by men for taking more than one wife is the labour requirement of livestock and land production; an additional wife provides an immediate additional labourer as well as being a source for future labourers. Some young women in the greener region bordering Syria work as paid employees of neighbouring farmers who were reported as paying JD1.5 per day plus some of the harvest.

I was surprised to learn that women also contribute their physical labour in building dry-wall livestock enclosures and dividing walls between houses and sections of land.

Child-rearing, cooking, cleaning, craft-making and sewing are time-intensive services women provide even when they, in albeit unusual cases, function as the primary financial support as confirmed by the 'new age man' I came across in one of the region's small villages. This husband left his job with the military because his wife was earning more as a head teacher in a school than he was in the army. He now looks after the children while she is at work and takes odd jobs driving neighbours to the market towns or working in 'trade'. However, despite the fact that his wife apparently provides the primary household income, she does her housework before she leaves for work in the morning and takes over responsibility for all on her return.

The rearing of children is, as partially suggested above, not solely within the realm of the woman's domain. Men appeared to be actively involved in the well-being of their children and even pre-school aged children sometimes accompanied their fathers, particularly while working with livestock. Children begin to contribute themselves to the household labour force at a very early age. It is not an unusual sight to see a six- or seven-year-old boy or girl wielding a heavy hoe chopping and gathering range plants for livestock consumption.

Boys generally tend livestock either on foot or on donkey, but girls may also do so if the tribal and/or family influence permit it. Children not only provide a source of labour for their parents but to other members of the family; a ready resource to be utilized as and where needed. The hands-on education they receive in this way could not be replaced by institutionalized education. Some might consider this exploitation of children's labour, but life in the Badia demands the contribution of each able member of the community regardless of age or sex and in many ways offers 'equal opportunities' in terms of being valued, respected members of society.

5.2. Modes of Production

The military is, of course, one of the primary sources of income for the people of the region but livestock and to a lesser extent cultivation are the other primary sources, which may be combined to increase income opportunities throughout the year. There were some complaints about 'lack of co-operation from the Ministry of Agriculture' and water limited the level of cultivation. Some of the inhabitants, primarily those who have moved from livestock production to cultivation, said that they did so because they find that they get more of a yield in less hard-working conditions. Those individuals had the capital to invest in irrigation systems making such a transition feasible. They also seemed more resolved about their decision to settle in an area than did those who have settled but still tend livestock even if they cultivate land as well.

CHAPTER 19

Women and Unpaid Work in Nayifa Village

CLAIRE NAYA
(University of Durham)

1. BACKGROUND

Over the centuries the Bedouin of Jordan have learnt to adapt and survive the numerous adversities and pressures of 'modernization' and 'progress', but, changes have occurred which have affected every aspect of Bedouin life. People external to the Badia have created international borders, built oil pipelines and pumping stations at which villages like Safawi and Ruwayshid have evolved, introduced vehicles, provided livestock subsidies, initiated health and education systems and developed water resources. The Badia people have in turn exploited the advantages of cross-border trading, replaced their camels with trucks, taken employment in the military and civil service and acquired a taste for material goods. As a consequence there has been an increasing trend of once nomadic families now becoming permanently settled (see Chapter 1). As a further consequence women too have suffered, becoming bored and secluded. The traditional activities that used to allow them to meet other women and socialize have been taken away and replaced by new technology. Nomadic tribes that have settled down mean that women don't have the experiences of setting down a tent, riding camels to the next place or collecting water from a community point. All these activities that women did together as a tribe allowed them to interact with each other and now it no longer happens.

2. THE INVISIBILITY OF WOMEN'S UNPAID WORK

2.1. The Baseline Survey

There are still considerable steps to be taken to fully document data on women's paid work, so accounting for women's *unpaid* work proves to be an even greater task. The BRDP carried out a Baseline Survey in 1994 to gather information on the population of the study area. This resulted in a huge data base of quantitative information. In general, the Survey revealed that 98.8 per cent of women's major activities in the village of Nayifa were not specified and that 97.6 per cent of women were not classed as earning - reasons for why this was so is unknown for 98.8 per cent of cases. Because of the nature of the survey techniques used, information on women's main activities and earning patterns in Nayifa was systematically excluded. This type of invisibility is serious because if these women are not recognized as workers they cannot have access to training, credit facilities or technology.

2.2. Failures in Survey Techniques

There are two main reasons for the Survey's failure to capture fully female activity. As stated by Papps (1992), the first failure with conventional survey techniques is conceptual and deals with the global definition of economic activity, which encompasses the production of a good or service that is bought, sold or exchanged on the market. This definition does not even begin to capture the unpaid, routine activities of women. The second failure is the problem of measurement. How do we allocate time spent on activities? How do we assign a monetary value to unpaid work? How do we measure the importance of certain activities over others?

2.3. Contributing Factors to the Invisibility of Women's Unpaid Work

Other factors also contribute towards making these unwaged activities invisible. The Badia women are not a homogeneous category. They are divided by class, age, educational level and so on and are therefore particularly vulnerable. They carry out an array of tasks that occur in and out of the home and this fragmentation makes it hard to measure their activities. When viewed as separate components, cultural assumptions make these activities (such as child care, cleaning and food preparation) appear of secondary importance - even to the women themselves. Women's unpaid work com-mands little importance in society. Further, attempts to break down women's work into separate components are, in reality, a difficult task. Women must have the ability to carry out multiple tasks simultaneously, like food preparation while looking after children. Added to this complexity is the fact that non-domestic tasks, such as carrying out con-tract work or making products for the market, are also included in these multiple tasks. Another factor is that unpaid work has no monetary value attached to it. As previously mentioned, value tends to be synonymous with market worth and yet many goods and services with economic value are not marketed. This impacts on the status of Nayifa women in their society and may lead the BRDP into making gender-blind development policies.

Rapid Rural Appraisal (RRA) was a technique implemented because of its simplicity and its ability to help recognize the complexities of rural life that traditional surveys keep hidden. In theory, if a researcher is experienced in the use of RRA, the information gathered can be returned to the people surveyed so that they too can learn from it, become aware of and be sensitized to the issues raised and therefore change things for the better. Qualitative information can be gathered at a rapid rate and used to support the quantitative data of official questionnaires and baseline surveys (Chambers 1992).

With these issues in mind, research was carried out in Nayifa in the form of two case studies using RRA techniques, namely the Daily Routine Method, through semi-structured interviewing and direct observation. This allowed more qualitative data to surface on women's unpaid work within the village, but these data cannot be used to generalize women's situations in Nayifa because they are not representative, although they do illustrate what type of information is being excluded.

2.4. Nayifa

Situated 50km west of Safawi, Nayifa is a very small village with a small population and approximately 100 households (Wintersgill 1994). Its terrain is mainly sand and rock with very sparse vegetation and few trees. Because of poor tree cover, constant winds erode surface nutrients making cultivation very difficult. Annual rainfall is low and water is a very precious commodity. Domestic water is brought to the village twice a week by tanker to fill home tanks.

3. CASE STUDIES

A Jordanian woman can anticipate a life expectancy at birth of 69 years, and has average age at marriage of 22. She can expect to have about 6 children during her life time and any daughters she has will have a 73 per cent chance of being enrolled in secondary education (ILO 1994). In Nayifa the oldest woman was 75 and the youngest married woman was 18. On average each woman had 6 children and primary school enrolment rates for Nayifa were unknown. Female educational enrolment rates and literacy rates have improved greatly in the Middle East, but this is not reflected in Nayifa's educational statistics where only 8.3 per cent have had primary schooling (Fig. 1) and over 76 per cent are illiterate. Also, enrolment into tertiary education and training in the sectors less traditional for women, such as the technical and scientific field, still remains low because of prevailing exclusionary attitudes and prejudice against women. Only 1.2 per cent of the Nayifa female population are in post-secondary education. This does not encourage Nayifa women to envisage a role for themselves in the sustainable development of the Badia.

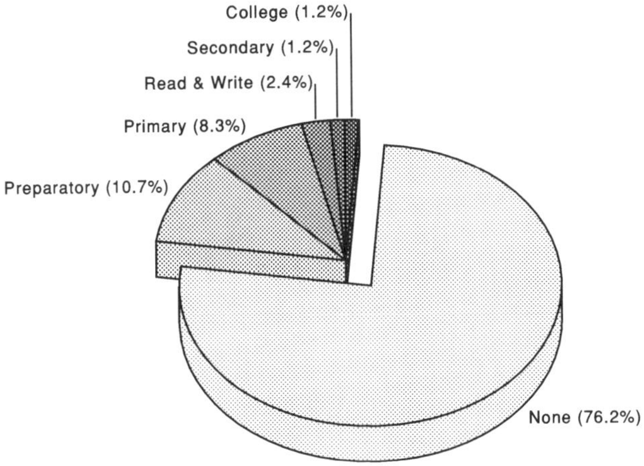

Fig. 1. Percentage of women at different education levels in Nayifa village, 1994.

The two case studies, Aisha and Meena, are women with wants and desires normal for women their age. However, many factors influence their ability to fulfil these desires. Because of the emphasis placed on the continuation of Islamic values by Muslim society, such as honour and modesty, women's participation in non-agricultural or paid employment carries a social stigma, therefore gainful employment is not perceived as part of their role. However, although it may be facile to blame their culture for their restricted access to paid employment, 'culture' must not be regarded as a constant. It is variable and the extent of its impact depends on other factors such as depth of development, state policy and class or social structure (Moghadam 1991, in Pearson, 1992).

The Daily Routine Method was used to log the average daily activities of Aisha and Meena, the results of which form part of Figures 3 and 4 respectively. A table was drawn and split into segments each representing the hours of the day that the women were awake. These segments were then shaded according to whether the women were working or had free time. Details of these major activities are stated beneath each grouped time segment.

3.1. Case Study 1 - Aisha

At 24 years of age Aisha originates from Umm al-Quttayn and is educated to college level. She was married at 21 after which she moved to Nayifa to live with her husband's family and at 22 had a son. Nayifa has few prospects for a woman whose ambition is to be a teacher. Travelling to another town to work was not a feasible option as far as her family were concerned, because she cannot travel on her own. They believed it was more important that she managed the responsibility of a home. Also she did not need to contribute to the household income as her husband considered his wages to be sufficient to support the family. In Nayifa the majority of household income sources are from wages and sales of livestock (Fig. 2).

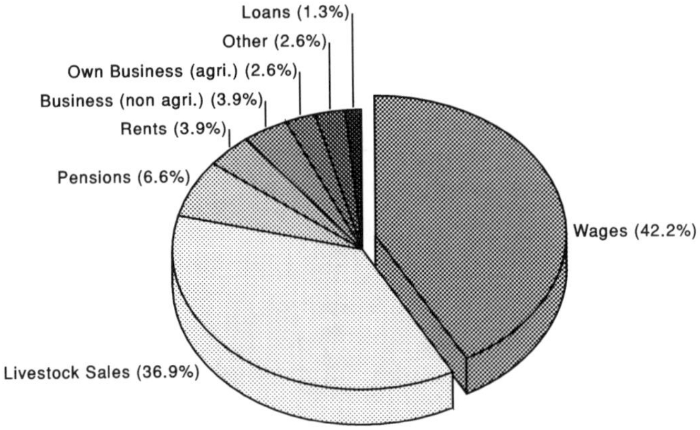

Fig. 2. Sources of income for households in Nayifa village, 1994.

Aisha's daily routine consists of a number of activities (Fig. 3). She rises about 8 am, says her prayers and then prepares breakfast. This generally consists of olives, *lubna* or goat's cheese, sometimes an omelette covered in *semna* (goat's milk butter), and plenty of *shrak* bread. She dresses her son before breakfast after which she clears away, washes up and cleans the kitchen. Once done, she makes the beds putting away blankets and mattresses, as well as everyone's clothes, generally tidying the bedrooms as she goes along. In the milking season, Aisha will milk the goats using the milk to make warm, sugary drinks for her son and the rest she turns into yoghurt. The majority of that is eventually dried in the sun and preserved until required. She also makes the dough for the *shrak* bread if it is necessary. This is baked later on when the dough has risen. Around 11am she rests if her chores are done and about 1pm she starts to prepare lunch. Most meals are vegetarian unless guests arrive or it is Ramadhan, then meat is eaten. Lunch consists of rice cooked with onion, *shrak* bread and salad. The process of clearing up, washing the dishes and tidying the kitchen are repeated. Around 3pm she has free time when she may sleep or visit her relatives next door. About 6pm Aisha prepares the evening meal following the same clearing up routine. Then she puts her son to bed and relaxes with her husband and family.

3.2. Case Study 2 - Meena

Born in Syria Meena is 35 years old and about to be a mother for the ninth time. Having had no education, Meena was illiterate and found the idea of 'adult education' an appealing thought but would not enrol unless her husband allowed it. She believed he would not as she was needed for the milking season. Like Aisha, Meena does not directly contribute to the household income. Her husband is a sheep farmer and so trading livestock is the main source of income. Still semi-nomadic, her and her family divide their time between their tented home out in the Badia desert, where the sheep are kept, and the permanent building set in Nayifa. Meena and her daughters are relied upon to milk between 150 to 300 sheep every day in the spring so the milk can be sold. Meena's second eldest daughter, who is only in her early teens, looks after the domestic routines in Nayifa while she's away in the desert.

Meena rises about 6am (Fig. 4), and her first chore in the morning is to prepare tea and then make *shrak* bread with dough prepared the previous night. Kindling is used to fire the metal dome the bread is cooked over which creates a very smoky atmosphere in her 'tent' kitchen. Every three days she boils water at this time to wash clothes - depending on water availability. After breakfast Meena tidies the house and then washes her clothes. She uses an electric washing tub to spin the clothes, an activity that sometimes her daughter does. From 10am to about 12pm she rests. She enjoys making cushions as well as knitting covers for them during this free time using various coloured wool purchased locally. This is done for pleasure and also to build up her daughter's dowry. Between 12pm and 2pm Meena prepares lunch cooking mainly vegetarian dishes of rice with a vegetable stew. After lunch she washes up at a tap located outside the house, unlike Aisha, she does not have a fully equipped modern kitchen. She also cleans the house in this time. From 2pm to 4pm she rests again maybe sleeping off the intense desert heat or she sews and makes cushion covers. About 6pm she will prepare the evening meal amongst doing other things and once they have eaten she washes up and clears away. After this she rests and watches TV, and then retires.

3.3. Summary

In summary both women appear to have very similar routines in the household. They both have the responsibility of preparing all the meals, caring for the children, cleaning dishes, washing clothes and tidying the house up. Meena rises much earlier than Aisha but their routines follow a similar pattern. It may be that Meena takes more time to carry out the same tasks because she doesn't have the labour saving devices and large working areas in her house like Aisha. Meena may suffer less boredom because she constantly occupies herself with sewing and knitting while Aisha prefers magazines and so on. Where their routines differ a lot is in the spring time when Meena has to milk her husband's large sheep flock. Aisha does not have to do this on such a large scale.

3.4. Hidden Issues

On further examination of these women's daily routines, several issues arise.

Women's unpaid work is taken for granted. These activities have low visibility and are deemed to be done out of love or motivated by feelings of duty. For example, Meena is expected to milk sheep, a very necessary activity if her husband's livestock business is to function profitably, and yet her unpaid work is not classed as work nor is it rewarded. The returns from unpaid work are critical to the household and the economy of a nation

Fig. 3. Time allocations for Aisha's daily activities

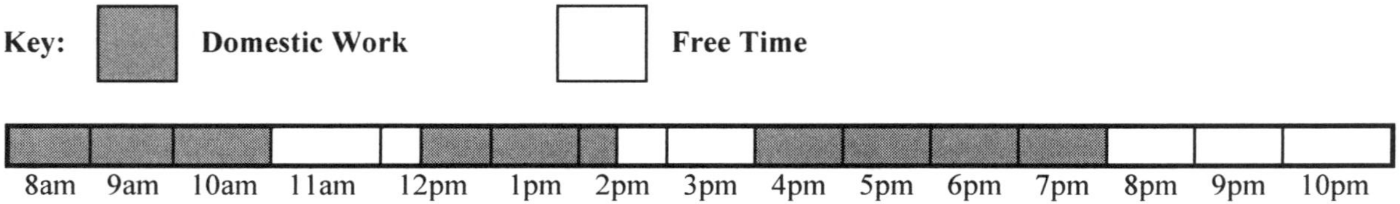

▪Makes breakfast ▪Dresses her son ▪Eats breakfast, clears and washes up ▪Makes the beds and tidies the rooms ▪Every other day she'll also make bread in this time slot	▪Free time - she may have a shower, pray or watch TV	▪Prepares lunch ▪Says her prayers ▪Eats, clears and washes up ▪Cleans the kitchen	▪Free time - she and her son may sleep ▪Maybe watch TV or visit family next door	▪Tidies up and clears rooms ▪Prepares evening meal ▪Eats, clears and washes up ▪Makes drinks (tea normally) for everyone ▪Washes and changes her son	▪Free time - watch TV, chat to family, spend time with her husband ▪Puts her son to bed ▪Shower and changes for bed ▪Says her prayers

Fig. 4. Time allocations for Meena's daily activities

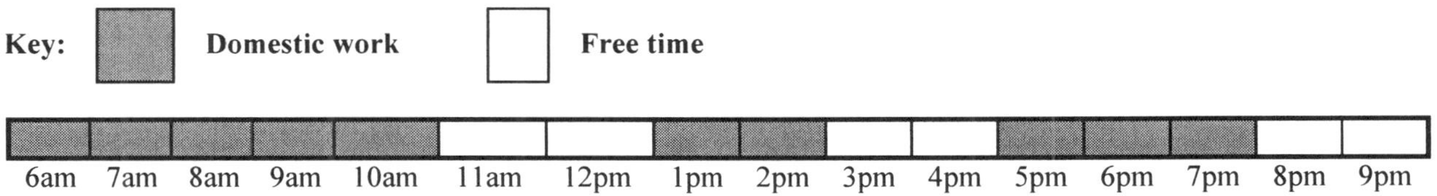

▪Prepares tea ▪Boils water to wash clothes (3 times a week if there's water) ▪Gets children up and dressed ▪Prepares breakfast ▪Tidies the house ▪Washes the clothes	▪Has a rest - she may lie down or knit cushion covers or sew	▪Prepares lunch ▪Washes up ▪Cleans the house	▪Rests - may sleep but usually sews	▪Prepares dinner ▪Washes up and clears away	▪Rests and watches TV then goes to bed

and should not only be viewed as housework but as a range of women's activities which are culturally valuable and involving women's time, energy and organizational skills.

Their daily activities may be considered mundane and routine allowing no scope for intellectual stimulation which an educated woman like Aisha needs, and a wasted potential that the BRDP must notice.

Both suffer from a lack of regular social interaction with other women, not only because of distance from close family but also because there are no communal activities or societies that women can interact together in.

There is the issue of 'adult education' schemes arising for people like Meena who expressed interest in learning. Such facilities should be available to those who wish to use them. Also the issue of health deserves attention. Both women work in environments that at times threaten their health, when they bake *shrak*. The smoke and fumes emitted affect their respiratory systems and eyes causing irritation and at times illness.

4. OBTAINING MORE DATA ON WOMEN'S UNPAID WORK

Conventional labour force methodologies are not designed to accommodate the issues contemporary inquiries into women's work are concerned with (Massiah 1990). National surveys and censuses fail to reflect the range of activities in which women engage. In order to respond to the labour force potential of women, policy-makers need to know what work women are doing in market and household spheres.

Arguably it is a revised definition of economic activity that is the first step needed towards improving measurement of women's work outside the 'production boundary'. A number of difficulties arise when trying to account for household production because a decision has to be made of what should and should not be counted. Should the economic value of a) all work done in the household or b) only economic services that correspond to market activities, such as food processing, be counted? Because household work merges with everyday living, systematic observation of the time spent on household activities is required (Evans 1992).

Massiah (1990) suggests techniques used by a Caribbean women's project. Using the women's definition of work, 'activity status' categories can be devised which are comparable with conventional categories and are more precise about what women are actually doing. Measurement of these activities are then possible in three ways: first to estimate the volume of work activity of the women, second to estimate the amount of time women spend in different types of work activity providing an estimate of input hours, and third to estimate the monetary value of the different types of work.

Time-Use surveys offer a viable way to collect accurate data on women's work. They can be used to measure the flow of labour between productive work, household chores and leisure on a daily, weekly and even seasonal basis. They identify primary, secondary and tertiary occupations, include productive activities on the borderline between economic and non-economic categories and permit analysis of gender and age specialization and trade-offs within the household unit (Evans 1992). There are methodological problems with Time-Use surveys, but they are resolved as such research methods develop. Aggregate information like the Baseline Survey can then be supplemented with more detailed regional or community studies of economic activity and time allocations.

Evans (1992) also suggests that women's expenditure patterns should be monitored to enhance the kind of data already available to policy-makers. It may be possible to determine the extent to which data on women's incomes reflect accurately their full contribution to the welfare of their households and families. Women make significant contributions to household income from their activities which aren't necessarily in monetary terms. Evans also states that women have been shown to allocate more of their income than men to the basic survival needs of their children. Therefore from a policy point of view, realistic efforts to raise the general standards of living of women have to deal with both the earning and the expenditure side of the income issue. Also the concept of income must be expanded.

Khoury and Moghadam (1995) suggest some possible policy measures that the BRDP might consider to promote female participation in the labour force: extensive media campaigns concentrating on the importance of women's roles in economic development; involvement of women in the decision-making framework; misinterpretations of Islam have hampered women's abilities to be active members of society and so an effort should be made to bring cultural traditions into line with Islam; more secondary and tertiary educational opportunities available to women who should be encouraged, via the media, to not remain content with a low level of education; an intensive campaign to promote family planning practice among married women giving them more control over their fertility and allowing them to remain in work; a system of part-time employment for those women with household responsibilities who cannot take full-time work but would like to work anyway; a more complete system of data compilation regarding women's activities; and programmes that will create employment for both women and men.

5. DISCUSSION AND CONCLUSIONS

The whole issue of women's work, paid and unpaid, is fraught with problems. To bring about change in the issue of underreporting, underestimating and undervaluing of women's work, there will have to be major transformations such as those suggested above. Improving women's opportunities can foster growth. Education improves a country's stock of human capital as the better educated spur a country's output and growth. Educating women reduces fertility (*The Economist* 1995) which increases the potential wages that women can command, raising the value of women's time. The most effective communicators of women's needs are the women themselves and they must be paid attention to if assistance is to be relevant and non-destructive. Women don't want special treatment, they want to be recognized for their contribution to the economy and be rewarded for it. It is a mistake to think that economic development will be successful without them.

Many issues arose from the investigation of Aisha and Meena's daily activities. Boredom was one of the main ones. Satisfying a woman's psychological needs is not given the importance it so deserves and yet it is crucial to the improvement of women's well-being. Women should be given every opportunity to develop their abilities in and, especially, out of the home. Allowing women to work need not be something men should feel threatened by because women are not out to 'get' men. Women want equality and a sense of fair play. Collins (1985) states quite strongly that men still have a fear that independent earnings will lead the wife to be more conscious of her personality and that therefore she will refuse to accept the demands she was subjected to before and reject an empty and indolent life at home.

The whole mechanism of co-operation has not developed very far because there are still planners and policy-makers who remain either unaware or choose to ignore the fact that there is a need to deal specifically with women's roles. As stated by Fetherolf Loufti (1980), women in general are already overburdened so it is not *just* a question of providing new employment opportunities or raising education and training standards - although urgent - nor is it a question of the law recognizing women as equal. It is the fundamental issue of women and men participating in the direction, the process and the benefits of development for reasons of justice and of development. It is because of this argument that the BRDP should make every effort possible to learn all they can about the women in its area range. This will allow policy decisions and development plans to fully account for women and reward them for their vital contributions.

ACKNOWLEDGEMENTS

My heartfelt thanks to Dr. Elizabeth Oughton for offering inspiration and guidance in times when I felt there was no hope, to Roderic Dutton of the Centre for Overseas Research and Development (CORD) and to Mohammed Shahbaz of BRDP for facilitating my research in Jordan, to Muna Salem Shiefat for her time and patience in being my translator, to Sandra Kelly for her companionship throughout the whole study and finally I would like to offer special thanks to 'Aisha' and 'Meena' for welcoming me and accommodating my intrusion into their lives.

REFERENCES

Badia Research and Development Programme (1994). *Baseline Survey*, Stored on SPSS at the Research Centre in Safawi, Jordan.

Chambers, Robert (1992) Rural Appraisal: Rapid, Relaxed and Participatory, Discussion Paper 311 Institute of Development Studies, Sussex University.

Collins, Sarah (1985) Women In Contemporary Arab Society as Reflected in the Writings of Nawal El-Saadawi, Dissertation, Middle East Centre, Durham University.

Evans, Alison (1992) Statistics. In Lise Ostergaard, *Gender and Development: a Practical Guide*, Routledge. pp.11-40.

Fetherolf Loufti, M. (1980) *Rural Women: Unequal Partners in Development.* International Labour Organization.

ILO (International Labour Office) (1994) *Women And Work In The Arab Region: A Perspective.* ILO, Geneva, Arab Labour Organization, Cairo and the Centre of Arab Women for Training and Research, Tunis.

Khoury, Nabil and Moghadam, V.M. (1995) *Gender and Development in the Arab World.* Zed Books Ltd. and United Nations University Press.

Massiah, Joycelin (1990) Defining women's work in the Commonwealth Caribbean. In Irene Tinker, *Persistent Inequalities: Women And World Development.* Oxford University Press.

Moghadam, V. (1991) Gender and 'Culture' in the Middle East. In Ruth Pearson (1992), Gender Matters In Development. In Tim Allen and Alan Thomas, *Poverty and Development in the 1990s.* Oxford University Press.

Papps, Ivy (1992) Women, work and well-being in the Middle East: an outline of the relevant literature. *The Journal of Development Studies*, 28(4):595-615, Frank Cass Publishers, London.

Pearson, Ruth (1992) Gender Matters In Development. In Tim Allen and Alan Thomas, *Poverty and Development in the 1990s.* Oxford University Press.

The Economist (1995) *Closing The Gap: Why Improving Women's Opportunities Makes Economic Sense,* 2 September, 112.

Wintersgill, Owen (1994) Economic Differences between Villages in the Badia Region of Jordan. Dissertation, CORD, Geography Department, Durham University.

CHAPTER 20

Preliminary Findings of a Health and Education Survey

CYNTHIA BRANDENBURG
(School of Oriental and African Studies, University of London)

1. INTRODUCTION

Following research work in development anthropology during the summer of 1963 (Ch. 18), I returned to Jordan in June 1994 with a colleague, Carole Rehfisch, to participate in a joint health and education rapid appraisal in which my colleague was to lead the former component and I the latter. Jointly, we conducted some 18 interviews between 13 June 1994 and 22 June 1994. At this point, due to unforeseen circumstances, my colleague terminated her work with the programme. At the request of the then Programme Director in Jordan, I agreed to proceed with the work that we had begun, adding the health component to my former agenda. My intention was to work until September but like my colleague before me, I was forced to terminate my work prior to its completion (on 16 August 1994).

1.1. Objectives

Development projects in developing countries have historically been approached from a top/down perspective with funding allocated to projects which are designed by executive administrators and so-called development experts with little or no consideration being given to the needs, aspirations and practices of the said 'beneficiaries' of such projects. Asking the individuals to be 'developed' and whose socio-economic circumstances are to be 'improved' about their own views and willingness to become active rather than passive participants in their continuing development seemed to be an unnecessary and costly exercise. Quite to the contrary, a disregard for the importance of the individual's contribution and participation in development programmes has proven to be the most costly mistake with failed project after project testing to the need for joint co-operation and communication between development planners and local inhabitants.

For this reason, the joint health and education appraisal in the Badia focused on the attitudes of the local inhabitants towards past, current, and future provision of these services and is biased towards those individuals participating in the reception rather than in the provision of services (although these are clearly not static, bounded categories). It may best be placed and considered as a complementary rapid appraisal and compared with other studies whose bias is towards the service provider[1]. However, such short-term

[1] Concurrent with the period of my own fieldwork during the summer of 1994, officials from the Ministry of Health conducted a rapid appraisal in the same region (of which I was presented a draft summary of their findings) which was based on questionnaires distributed to health care professionals and patients at 5 primary

appraisals must be viewed as indicators of contesting views to be further explored and not as final dictums.

1.2. Methodology

Field visits were conducted on a daily basis when adequate resources (i.e. transportation, driver and Jordanian counterpart) were available to the researcher. Between four and six interviews per day were conducted and included as many representative members of the community as possible, including household members (men, women and children), health care and education service providers who resided both within and outside the region, and village administrative officials and employees. While the sampling of individual villages was so small as to be statistically insignificant, I was able to cover 30 out of 35 villages in the region and the work provided a sound basis for gaining an understanding of some of the principal concerns of the inhabitants across the region. The concentration of interviews occurred within households (with as many members of the household included in the interview as possible), followed by interviews with health care and education staff, and then by interviews with village administrators.

The interviews were as unstructured as possible while covering the relevant areas as outlined in the proposal (below). Structured interviews may be used to obtain certain, rudimentary information for statistical analysis but this method alone is an ineffective means of obtaining the kind of information on which development policy may be established and implemented. The focus must rather be on unrestricted dialogue; including the discussion of other areas which influence attitudes towards health and education. Denying the inhabitants the opportunity *fully* to participate in the decision-making process only results in a top/down implementation which the inhabitants will justifiably reject.

The research topics on which the dialogue was focused are as follows:

1.2.1. Health component:

1. Attitudes of local inhabitants towards:
   a. National Schemes: Immunization Programme.
   b. Traditional medicine.
2. Reasons for little uptake of services available in health centres.

1.2.2. Education component:

1. Reasons for children leaving school prior to completion of secondary education.
2. Reasons for children not enrolling or attending school.
3. Uptake of adult education programmes.

---

health centres in the project area and 1 comprehensive centre servicing the region. Given the opportunity to work together, mutual benefit could have been gained.

2. PRELIMINARY FINDINGS

2.1. Uptake of Health Services

The people of the area are by no means to be viewed as part of a homogeneous community with like interests and needs. Attitudes towards health care choices are influenced by a multitude of factors including one's sex, age, family, geographical location, religion, education, employment, and exposure to urban areas. These factors must be taken into consideration in developing health and education services tailored to meet individual communities' needs.

2.1.1. Prejudices retarding development

One of the key concerns identified was the seeming opposition between the position of health care providers as possessors of 'expert knowledge' and the perceived ignorance and inability of the people to make appropriate health care choices for themselves. From the point of view of many of the inhabitants, health care professionals are dictatorial outsiders who do not understand the way of the people and who instil fear or mistrust.

Some inhabitants refused initially to talk about certain alternative medicine practices fearing that I would reprimand them as their doctors had done. Women who did not wish to be examined by male doctors were often viewed as ignorant and such refusal was seen to be irrational. No female doctors were employed in the region when I conducted my research, and this was one of the primary factors for greater uptake of health care services by women outside the region.

Health care provision should be an empowering process rather than a disempowering process whereby control over one's own body is fully relegated to someone else. Unless the disparity in understanding is redressed, 'development' in terms of health care choices will remain retarded. A true development process would be facilitated with greater communication, collaboration and co-operation between service providers and beneficiaries. This may not be achieved through health care education programmes 'targeted' only at the local inhabitants, when equal targeting needs to be aimed at the health care professionals themselves.

2.1.2. Women and children as principal users of health care services

Women and children were said by both health care providers and the inhabitants themselves as being the principal users of health care services - be they services on a 'Western' model or services on a 'traditional' model or somewhere in between these two abstractions. As such, it would seem logical that their needs should take precedence in a delivery system. Clearly, as was above alluded, this is not the case. Furthermore, the health care that is available locally is socially alienating, difficult to get to, and inadequate in terms of facilities and services. The only comprehensive centre is located on the north-western border of the project area in Umm al-Quttayn. However, the comprehensive centre most used by the inhabitants of the project area actually lies outside the project area in Muthalath Sabhah, which was seen as better equipped, staffed, and more easily reached using public transportation. The same doctors from this centre work part-time in Umm al-Quttayn but the people said that they were more likely to be found at Muthalath Sabhah.

Women (with exceptions) have to rely on a cursory examination by a male doctor who makes his diagnosis based on answers by the patient to diagnostic questioning and less often by actual physical examination. Women and men alike complained of the lack of responsiveness to women's particular needs. The professional midwife is slowly replacing the role of the traditional midwife and there is heated debate over this process. On one occasion when I was discussing the availability of midwifery services (and used the term for traditional midwife) with one of the local doctors he retorted vehemently that such people are a remnant of past backwardness and that only the professional midwife has a place in 'modern society'. Birth, a marked and celebrated social event, is becoming an event isolated from family and friends occurring in a hostile environment rather than in the familiarity and comfort of the home and treated as an illness instead of a natural process. Only the experts are qualified to take control of the process of birth. While many women continue to use a traditional midwife in preference to hospital birth, younger women particularly are becoming indoctrinated into the ideology that 'West is best' and fear their own death or that of their child if they opt for home birth.

Clearly, there are cases where hospital birth is essential for the well-being of mother and child but women's life choices with regard to health care should be respected and observed. Facilities could be made less hostile and less restrictive for 'normal' births: i.e., allow greater freedom of movement for the women during delivery as opposed to being left in an uncomfortable, undignified posture for hours on end; have birthing rooms which are less 'clinical' and allow for the presence and perhaps participation of a traditional midwife or a close female relative. Traditional birthing practice is empowering in contrast to current health centre/hospital birthing practice. If institutionalized health care service is to 'develop', it needs to incorporate means within its delivery system that empower the patient. Both forms of practice are viable and necessary, and development plans with regard to health care provision should focus on the complementarity of these practices - trying to close the gap between them rather than trying to condemn and ultimately annihilate one or the other.

Paediatric services are limited in the programme region, with most children being seen by general practitioners who may then refer children to the paediatrician who is available on a part-time basis at one of the comprehensive centres. Children often have only blackened shards remaining in place of former healthy teeth and gums. With the heavy reliance on consumer luxury items such as tinned colas, sweets, and other sugar-laden items, dental hygiene services should become a priority in the overall health care provision to the region. At the time of fieldwork, I visited the beautiful, newly furnished dental room at the Safawi primary centre - the only thing missing was the dentist himself. As I found consistently throughout the programme area, health care professionals from urban areas look upon the Badia as a kind of Jordanian Siberia - no one wants to work there. If an opportunity arises to move to private practice or to 'return to civilization' the professional moves on. The people of the area are by no means oblivious to this divide. They would much prefer to have members from their own communities involved in the delivery of services.

Again, it is the people themselves who lose out. Some families simply will not take their children the great distance to the next closest dental facility at one of the comprehensive centres for treatment of a toothache - no matter how painful. Sadly, development (in terms of progressive change) often means an increased hardship or a worsened condition unless steps are taken to provide adequate means to accommodate such changes. Adults in their 50s and 60s often had their own teeth whereas many of the children that I met will not have their own teeth when they are in their teens.

2.1.3. The Vaccination Programme - a success story?

None of the inhabitants questioned about the vaccination programme faulted it in any way. All said that they approved of it and that their children were immunized at school. Only the benefits of immunization had been explained to them by either health care professionals or education professionals. They all saw it as mandatory and beneficial with no possibility that anything might go wrong. In contrast, more concerns were expressed over the unknown substances present in pills or other manufactured medicines than an injection of an unknown substance. Virtually all health care professionals queried on this subject confirmed the success story believing that all children are being immunized. However, the success of the vaccination programme presupposes a number of factors:

(1) All children of school age in the region attend school on a regular basis;
(2) Those children who do not attend school at all (or who do not attend regularly) are brought into a local health care centre for immunization; and
(3) These same school-leavers are returned to the health centre for boosters as scheduled.

Clearly, some children are falling through the cracks in the system; often the same children that are falling through the cracks in the education system. I understand that there was a mobile immunization programme in operation in the recent past actually taking the service to the people - particularly to tent-dwellers who were less likely to have children in the school system on a regular basis, if at all. The service was discontinued due to lack of perceived need because of greater numbers of children attending school.

I see the vaccination programme as a success, with the caveat that improved educational opportunities will create an increased demand for it and, at the same time, improve the immunization programme record by aiding those children currently falling through the cracks. I would also hope to see a more balanced information dissemination by health care providers with regard to the risks as well as the benefits of preventative and curative treatments.

2.1.4. Attitudes towards 'traditional' medicine - fire, honey and magic

The subject of Arabic Medicine is such a vast and complicated one that I willingly acknowledge my inability to do justice to the magnitude of this fascinating encyclopaedia of local knowledge. I can only be amused at my naivety in the summer of 1993 when I thought that Arabic Medicine principally comprised what 'we' might term, herbal medicine. Now, I am convinced that even in the subsequent summer - when my remit included an investigation into attitudes towards the use of Arabic Medicine - I have only scraped the surface. In addition to the magnitude of the subject with its roots in Islamic and pre-Islamic practice there is the further contemporary complication of the practice being discouraged in modern Jordan and thus veiled in secrecy and denial.

It is impossible to provide a generalization for the attitudes by local inhabitants towards Arabic Medicine as beliefs, practices and evaluations differed widely. However, it is practised in one form or another by the majority of people that I spoke with, including those who denied using it. As such, it cannot merely be dismissed as a remnant of 'the past'. An indicator of the belief in and use of such practices was often the most emphatic denial, often a result of fear of reprimand or reproof.

No one that I spoke with suggested that Arabic Medicine was 'better' than 'Western medicine'. Rather, Arabic Medicine was viewed as offering greater opportunities for choice - as being appropriate in certain circumstances and not necessarily in others. It might be used on its own, in conjunction with 'Western medicine' or in alternation with 'Western medicine'. The greatest problem attributed to its use was the decline in 'knowledgeable' practitioners. This was directly linked by the inhabitants to the privileged position of Western medicine.

While both men and women may be practitioners of Arabic Medicine, it seemed to me that women have a privileged position as practitioners and are more numerous. This may be linked to the fact that women may offer their services to women, men, and children whereas male practitioners tend to be restricted to men's afflictions. Women practitioners as professionals with external clients were female elders of the community. Women in their 30s and 40s often had a wealth of knowledge but none admitted to providing services to others. To contradict an elder about a particular prescription is not tolerated though sometimes attempted. One such practitioner flung the mortar and pestle to the ground spilling all of the contents of the preparation she was making for me when one such upstart tried to contradict her on the requisite ingredients.

Again, there is an existing complementarity which exists between notions of 'Western medicine' and 'Arabic Medicine'. If there is to be any effectual development of the region, Arabic Medicine should be given its respected place and not subjugated to levels which will only succeed in reducing the opportunities for informed choice. 'Western' and 'Arabic' medicines are not mutually exclusive options, as has been so clearly recognized by the inhabitants themselves. But will the 'experts' gain such insights, or will the dominance of the ideology of modernization continue unabated?

2.2. Uptake of Education Services

2.2.1. Équal access? - disparity between the sexes

After raising the question of April of 1994 as to the usefulness of education for the people of the region, this year I visited local schools, interviewed local teachers from within the community and outside it, talked to male and female school-aged children of all ages about their experiences, conferred with parents about their children's experiences and their own experiences, and spoke with past or present male and female attendees of adult education programmes in the region.

After all this, I have to conclude that institutionalized education in its current form in the Badia is of little use to the majority of people 'eligible' for such education. How can I possibly suggest this when this same majority perceived institutionalized education to be a good thing? To begin with, not everyone has equal access to educational opportunities.

If you are a boy, you are more likely than a girl to be encouraged by your parents (I should say encouraged by your mother and given permission by your father) to continue to the highest level of secondary school - even if you have to travel great distances to do so. There are exceptive cases for girls but these currently represent a minority. If you have been born a girl and the only school in your village is co-educational you will almost certainly be removed from this threatening environment before the onset of puberty. Sex is not the only divider, for where would sexual differentiation be without local politics! If your particular village does not have a school, you may not be allowed to go school in a neighbouring village - even if you are a boy - if that village is inhabited by a 'rival' tribal section[2]. The village elders demand to know why this rival village should have a school when they do not. Instead, you may have to travel a great distance to a village with established ties between your own - your sisters in such a situation will remain at home to protect them from members of the other tribal section who may even have a predisposition to treat you with less respect than girls from their own community.

What is the benefit of education when it often means sitting in a one-room school building with teachers who sometimes show up and sometimes do not, teachers who are consistently late because of the long distances they may have to travel due to lack of suitable staff accommodation, teachers who (like many of the doctors) don't really want to be in the Badia in the first place and are not interested in the success or failure of their students whose condition they may also deprecate? Your teacher may even be from the same region (although this is much less likely) but he or she may still be pressed by the constraints of insufficient facilities and resources. What kind of quality of education can you have when classes are often combined together with vast age difference and abilities and where each class division may only receive a quarter-of-an-hour 'teaching time' per lesson per day? This situation was confirmed by both parents and teachers. As one teacher sighed resignedly, 'this is a school in name only'.

Clearly, it is impossible to have segregated, secondary (or even primary) schools in every village and the inhabitants for the most part are generally aware of this fact - although many will continue to insist that there should be a school in every village. However, the inhabitants themselves proposed their own solution to their problem with the existing school system: a segregated bus service. Bring the students to the schools instead of attempting to bring the schools to the students. In this way, a concentration of available resources could be focused on the existing, segregated secondary schools (which include primary education). Without such a system, girls will continue to be denied access to formalized education and a large number of boys will continue to leave school before they complete secondary education. If the school system is improved in this way, and with the incentive of the new university in Mafraq, greater numbers of young people from the region will have the foundation needed to pursue higher education. They in turn will be more likely to reinvest their talents and skills in the region.

---

[2] The use of the term 'rival' is my own construction and I do not mean that all members of the different tribal section are necessarily in a rival position all the time. However, a tribal idiom was invoked by the inhabitants in such circumstances to explain their reasons for refusing to send their children to the neighbouring village. Such rival splits also occurred between leading families of the same tribal section with a subsequent village split (east/west or north/south). It is also worth mentioning that there is an increasing trend to deny that any such disparities exist but to ignore them would, in my view, result in development plans that could be destined to fail before they even got off the ground.

3. CONCLUSIONS - WAY FORWARD

3.1. Communications

A publication outlining the aims of and the opportunities for individual involvement in the Programme should be composed and distributed to inhabitants. It should include contact names and addresses so that people know who they may put their concerns to and where they may do so. This publication should be in straightforward language in Arabic and should include as many visual images to support the information being conveyed as is possible. Regular updates of Programme activities/successes should also be communicated in printed form. While printed material, on its own, is an inadequate means of communicating with the people of the area, it certainly represents a step in the right direction and could be effected expediently. The inhabitants would then see for themselves that their concerns are going to be taken seriously and that they are valued participants of an ongoing process.

3.2. Reasons for Little Uptake of Health Services within the Region

The health services in the region are widely varied in terms of staffing, facilities, treatment provided, specialist services etc. but the following gives an indication of the primary complaints discussed:

* Lack of availability of women doctors in the region - women will even refuse IM (intra-muscular) injections by a male doctor.
* Lack of provision of quality health services for women.
* Insufficient opening hours with limited access to doctor (who may or not be there at scheduled times) in the smaller, primary centres.
* Limited facilities, treatment, and medicines at primary centres.
* Insufficient examination time/methodology for both men and women.
* Preference for drugs manufactured outside Jordan (available outside region), as these were viewed as more 'effective'.
* Belief that you pay for what you get - private practice was often preferred because staff, facilities, and treatment were said to be of better quality.
* Insufficient public transport for access to health centres (primary and comprehensive).
* Availability of alternative forms of medicine, i.e. Arabic Medicine.

3.3. National Schemes: Vaccination Programme

The national immunization programme has been a resounding success in terms of reaching the majority of the population. Improvements to enable greater access to education will also increase the success rate of this programme. Unlike some of the health care professionals in the area who stated that all the inhabitants come to the health centres now, there is still a need to provide services for the children of tent-dwellers. A mobile 'well-child' service would have positive effects on this increasingly marginalized community.

I would like to see patients treated as equals capable of making informed decisions rather than as ignorant clients who must be shielded from too much information.

3.4. Attitudes Towards Arabic Medicine

Categories of medical practice are not the dichotomous oppositions that they may at first appear. For while one doctor may harshly scold a patient for using one form of alternative medicine, another doctor may 'prescribe' a herbal remedy or at least suggest that this might be tried for minor ailments. More needs to be learned of existing and former practices before making an all-encompassing denouncement of the efficacy of such treatment.

3.5. Three Points for Immediate Consideration

Teeth: The standard of dental care seems to be getting worse, probably as a result of easier access to sweet foods and drinks. There may be a need for an awareness campaign directed to the people of the Badia. Some of the problems might be reduced simply by the use of more fluoride toothpaste.

Special female needs: As with education considerations, the specific needs of the female members of the society should be taken into account. If there is to be a development for the Badia it should be a development for all and not merely for the few who are currently economically and politically advantaged.

Community participation: In general terms, there needs to be much greater flow of information between parents and teachers with a focus on communities solving their own problems through joint participation in developing educational programmes. A top/down linear educational delivery system is less likely to be effective than one that includes the individuals within the system in its delivery: learning from these individuals as well as teaching them.

3.6. Education Uptake

Itemizing the local inhabitants' concerns regarding the provision of education services to the area may only serve to trivialize the complex reasons for the little uptake of education services. But if the list of points below is accepted as a shorthand form only and not a detailed analysis then it may serve some use.

3.6.1. Reasons for children of various ages and both sexes leaving school before completion of secondary education.

Boys

* Family income dependent on labour contribution of sons. If education services in the area involve great hardship for access and/or are of questionable merit, these factors often provide additional incentives to keep sons at home.
* Hardship factors:
  - Inadequate transportation;
  - Cost (in relation to return on investment with reference to quality of services and future employment potential).
* Educational facilities/services:
  - Combining of classes because of insufficient space;
  - Teachers lack of interest in region and inadequate training: this region was often referred to as the 'training centre' for teachers where the students were viewed as expendable victims of this training;

- Lack of educational materials and equipment: lab facilities/equipment; library materials; and general teaching aids;
- School day inconsistent throughout region: due to transportation difficulties, teachers leaving early and arriving late, etc.; and
- Insufficient provision of teacher/staff accommodation: often provisions are provided by the community themselves and so may be very rudimentary; teachers generally complained that there were not enough incentives in terms of comfortable accommodation, salary etc. to make their tenure in the region a pleasant one.

* Lack of encouragement by parents; parents of school-leavers often lacked an adequate understanding of how to encourage their children in their education. Those (fathers) that were said by teachers or themselves to encourage their children were generally educated themselves: either through the school system or more often through voluntary educational opportunities associated with military service. This supports the case for extending adult education opportunities in the region. While mothers often seemed to want more in terms of education for their children (both boys and girls), they also were least equipped to provide practical encouragement in the form of help with homework, timetabling schoolwork, etc.
* Failure of student: teachers often referred to the significant numbers of students (both males and females) who were allowed to remain within the school system although they consistently failed their year-end exams. They felt it was better to keep the child within the system in the hope that they might benefit in some way. Failure is an often invisible problem within the school system - concealed by all parties to it. It may result in the termination of the child's education or, perhaps more insidious, his or her continuation in the system with the result that this child may miss out on the vocational education they might be receiving at home. Ignoring the problem appears to be more popular in the Badia than tackling it head-on. The consequences of such ignorance could well be the creation of economically dependent individuals rather than income-generating ones.

Girls:

* Limited access to education because of sex differential.
* Labour input required at home. This is linked to lack of access to continuing education and training period at home for future marriage.
* Precedence of early marriage for girls in the region. Many girls are said by the inhabitants to marry only after the age of 14 or 15, but many of the inhabitants conceal the fact that much younger girls than this are marrying for fear of recrimination by 'urbanized experts' who, the inhabitants clearly understand condemn such an event. Many mothers that I spoke with would actually prefer later marriage for their daughters because of the hardships they themselves incurred: pregnancy rate, labour and responsibility of marriage, lack of formalized educational opportunities, etc. Early marriage is also influenced by the lack of educational opportunities in the region because of planners' refusal to acknowledge and respond to the sex differential. Instead, the inhabitants are just labelled, 'ignorant' because of their refusal to allow co-education beyond a certain age. Allowing co-education up to a certain point (which varies from family to family, village to village, etc.) actually represents a compromise on the part of the inhabitants because most prefer segregation from year one.

* Lack of adequate transportation played a more crucial determinant in access by girls to education than it did for boys. Investment in education for boys was also perceived by most fathers as having a greater potential return on investment than the same investment in a girl's education.
* Lack of educational facilities/services.
* Lack of encouragement by parents.
* Failure of student. However, girls were not perceived by either themselves, parents or teachers as having a higher failure rate than boys.

3.6.2. Reason for children not enrolling or attending school

Study of this problem was incomplete, as few tent-dwellers were interviewed and they tended to be those in close proximity to the villages. It is the more remote group which would most likely provide indicators for the problems associated with children who simply never enrol in school and have no access to formalized education whatsoever.

In the villages, few actually stated that lack of enrolment was a problem. If it was a problem, it was a problem for a very few - the few that remain invisible. Tent-dwelling groups are referred to within anthropological literature on the region and often by government officials and development planners to be so few as to be insignificant. Is it any wonder, therefore, that such groups are seen as insignificant in terms of need for the delivery of health care and education services?

My contention is that while such populations may represent a low percentage nation-wide, they represent a significant number of the population locally. I would also suggest that the baseline survey does not give an adequate representation of the numbers of people who are classed as tent-dwellers which is in itself an elusive category and tells us little of people's actual residence/vocational practices.

3.6.3. Uptake of adult education services

If you ask an adult male or female inhabitant whether they think that adult education is beneficial or would be beneficial to them, they are most likely to respond in the affirmative. However, implementation of such programmes has proved less than effective, with frequent termination of such programmes occurring within a short space of time. Principal reasons given for low uptake of services were as follows:

* Elderly men and women join for the social interaction and then lose interest. Problem with the disparity in age groups studying together.
* Young women with children have difficulties in attending.
* Lack or sufficient numbers to justify an adult education class in area, keeping in mind that women and men must be segregated.
* Lack of teachers willing to run adult education classes. The running of adult education classes is taken up on a voluntary basis on the part of the headmaster/headmistress of the local school. A part-time adult education teacher takes home a salary of approximately JD 40 per month which for some is hardly worth the time or effort.
* Adult education books are not tailored to a particular region's interests and so the students may find it difficult to relate the benefits of such learning to their particular situation and environment.

3.7. Four Points for Immediate Consideration

Educational cluster environments: Many of the interrelated problems preventing access to education could be overcome with a concentration of educational resources in cluster environments with regular transportation of students to these centres rather than spreading the resources too diffusely throughout the region. This proposition is one made by the inhabitants themselves across the region and I believe it to be the most sensitive way of providing greater numbers of individuals with access to education, which would prepare them to reinvest their labour, talent and skills back into the community.

Special female needs: As with health considerations, the specific needs of the female members of the society should be taken into account and not simply marginalized. Development for the Badia should be a development for all and not merely for the few who are currently economically and politically advantaged.

Remedial education: The currently tolerated failure rate of students within the school system should be treated by remedial education and more interaction between parents and teachers.

Community participation: In general, there needs to be much greater flow of information between parents and teachers with a focus on communities solving their own problems through joint-participation in developing educational programmes. A top/down, linear educational delivery system is less likely to be effective than one that includes the individuals within the system in its delivery: learning from these individuals as well as teaching them.

CHAPTER 21

Improvement of the Health Services

REYAD AL-AKOUR
(*Ministry of Health*)

1. HEALTH STATUS AND HEALTH SERVICES

1.1. In the Kingdom

Since its establishment, the Ministry of Health in Jordan has been committed to the principle that health is the right of every citizen, and despite its limited available resources it is trying its best to improve the health services, both quantitatively and qualitatively.

The infant mortality rate (IMR) is considered to be a sensitive health and socio-economic indicator in any society, and by 1993 it had declined in Jordan from a previously high level to 34.6 per thousand live babies. Mortality due to communicable diseases had declined to 0.3 per cent of all causes of mortality. Malaria had been eradicated for more than 25 years. The immunization ratio against the communicable diseases of polio, pertussis and tetanus had increased remarkably to reach 98 per cent in 1993, for pregnant tetanus 53.5 per cent and for measles 87 per cent. Improved sanitation has also become a major objective.

Manpower for health has increased greatly: there were 7,003 physicians in 1993 at a ratio of 17.4 per thousand inhabitants, and a similar growth for other health personnel (except for midwives, where there is a severe shortage). Jordan also had 63 hospitals, with a capacity of 6,398 beds, and with a ratio of 1.6 beds per thousand inhabitants.

1.2. In Mafraq Governorate

Primary and secondary health care are provided through 64 health care centres, which are connected and accessible by a net of roads, and distributed all over the Mafraq Governorate, with a distance of about 10km between them. Mafraq general hospital has 120 beds and there is an emergency hospital in al-Ruwayshid. As for manpower in the public sector, general practitioners number 70, specialists 29, and dentists 12. The private sector also contributes to health care in the area, with 15 general practitioners, clinics, one specialist and a 40-bed pulmonary hospital.

Manpower of other health personnel includes 42 registered nurses, 32 midwives, 117 assistant nurses, 192 practical nurses, 2 pharmacists, 40 assistant pharmacists, 3 laboratory technicians, 21 assistant laboratory technicians and 13 x-ray technicians.

Thus there is one physician per 650 inhabitants, one registered nurse per 4,000 and one midwife per 5,300.

1.3. In the Jordan Badia Programme Area

Health services in the Programme Area are provided by the comprehensive health centre (CHC) at Umm al-Quttayn which is open for 24 hours, 5 primary health centres (PHC) at Dayr al-Kahf, Safawi, ar-Rifa'iyyat, al-Mukayfita and al-Ashrafiyya, and 8 peripheral clinics at al-Bishriyya, Hulaywat al-Masariha, Hamra Suhaym, al-Manara, Abu al-Farth, al-Jubaya, Tall ar-Rimah and Dayr al-Qinn.

The area is suffering from a severe shortage of health manpower, especially of specialists (see Tables 1 and 2). It is served by just 10 physicians (one family, 9 general), one per 1600 inhabitants, much less than in Mafraq Governorate and less still than in Jordan as a whole. There is only one registered nurse in the Programme Area and one midwife, very much less than the regional and national averages.

2. THE COMMUNITY'S HEALTH NEEDS AND APPROACHES TO IMPROVE HEALTH CARE SERVICES IN THE PROGRAMME AREA

2.1. Objectives

The general objective of this study is to accelerate development and to improve the living and health conditions of the population in the Programme Area. Specific objectives are:

- assessment of the health status of citizens in terms of coverage, accessibility, acceptability and utilization.
- study of the available health resources.
- suggesting the best alternatives to improve the health services and upgrade them in line with demographic changes.
- ensuring and encouraging community involvement and teamwork in the provision of health services delivery.

The study was undertaken by the author as principal researcher, Qais al-Saket as study moderator, Taissir Fardous as principal researcher's assistant, Bassam Khasawneh as field supervisor, with Abdel-Rahman Fataftah the Programme Director as study consultant. Data analysis was assisted by Mohammed al-Khatib and Nasser al-Satari.

2.2. Importance of the Study

Jordan has been committed to the principle since 1979, since the al-Ma'ata Declaration, that health is a right of every citizen, and has relied on the Primary Health Care approach and method to achieve the target of health for all, and make these services available within appropriate costs and adequate technology.

The Ministry of Health as the main health provider started to establish health centres for every population group with 2,000-3,000 inhabitants, except for remote and unprivileged areas where centres were established for fewer inhabitants. Our aim was to let the beneficiaries of these services engage in their evaluation, and to use the data generated in a holistic approach to the development of the Badia region.

Table 1. Main health personnel (available and needed) in the Jordan Badia Research and Development Programme area.

Description	Specialist physicians		General physicians		Dentists		Registered nurses		Registered midwives	
	available	needed	available	needed	available	needed	available	needed	available	needed
Safawi	-	3	2	3	1	1	1	1	1	2
Dayr al-Kahf	-	-	2	1	-	1	-	1	-	1
Umm al-Quttayn	1	3	2	2	1	1	-	2	-	2
ar-Rifa'iyyat	-	-	1	1	-	1	-	1	-	1
al-Mukayfita	-	-	1	1	-	1	-	1	-	1
al-Ashrafiyya	-	-	1	1	-	1	-	1	-	1
al-Bishriyya	-	-	-	1	-	-	-	-	-	-
Hulaywat al-Masariha	-	-	-	1	-	-	-	-	-	-
Hamra Suhaym	-	-	1	-	-	-	-	-	-	-
al-Manara	-	-	-	1	-	1	-	1	-	1
Abu al-Farth	-	-	-	1	-	-	-	-	-	-
al-Jubaya	-	-	-	1	-	-	-	-	-	-
Tall ar-Rimah	-	-	-	1	-	-	-	-	-	-
Dayr al-Qinn	-	-	-	1	-	-	-	-	-	-
TOTAL	1	6	10	16	2	7	1	8	1	9

Table 2. Auxiliary health personnel (available and needed) in the Jordan Badia Research and Development Programme area.

Description	Laboratory Technicians		Pharmacist		Pharmacist		Health Inspector		Assistant Nurse	
	available	needed	available	needed	available	needed	available	needed	available	needed
Safawi	-	1	-	1	1	1	1	1	1	2
Dayr al-Kahf	-	1	-	-	1	1	-	1	1	1
Umm al-Quttayn	-	1	-	1	2	2	1	1	7	-
ar-Rifa'iyyat	-	1	-	-	-	1	-	1	1	1
al-Mukayfita	-	1	-	-	-	1	-	1	1	-
al-Ashrafiyya	-	1	-	-	-	1	-	1	2	-
al-Bishriyya	-	-	-	-	-	-	-	-	-	1
Hulaywat al-Masariha	-	-	-	-	-	-	-	-	-	1
Hamra Suhaym	-	-	-	-	-	-	-	-	-	1
al-Manara	-	1	-	-	-	1	-	-	-	2
Abu al-Farth	-	-	-	-	-	-	-	-	-	1
al-Jubaya	-	-	-	-	-	-	-	-	-	1
Tall ar-Rimah	-	-	-	-	-	-	-	-	-	1
Dayr al-Qinn	-	-	-	-	-	-	-	-	-	1
TOTAL	-	7	-	2	4	8	2	6	13	13

2.3. Methodology

The researchers believed that the best approach to study these interrelated factors and variables was (a) by studying the population structure of the Programme Area and the geographical distribution of the target population in relation to the health centres serving the community; and (b) by studying the health needs of the population through:

- distribution of a questionnaire to the people treated in the CHC and PHCs during two months;
- distribution of another questionnaire to the maternity and child health (MCH) beneficiaries during two months;
- survey and crude age classification of the cases visiting the selected centres during two months, considering their diagnosis ;
- provision of statistical data and information regarding the available health services in terms of their adequacy, equity of distribution, and manpower development through 1993; and upon completion of the study recommend that all health care units be upgraded in line with local, regional and national standards.

2.4. Analysis and Discussion

The study team has reviewed the available and collected data in line with Ministerial instructions about evaluating health services provided, and has gathered information through field trips to the Programme Area and meetings with key officials. The study revealed the following findings with regard to the general evaluation of services in 4 PHCs.

Population: The sample reached 1,250 persons, 516 (41.3 per cent) males and 734 (58.7 per cent) females.

Labour status: 308 (24.6 per cent) were employees, 581 (46.5 per cent) householders, 178 (14.2 per cent) free labourers, 130 (10.4 per cent) unemployed and there were 54 (4.3 per cent) others.

Education: 620 (49.6 per cent) were illiterate, of whom 31 per cent were males and 69 per cent were females. Of the remaining 630, 198 had primary education, 249 intermediate, 135 secondary, 31 diploma and 18 bachelor. The study illustrated the urgent necessity of examining educational efforts in the region.

Fertility: The average number of siblings was 5.7, close to national average fertility.

Socio-economic status: 536 (45.0 per cent) of the total sample had a monthly income of less than JD100, 484 (38.7 per cent) had JD100-200, 10 (0.8 per cent) had JD200-300, 4 (0.3 per cent) had more than 300, and 190 (15.2 per cent) were unknown. It is evident that the majority of the sample are well below the absolute poverty index, and more attention should be given to this very important stratum of the Jordanian population and how to improve it.

Housing: 1,190 persons (95.2 per cent) were permanent dwellers in their own houses, 49 (3.9 per cent) had rented homes, and less than 1 per cent were nomadic inhabitants. Therefore almost all the population sampled were settled, highlighting the reduction in the unstable nomadic life.

Primary Health Centres: The main features of the services delivered are:

- 32,508 inhabitants are served, but that includes people living outside the area;
- no specialists are available in the Programme Area, even in the CHC, so critical cases have to be referred to the nearest hospital at a distance of 40-80km - there is an urgent

need for specialists in internal medicine, ophthalmology, dermatology, obstetrics and gynaecology, orthopaedics and dentistry;

- health centres are usually covered by one or two general practitioners;
- nursing services are mainly covered by one registered nurse (male) and 2-3 practical nurses, and there is a shortage of female nurses;
- midwifery services are lacking quantitatively and qualitatively. There is one midwife in Safawi with poor acceptance and accessibility, and one in the CHC. This problem has to be solved, as it is crucial for maternal and child care;
- there are no x-ray or laboratory technicians, so these essential services are not available in PHCs;
- there is a relatively satisfactory supply of other services such as public health supervisors, drivers, clerks and porters;
- follow-up services and outreach activities are lacking;
- while medical supplies and equipment are generally available in all centres, they are insufficient and the study indicated that medicines are not always available;
- government cars were insufficient in two of the four centres;
- safe water supply is available, but no reliable information was available on sanitation;
- there is not much activity in health education, and attendance is unsatisfactory, so there is an urgent need to revive this important aspect of health care provision and development;
- a minority had to walk a long way for care - 26 (3.7 per cent) more than 4km - but 930 (74.5 per cent) reported that they had access to transport;
- statistics of the prevalent diseases among the sampled population aged 15-49 indicate that 752 cases concerned the respiratory tract (most common in summer months), 306 the digestive system, 135 opthalmic diseases and 183 skin and orthopaedic ailments.

Maternity and Child Health Services: 147 females of child-bearing age completed questionnaires, 145 of whom visited the MCH centre one or more times during the year. Unfortunately, this accounts for only 4-5 per cent of women of child-bearing age. Attendance is poor because of staff shortage, traditional and cultural factors, and local ignorance of services. Questionnaire data revealed that 72.1 per cent attended for treatment, 53.1 per cent for immunization, 71.4 per cent for examination during pregnancy, 83.0 per cent for baby care and 87.7 per cent for family planning.

- only 19 (12.9 per cent) of the sampled population said that they used contraceptive methods, while 86.1 per cent used no contraception at all, indicating a low rate of family planning awareness;
- 95.2 per cent of the sample had not been visited by a midwife, while 4.8 per cent had been visited once for different purposes: baby care or mainly labour;
- 23 per cent travelled 6-10km for health care, and 33 per cent 3-5km;
- 20.4 per cent said that they had been examined once during pregnancy, 53.7 per cent twice, 16.3 per cent three times and 8.8 per cent four, so maternity care is less than desirable;
- 21.1 per cent reported that they had been examined once during the post-partum stage, 12.9 per cent twice, but 65 per cent had not been examined during this important stage;
- 91 per cent of children are immunized against measles, polio and diphtheria, and 87per cent are vaccinated regularly, revealing that the expanded immunization programme has been conducted successfully in the Programme Area; and 15.6 per

cent of patients sampled reported that they had been delivered by traditional birth attendants, 66 per cent by a midwife and 3.4 per cent by a doctor.

Comprehensive Health Centre at Umm al-Quttayn: Services lack specialized care, as no permanent specialist works there. The main ones needed are internal, obstetric and paediatric. No space is available for x-ray machines, and there are no technicians to operate them. The laboratory service is not available for 24 hours, owing to staff shortage, and the midwife works from only 08.00 to 14.00 hours. Only one dentist works in the Programme Area, with no replacement during his absence. No emergency services are available.

3. RECOMMENDATIONS AND SUGGESTIONS

The study undertaken may be considered as a basis for further detailed studies, projects and research activities, with the aim of contributing to sustainable development of the desertified Badia environment and to improvement of the standard of living of its inhabitants. With the understanding that the improvement of health status is an essential part of total socio-economic development, it is crucial to think holistically and initiate plans of action dealing with improvement of labour status, education, housing, water supply, environment and other aspects of daily life, as well as forecasting the future. In this light, the following recommendations and suggestions are made:

1. *Improvement of educational status*: by continuing courses for illiterates, especially women, and by better working conditions for teaching staff and learners.
2. *Improvement of the socio-economic conditions*: by creating fund-raising projects, mainly agricultural but also small handicraft and family projects similar to those successful Queen Noor al-Hussein, Save the Children and Queen Alia projects, which would contribute to population stabilization and development.
3. *Health care and health services delivery*:

- upgrade the peripheral health centre at al-Manara to PHC status by the Ministry of Health.
- upgrade Safawi PHC into a model CHC to serve the Programme Area, as it is a long way from Mafraq and al-Ruwayshid hospitals. Funding might be sponsored by the Programme administration and co-ordinated with the Ministry of Health.
- supply all PHC and CHCs with the required manpower specialists, and dentists.
- improve maternal and child health care through increasing the number of MCH centres to 5, equally distributed and equipped and improved through various components of international programmes such as USAID and UNFPA.
- constructing a small hospital in the Programme Area with a capacity of 25-30 beds in the coming 5 years is merely a suggestion if funds are available, but for the time being it is sufficient to improve and upgrade the existing services and facilities.
- raise the performance of health staff, particularly PHC and MCH centre staff, through training courses, continuing education and tangible incentives such as housing, fringe and field allowances, and special allowances for remote areas.
- activate health educational programmes by establishing incentive plans for community health committees, clubs and voluntary associations in collaboration with other NGOs and governmental institutions.

- perform studies, field research and health surveys aiming to screen the most common endemic, hereditary, communicable and non-communicable diseases.
- as theBadia CHC is considered to be the main provider of health services to the inhabitants of the Programme Area, it is suggested that it is attached directly to the development programme centre at Safawi by establishing an advanced emergency unit and gyno-obstetric unit with a capacity of 4 beds and appropriate staff.

CHAPTER 22

Sedentarization and Accessibility to Health Services

N.J. SPICER
(University of Glasgow)

1. INTRODUCTION

The population of the Badia of Jordan continues to shift from being primarily nomadic-pastoralist to being settled and ultimately urbanized. On the face of it, this is a geographical change; a tradition of substantial mobility being replaced by a sedentary existence in villages and towns. However, changes also include important economic, social and political shifts.

These changes reflect the process by which the population of the Badia is in the ongoing transition of becoming more and more integrated within the modern state of Jordan, both economically and socially, and less imbued in the Bedouin tribal system.

The dichotomous institutions of state and tribe both remain significant in almost every part of life in the Badia. For example, the urban government employee who resolutely identifies himself as a Bedu and the family with some members working in the government or in the military and others working with the family livestock. Nomadic Bedouin families use government services such as medical and health centres, and the Bedouin pastoralist economy continues to be integrated within the state sector.

Although both state and tribe coexist, the state is replacing the tribe as the dominant institution; economic and social changes are profound and the synchronous process of sedentarization is reducing the nomadic people to very small numbers.

This chapter illustrates how people of the Badia in different stages of the process of settlement exist both within the tribal system and within the modern state system. This is done by focusing on one important element of everyday life, access to and the use of 'modern' state provided health care services and alternatives such as 'traditional', tribal medicines.

Views of the users of health services regarding different options and accessibility are compared to those of the suppliers of health services, viz. the clinicians. The primary concern of the chapter is the health, accessibility and utility of health care for children.

2. POPULATION: A BRIEF OVERVIEW

The total population of the Badia Programme Area was 15,318 in 1993, of which the majority are settled in 34 villages in the north-eastern corner. The villages range in size from the smallest, such as al-Mansura with 4 houses and a population of 45 (1993), to the largest, Umm al-Quttayn with a population of 2,504 in 1993, increasing to 3,064 in 1996. To the east of this cluster is Safawi village with a population of 1,356 in 1996.

The population of the area consists of both settled and nomadic people as well as intermediate 'semi-nomadic' families. Table 1 shows the number of households with either a permanent house, house and *bayt sha'r* (tent) or just a *bayt sha'r*, and therefore provides an approximation of the numbers falling into settled, semi-nomadic and nomadic groups.

These figures are an indication only, since many fully settled households in the villages retain their *bayt sha'r* alongside their permanent house.

Table 1. Number of households in the Programme Area by type of house

Permanent house	1,625
House and *Bayt Sha'r*	520
Bayt Sha'r only	227

Source: Summary of Demographic and socio-economic survey for the Programme Area, 1993.

The state of Jordan granted a concession to the tribal sheikhs in the 1930s of areas of land which subsequently became the first villages. The older villages include Umm al-Quttayn, Dayr al-Kahf, Rifa'iyyat and al-Mukayfita. In the 1930s the 'villages' consisted of a well and a number of tents. Permanent houses have been built over the years so that now, only a few tents remain in the villages, and these are often those pitched adjacent to a permanent house. Services such as schools and health centres are a relatively recent addition, such as the Umm al-Quttayn health centre opened in 1980.

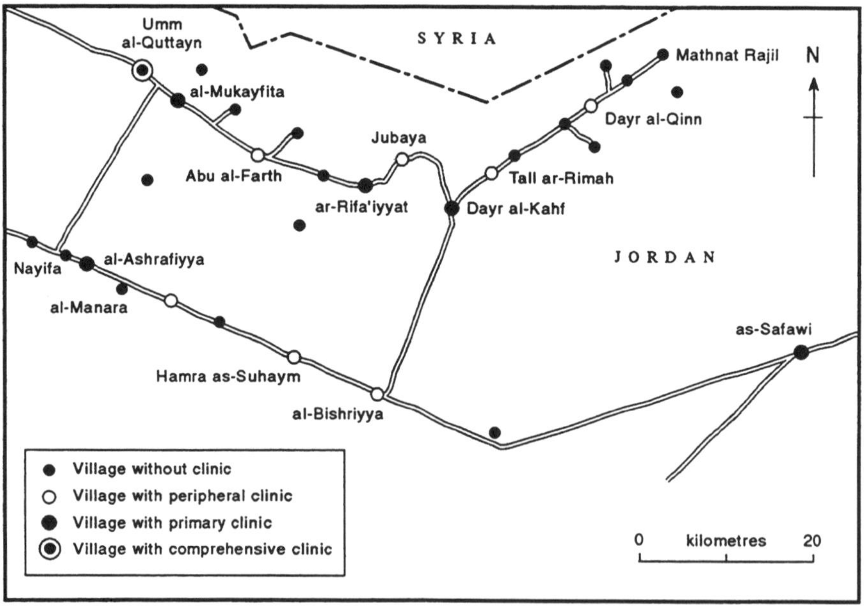

Fig. 1. Settlements and health care provision.

Before considering differences in the accessibility and use of health services between groups, it is first necessary define those groups. The contemporary meaning of 'Bedu' is more about 'identity', derived from strong sense of family and tribal roots than an indication that a person is currently a nomadic pastoralist. No doubt those identifying themselves as Bedouin have a nomadic heritage, but the vast majority now in the Programme Area have different occupations and circumstances. As an indication, Table 2 suggests that occupations are diverse in the Programme Area.

Table 2. The main income source of the head of household amongst participating families in the interview survey

Income source	Number
Farmer	89
Retired military (pension)	55
None	25
Military	11
Shop owner	9
Government employee	7
Employee (private sector)	6
Trader (unspecified)	4
Retired government employee	1
Retired Other employee	1

This study therefore concerns itself with the comparison of groups in different stages of sedentarization, rather than looking at the more subjective identification of Bedouin. As well as being primarily a geographic state, nomadism is intimately associated with pastoralism, mostly sheep in the present day, but also goats and camels. Not only does this represent a means of subsistence and/or income, but also a way of life. Defining a family or individual as either 'settled' or 'nomadic' or even 'semi-nomadic' is not easy since definitions of these groups must account for the geographical as well as economic and social characteristics of the families.

The follow working definitions are imperfect in representing the spectrum of social and economic circumstances of the Badia population.

However, they are useful in allowing households to be categorized for the purpose of comparison, as in Table 3.

Many families graze sheep outside the village, but within the locale. These households are considered settled because members of the family involved in the care of the sheep live permanently in the village throughout the year and return home most days, since the sheep are kept no more than a few kilometres away. Some settled families own livestock which is grazed and mobile within the Badia, but employ a shepherd to look after the sheep. It is often the case with semi-nomadic households that some members of the family remain in the village throughout the year, whilst others look after the animals in the Badia. Commonly, children remain with the parents in the Badia until school age, then move more permanently to the villages to receive schooling.

3. HEALTH AND HEALTH SERVICES IN THE AREA

A disparate population is one of the main difficulties in providing health services in the Badia. Although most of the population live in the villages clustered in the north-west of the Programme Area, a substantial number are nomadic or semi-nomadic. In addition, most of the villages are small, making it difficult to justify service outlets in each settlement. Centralization and a hierarchical system is the inevitable result. The Badia is also an area of low economic status compared to many other parts of the country, around 50 per cent of the Programme Area families reporting an income of less than JD100 per month (Note: 1 JD ≈ UK £1). Indicators of health lag behind most of the rest of Jordan, for example the Zarqa and Mafraq regions have an infant mortality rate of 38.5 per 1,000 births, compared to Amman with 36.0 (Jordan Population and Family Health Survey 1990, 1992). Jordan does, however, compare favourably with other Arab countries; Syria, with a similar GNP per capita had an infant mortality rate of 37 per 1,000 births in 1991, compared to 29 for Jordan in that year (Dejong 1995).

Table 3. Categories and description of sedentarization groups

Category	Description
Nomadic	Those who permanently use a tent and spend the whole year in the Badia with livestock, usually with significant seasonal mobility.
Semi-nomadic	A transitional state in which a family graze their livestock in the Badia area in a similar way to those who are fully nomadic, but also spend a significant part of the year based in a village, with a permanent dwelling. Often some members of the family are permanently based in a village whilst others spend much of the year grazing the sheep.
Settled	Those who have a permanent house in a village and are not occupied with pastoral activities outside the village.

There are 13 government health clinics of varying size in the villages of the north-east of the Programme Area (see Fig. 1). The village clinics are hierarchically organized, with Umm al-Quttayn village having the largest clinic in the Programme Area; a comprehensive level clinic that is open 24 hours a day, 7 days a week. There are 5 intermediate primary centres shown, open full time with lower order facilities for basic treatment. There are also 7 peripheral health centres in some the smaller villages. These are staffed around 1 or 2 days per week by a qualified doctor, but often have a nurse trained to a basic level for more of the week.

The area is within the Mafraq Governorate jurisdiction, and is accordingly part of the Jordanian hierarchy. There is a system of referral from the bottom to the top of the hierarchy; the peripheral clinics refer patients to the primary clinics, then to the comprehensive centre and ultimately to Mafraq city hospital, around 50km from the western border of the Programme Area, or even the hospitals in Amman if necessary. Other services include an extensive national vaccination programme which is administered through the health clinics in the area as well as through the local schools. Maternal and child health services and family planning services are also provided through the comprehensive centre and more modestly through the primary centres.

Costs of medicines and treatments vary depending on whether a government concession card is held. These are available to military and government employees and their families. Many Jordanian nationals also have concession cards, and special poverty cards are issued in a very limited number of cases. Military employees receive most medicines free, most government employees pay around JD1/2, nationals pay JD1 and non-nationals pay JD2. This could arguably be considered quite expensive since many households earn less than JD100 per month. Medicines are dispensed through the pharmacists at the health clinics in the Programme Area, although relatively few are available in the smaller clinics. A number of private pharmacists in the city of Mafraq are often used by those in the Programme Area. Supporting the main health service there are also a number of military hospitals in Jordan, as well as 'civil-defence' stations which provide ambulance and fire services. Both of these services are available in As-Safawi. There are also a number of private doctors and clinics within Mafraq.

The modern government services are not the only option; an important alternative, traditional medicine, referred to as 'Arabic Medicine' is widely used by the people of the Programme Area. The most commonly mentioned medicines were the 'famous' Arabic herbs, used for relatively minor conditions such as colds and coughs, headaches abdominal pains and the like. Other methods mentioned include the setting of bones using flour, eggs and water and a splint. Branding by fire on various parts of the body was also mentioned although no one indicated that they currently used this method with their children.

The vast majority of families interviewed make some use of Arabic herbs as a tea, both on a curative basis as well as more habitually. Very often these herbs are obtained directly by the families as they grow in or around the villages or in the Badia itself. Some also mentioned that they buy herbs from shops in Mafraq. Use of non-Western medicine would appear to be mostly confined to the use of herbs administered within the family. A few 'practitioners' of Arabic medicine were interviewed directly and referred to by others, and these are mostly older members of society and are rarely sought for the purpose of medical advise now.

4. ACCESSIBILITY AND UTILIZATION

The main focus is to consider problems of accessibility derived from geographical, economic, organizational and attitudinal aspects and to consider how these reflect patterns of differential utility between nomadic, semi-nomadic and settled populations. This chapter will focus on three detailed case studies of families in the area; one nomadic, one semi-nomadic and one settled. Whilst it is recognized this does not fully represent the population, it does illustrate the breadth of circumstances and problems which affect accessibility and utilization of health care. It also reflects many of the comments recapitulated within the overall survey of 200 interviews carried out in many of the Programme Area villages.

4.1. Case 1: Nomadic Family: Abu Abdullah

Abu Abdullah has two wives, one his age (43) and one substantially younger, aged 20, and three generations live within the household. The family has a substantial flock of sheep, from which it derives most of its income. They are grazed throughout the Badia and moved seasonally, the position depending on annual and seasonal opportunity for grazing, although the southern part of the Programme Area, such as towards al-Qutafi Well is frequented. The family live in the *bayt sha'r*, although the children aged between 10 and 17 receive education at the school in al-Bishriyya village, staying during the term time with relatives who own a permanent house in that village, and with their immediate family during the school holidays. The mother of the children is often now in the village with the children. The family owns a truck for transporting the flock as well as for personal use.

Use of modern health services depends on whether the children are in al-Bishriyya village, that is during school term or with the rest of the family, away from the villages at other times of the year. The al-Bishriyya peripheral clinic is used for minor illness when, and only when the children are there. During other times of the year, the children are not taken to the clinics for minor problems. Two of the children are under the age of six, that is before school age. They are away from the village permanently, so have poor geographical accessibility to the village clinics. They consequently have poor accessibility to the government vaccination programmes, such as the current one by the clinics aimed at the eradication of polio among those under school age. No use is made of the maternity and child health services available at the comprehensive and primary clinics in the Programme Area.

Distance, availability of transport and journey time were stated as major factors affecting the ability of the family to use the village clinics. Indeed, when the children are away from the villages, these factors stop the use of the clinics for minor illness. The vehicle is intensively used for the transportation of the sheep, as well as water and animal feeds, so availability for personal transport to the villages is limited. Whilst in the village the lack of private transport is not a problem since neighbours with vehicles are usually willing to help. There is considerably less opportunity for this kind of assistance whilst camping in the Badia since there are few or no immediate neighbours.

Waiting time at the clinics and hours of opening exacerbate the problem of geographical inaccessibility; there is little point in travelling a substantial distance to find the clinic closed, contributing to the non-use of the smaller clinics.

Abu Abdullah felt that those families who are well known at the health centres often receive better care. It was felt that the family often received a far more perfunctory service than they would have hoped for, and it was believed that if they were non-nomadic or better dressed they would be better received. In general, trust for, and confidence in the clinic at al-Bishriyya is low. The hospital in Mafraq is considered much better with this regard.

The family is not in the possession of any form of government health concession card, and stated that this affects their use of the health services considerably. Private medicine was rated more highly in terms of quality, but considered even more problematic due to being more expensive than government medicine.

I was told that the family use Arabic medicine only use in a parsimonious way with the children, usually in the form of herbal tea and on a curative basis only. These herbs are picked by the family from a number of wadi in the Badia.

4.2. Case 2: Semi-Nomadic Family: Abu Hussein

The family owns around 300 sheep, and additionally derives income from employment within the government and military, two adult brothers are occupied in this way and another grazes the family sheep. The semi-nomadic brother has use of a truck to transport the sheep, mostly within the Programme Area, and frequently east towards the Shubayka area. There are 3 generations living in the household.

Until 20 years ago the family was fully nomadic, with a flock of 500 sheep. The stated reason for beginning the process of settlement was the education of the children. Initially a few members of the family stayed in the village, particularly the mothers and the children for the purpose of education, whilst the remainder grazed the sheep in the

Badia. The situation 20 years ago is therefore similar to the present circumstances of case study 1. The Abu Hussein family settled in Abu Farth village because of connections with their wider family and tribal group. Ten years ago the permanent house was built in which most of the family now stay throughout the year. Much of the flock was sold to pay for the house. As the sons grew up they gained employment in the government and the military. One of the adult sons, his wife and two sons took over the responsibility of the flock, which is still seasonally mobile throughout the Badia. The father and mother, aged 65 and 55 respectively, are now permanently settled. Both economically and geographically the family has passed from being nomadic to semi-nomadic and is in a state of reviewing its current position with regards livestock production. Large scale pastoralism is becoming less and less tenable and the family recently sold a substantial proportion of the flock to pay for feed to sustain the remaining sheep.

The family use the local peripheral clinic as a first stop for children's minor illnesses and vaccinations. It is within easy walking distance in the village of Abu Farth. For more serious incidents the Umm al-Quttayn comprehensive clinic is used initially, although the family will use the Saidia clinic around 30km to the west, or a private doctor in Mafraq if considered necessary. The family have a pickup car in addition to the lorry used for transportation of sheep, making transport of people not overly problematic.

The big difference in accessibility and use of health services is between those children who spend the majority of their time away from the villages tending the sheep flocks in the Badia and those who remain in the village all year round. Those in the Badia always use herbal tea as a first option when illness is detected or suspected, waiting one or two days to see what the outcome is before going to a modern clinic.

The specific clinic used depends very much on the location of the people at the time, and will tend to be the larger clinics, where they can guarantee it will be open so as to not to waste a long journey. If a major illness occurs, those in the Badia will usually try to go straight to a higher order clinic, depending on the location at the time. Modern health services are rarely used for minor complaints by the adult brother, his wife and their sons who tend the sheep. Illnesses may become more serious by the time the journey is made to seek medical attention.

Those children in the village have very good geographical access to the local clinic. This becomes a little more difficult if one of the larger clinics outside Abu Farth is required. Although the family have a pickup car, often the women of the family, who are precluded from driving, need to take their children to a clinic, and have to rely on public buses. The limited number of buses that stop at Abu Farth make use of public transport difficult, although neighbours are often willing to assist if necessary.

The cost of treatments and medicine is an issue which affects accessibility differently with different members of the family. The respondent's brother is in the army and has a health concession card, which covers his wife and children as well as the parents of the house. However, the remainder of the family is excluded, so the family has to pay for medicines and treatment.

The opening times of the local Abu Farth clinic are considered a major problem. The clinic is only open one or two days of the week, and often for incomplete days. This means those in the village often need to seek a clinic in another village, for example the comprehensive centre at Umm al-Quttayn which is always open. The staff there often suggest the family return to Abu Farth and wait for the Abu Farth clinic to open to deal with the problem, although if the illness is clearly of a serious and urgent nature, patients are accepted. Referral through the system from lower order clinics to higher order ones is strongly encouraged by the providers of the health services. The children who are away for much of the year do not receive the government programme vaccines administered through the schools or clinics. Communication was also raised as a problem, resulting in not knowing when the vaccination programmes are to take place.

In terms of the relationship between the family and the doctor of the village clinic, I was told that this did not really have any particular affect on the use of the service above and beyond the other issues raised. However, a number of negative aspects were pointed out. For example, the doctors are predominantly from the cities of Jordan, with little local interest in the area and little empathy for the people of the Badia. Problems raised include the doctor distancing himself from the patient, rarely performing even a limited, perfunctory examination, usually just asking a few questions and prescribing medicines. It was claimed that those patients known or liked by the doctor are given priority.

4.3. Case 3: Settled Family: Umm Abdullah

This family is a three generation family living in Nayifa village. The father of the household is dead, leaving a widow, Umm Abdullah who is about 60 years old. The eldest son is established as the head of the household. The youngest member of the household is 3 years old, son of the head of the household. The family have lived in the village of Nayifa since 1963, and before that in Zarqa, a suburb of Amman. Last year the family sold their flock of around 70 sheep and goats, which were grazed in the immediate vicinity of the village. Family income is now mostly derived from the adult sons' employment in the government, but it owns a pickup car.

Use of health services is a lot simpler for this family than with the nomadic and semi-nomadic families because of proximity to the villages. The primary clinic, in the neighbouring village of al-Ashrafiyya, is used for almost everything and the Subha comprehensive clinic, a few kilometres to the west of the Programme is used in case of serious illness. The 3-year-old child receives vaccinations from the al-Ashrafiyya clinic, but no maternity services have been used by the family.

No geographical problems were raised as affecting the use of both the al-Ashrafiyya and Subha clinic, and being on the main road through the area, public transport is comparatively easy, for example a bus passes every hour to Mafraq, the journey taking about half an hour. I was told that costs are not a problem and the whole family has access to government rates of concession for treatments. Waiting to be seen by the doctor was raised as a major cause of inaccessibility as is the poor relationship with the staff of the local centre and can therefore reduce use for the more minor needs. Herbal medicines are used by the family on a regular basis, but modern medicine is rated as a more effective overall option for the children.

5. CLINICIAN'S PERSPECTIVES: UMM AL-QUTTAYN CLINIC

The issue of accessibility to medical services was discussed with a senior doctor of the Umm al-Quttayn comprehensive clinic by way of comparison with the views of the families interviewed. The clinician asserted that, although the people complain about the costs of treatments and medicines, these costs are very low. I was told that those without concession cards have to only pay a little more than those with them and that most people complain unnecessarily about the cost.

People's knowledge of health was raised as a core problem by the clinician. It was suggested that their perceptions of 'need' differ significantly from the formal modern ones of the service providers. The clinician complained about the people using the health clinic too often for minor, unnecessary cases. It was suggested that this is a greater problem than those who stayed at home when there was need to visit the clinic. Education was seen as an important way forward in improving the situation of the area, that is, the people being instructed what to do in certain situations.

The clinician agreed that many of the people in the area may have geographical problems with using the health centres, especially those who are nomadic. Regarding the relationship between the doctor and health services users, no specific problems were raised, although the problems of lack of knowledge and over-use of services were stressed. In terms of Arabic medicine, this was implicitly appraised as a placebo by the clinician and the overall attitude was to discourage the use of Arabic medicine in the area, although herbs are far less disparaged than other forms of non-Western medicine. It was stated that no popular medicine practitioners exist within the area.

6. DISCUSSION

The results of the household field survey have shown that patterns of accessibility and utility vary depending on specific family circumstances, whether they are nomadic, semi-nomadic or settled. Additionally, the type of service provided in the village influences their effective use; for example the smaller clinics, providing good geographical accessibility, but limited services, create problems. It is also clear that the factors are complex and affect each family differently. Often individuals have different circumstances within the family, for example individuals covered by a health concession card or not. I was assured, however, that there is no particular difference between girls and boys in terms of the way they were treated with regard to use of health services.

The process of sedentarization is more complex than simply a geographical transition, and this is reflected in the various problems of accessibility suggested in the case studies. The process is imbued within the economic, social, political and geographical integration of the population of the Badia into the modern state and economy. Differential accessibility reflects this state of transition, for example the nomadic family (case study 1) has poor geographical accessibility, having to travel considerably longer distances than those in the villages area, and are additionally hindered by not being fully 'within' the state system by not qualifying for the various government medical concession cards. These cards would appear to be easily obtainable by those who can perhaps better afford medicines and treatments, such as government employees and the military. Integration further into the state system, geographically, economically and socially would seem to increase the chances of good accessibility.

Conflict of opinion and interest would appear to be most pronounced between those at the opposite ends of spectrum such as the service providers and the nomadic users, manifested in imperfect relationships between those groups. The question is whether there are other ways to ensure that the people of the Badia, who are marginal within the state system, receive good access to fundamental services.

Perceptions of accessibility and inaccessibility and the consequential patterns of use of medicine and medical facilities vary between families. This is conceivably explained by each family, or even each individual, having a different set of circumstances which will reflect differential accessibility and use. However, considering comments made by the clinician, to what extent does accessibility to a service depend on an individual's perception of the need for the services, and how much does accessibility depend on difficulties in reaching it? Clinicians' views and those of the households interviewed seem to differ when it comes to considering some aspects of accessibility, as well as identifying need to use a service; for example those users who state that the cost of medicines adversely affects their use of the services, a view not shared by the service providers. The definitions of need for health care would also appear to differ between the people and the clinicians. Those interviewed stated that they took children to a health clinic when a need is perceived, particularly when a child was seen as suffering, however the clinician may not have agreed, suggesting the illness was not sufficiently important to be dealt with at the clinic. Clearly the different parties view need differently.

The clinician is the professional with regard to the administration of appropriate treatments or medicines for a certain classifiable illness. This assumes a previously defined notion of what health is and therefore what can be done about it, that is an identifiable disease with a certain cure. But is 'health' a broader concept and is it defined differently by different people? An important piece of consolidation research would be to tackle the question of differential views of 'what is health?' between those who provide the services and those who use them. It should then be possible to tackle the subjectivity of health as an influence on identifying need, accessibility to and use of health services. Perhaps a solution would be to help the people of the Badia to understand the scope and extent of the doctors ability to help them, and therefore more able to recognize the need to seek medical attention at the appropriate time.

REFERENCES

Dejong, J. (1995) Households, health and crises: Coping with economic upheaval in Jordan, 1988-1991. *Journal of International Development,* 7 (3) 443-65.

Jordan Population and Family Health Survey 1990 (1992).

Jordan Badia Research and Development Programme (1993) Summary of Demographic and Socio-Economic survey of the Jordan Badia Research and Development Programme Area.

ABSTRACTS

The following Arabic abstracts, and their English translations, summarize other work undertaken during phase one of the Badia Programme.

إدارة الموارد الأرضية في منطقة الصفاوي في البادية الشمالية الأردنية

بلال احمد علي أبو سليم

أجريت الدراسة بهدف تقييم الموارد الأرضية وتحديد مدى تدهورها. كما تم وضع خارطة وحدات أراضي بمقياس ١:١٠٠,٠٠٠ وخارطة تفصيلية لقاعي الشبيكة والبيقيعاوية بمقياس ١:١٢,٥٠٠. وتم اختيار نموذج لإدارة موارد الأرض والمياه والنبات في قيعان صحراوية أخرى، إضافة لقاعي الشبيكة والبقيعاوية.

لقد تم التوصل، باستعمال المنهج اللاندسكيبي في المسح والذي وضعه المعهد الدولي لمسوح الفضاء وعلوم الأراضي(ITC) ، إن منطقة الصفاوي وقاعي الشبيكة والبقيعاوية تعاني من تدهور الموارد الأرضية. كما أن قلة الاستفادة من مياه الأمطار واستنزاف المياه الجوفية أديا إلى تملحها.

أما التربة، فقد عانت كثيرا من الانجراف مما أدى إلى ضعف خواصها الطبيعية والغذائية. وأصبحت غير ذات فائدة رعوية.

وخلصت الدراسة إلى ضرورة الانتباه للتدهور في المنطقة وذلك من خلال تحسين مواردها من ارض ومياه، وتطبيق الحصاد المائي وزراعة النباتات المقاومة للجفاف ذات المنفعة البيئية والرعوية، وإنشاء محميات فيها. كما انه لا بد من توفر نظام معلومات جغرافي (GIS) خاص بالمنطقة لاستعماله وتحديثه باستمرار ضمانا لتحسن إدارة الموارد في المنطقة.

Land Resources Management in Safawi

BELAL AHMAD ABU SALIM
(*University of Jordan*)

Land resources were evaluated, their degree of deterioration determined and a land units map (1:100,000) prepared. A detailed map (1:12,500) for Shubayka and Bagi'awiyya was also prepared. A programme was selected for the management of the land, water and plant resources, also for other *qa'as* in addition to Shubayka and Bagi'awiyya.

By using the landscape approach for the survey, as devised by ITC, it was found that lands in Safawi and the *qa'as* of Shubayka and Bagi'awiyya have deteriorated. Also, the misuse of surface water and the over exploitation of ground water is the cause of some soil salinity.

Soil erosion is leading to the destruction of its physical properties and fertility, making it unsuitable for grazing. Therefore further deterioration of the soil and water should be prevented by proper management including the adoption of water harvesting measures and the creation of protected areas. A GIS could be a useful tool to assist with resource management.

الأشكال الأرضية البازلتية في منطقة الصفاوي في البادية الشمالية الأردنية

إبراهيم عبد المحسن علي الخلايلة

هدفت هذه الدراسة إلى حصر الأشكال الأرضية البازلتية، الموجودة في منطقة الدراسة، ونشأتها وتطورها وتصنيفها. كما هدفت إلى معرفة العمليات إلى أدت لنشأتها كالحت المائي والريحي وعمليات التجويه، وقد تلخص الهدف النهائي لهذه الدراسة في تحديد إمكانية استغلال أراضي المنطقة.

خلصت الدراسة إلى تحديد خمس وحدات جيومورفولوجية هي: المارل والكلس الطباشيري والمارلي؛ السطح التحاتي الاوليجوسيني؛ الوحدات الجيومورفولوجيه البركانية؛ وحدات الرواسب الرباعي، وأسطح الرواسب المائية والهوائية. كما تم تصنيف الأشكال البركانية في منطقة الدراسة٠

وقد وجد أن هناك ٣٢٠ بركانا صنفت على إنها من اندفاعات انسيابية (لابا قاعدية بازلتية) أخذت أشكال الدروع والقباب والربوات اللابية، واندفاعات انفجارية (فتاتية) أخذت أشكال مخاريط سكوريا وربوات سكوريا والمار، واندفاعات مختلطة أخذت أشكال البركان الطبقي والمخروط المشدوق، ووجدت أشكال أخرى كالموائد الصحراوية الناتجة عن الحت في اللابات، والقصبة البركانية الناتجة عن عمليات الحت في جسم المخاريط البركانية، كما تم تحديد أنماط شبكات التصريف المائي للأودية.

وعند دراسة الخصائص المورفومترية للمواد وعلاقتها بالعمليات الجيومورفولوجية وجد أن عمليات الحت كان لها نشاط مميز في تفتيت الصخور البازلتية وتشكيلها.

وخلصت الدراسة أخيرا إلى تحديد مراحل التطور الجيومورفولوجي البنائي للمنطقة وخاصة لاجسام البراكين وللآبار البركانية. كما تم وضع خارطتين جيومورفولوجية واستعمالات أراض بمقياس ١: ١٠٠,٠٠٠ وكيفية الاستخدام الاقتصادي الأمثل للبازلت ومنتجات البراكين.

Basalt Landforms in the Safawi Area of the Northern Badia of Jordan

IBRAHIM A. KHALAYLIH
(*University of Jordan*)

The origins of the basaltic landforms are outlined, as are the water and wind erosion processes and the potential for landuse development. Five geomorphological units are identified: marl, chalk and marly limestone; Oligocene eroded surface; volcanic unit; quaternary deposits; and wind and water deposits.

Volcanic structures are also classified and 320 volcanoes identified. The classification includes quiet eruptions (basic basaltic lava) that took the shapes of lava shields, lava domes and lava mounds. Shapes include pyroclastic ores formed like scoria cones, scoria mounds and maars, and mixed eruption volcanoes such as strata volcanoes and breached cones. Also found are classic desert landforms such as lava capped mesas (caused by erosion), lava fields, and volcanic necks that resulted from erosion. Drainage patterns in wadis were also studied. Morphometric analyses of volcanic landforms and their relationships to geomorphological processes showed that erosion has played an important role in rock weathering and formation. The structural geomorphological development of the area has been identified, especially for volcanic bodies and wells.

Two geomorphological and land use maps (1:100,000) have been also prepared. They show the current and suggested ideal economic use of basalt and other volcanic products in the area.

خصائص شبكة الطرق في البادية الشمالية واحتياجاتها المستقبلية

هاشم رفه المساعيد وتركي سليمان عبيدات

هدفت الدراسة إلى تحديد خصائص الطرق في منطقة المشروع من حيث صفاتها وتصاميمها وحجم المرور عليها. وقد تم القيام بمسوحات ميدانية عليها وتمت دراسة التوزيع السكاني في المنطقة أيضا.

ونتيجة لتزايد السكان والنشاطات البشرية المختلفة تم تطوير نماذج رياضية للتنبؤ بالمتطلبات المرورية على الشبكة.

بينت النتائج كفاية هذه الشبكة من حيث المواصفات والمسافات باستثناء الطريق الثانوي رقم ٢١٢ حيث يحتاج إلى تحسين مساره الأفقي والرأسي ليخدم سكان المنطقة. كما أن الطريق الرئيسي رقم ١٠ بحاجة للتوسيع ليكون بأربعة مسارب مع جزيرة وسطية إضافة لتحسين ثلاث طرق فرعية أخرى.

وبينت الدراسة كذلك ضرورة تحسين مستوى الصيانة للشبكة.

The Present Road Network and Suggested Improvements

HASHEM RAFFAH MASSA'ID
TURKI SULEIMAN OBEIDAT
(*Jordan University for Science and Technology*)

Road characteristics are identified including grades, designs and traffic flow. The population in relation to the road network is estimated. Future levels of traffic are predicted based on estimated population increases and changing patterns of activities.

The present roads meet current demands and were built according to specifications with the exception of road 212 which needs improvement to meet local needs. The main highway, no. 10, also needs to be enlarged to four lanes. Three other side roads need improvement.

In general, the maintenance programme for the whole road network needs to be upgraded.

الواقع الإداري
لمنطقة برنامج بحث وتطوير البادية الأردنية

عوني الهلسا و رجاء الفياض

هدفت الدراسة إلى التعرف على واقع الإدارة في منطقة المشروع وعلى دورها في تطوير المجتمع المحلي، وعلى المشكلات والمعوقات فيها التي تحول دون كونها منطقة جذب اقتصادي وسكاني، وانتهت الدراسة إلى وضع تطور تنظيمي إداري لتحقيق التنمية الاقتصادية والاجتماعية في المنطقة.

تخلل الدراسة زيارات ميدانية للمجالس القروية والقرى المتواجدة في المنطقة حتى الحدود السورية والعراقية.

احتوت الدراسة على وصف مفصل للواقع الإداري في المنطقة حيث بينت صلاحيات الحاكم الإداري بموجب الأنظمة المختلفة المعمول بها، بموجب نظام التشكيلات الإداري وقانون البلديات وقانون القرى ونظام إنشاء وصيانة الطرق والمشاريع القروية، ونظام مجالس الخدمات المشتركة. وبينت أيضا صلاحيات مجالس إدارة المحافظة والمجلس التنفيذي فيها والمجلس الاستشاري والبلديات من خلال أدارتها ومجالس البلدية ومجلس الخدمات المشترك.

كما بينت الدراسة مشكلات إدارة منطقة البرنامج من خلال الإطار التشريعي وصلاحيات واختصاصات المجلس القروي والمجلس البلدي في تحديد استعمالات الأراضي والخدمات العامة

والخدمات الصحية والصلاحيات الثقافية والمحافظة على البيئة الاقتصادية والمالية.

وعالجت الدراسة موضوع الرقابة على إدارة البلديات والمجالس القروية ووزارة الشؤون البلدية والقروية والبيئة، وعلى الأشخاص وأعمال المجلس البلدي ورقابة الحاكم الإداري وديوان المحاسبة وبنك تنمية المدن والقرى. وبحثت الدراسة في موارد المجالس المالية ومشكلاتها وضعفها واعتمادها على القروض نتيجة لتدني التحصيل وقلة المخصصات.

وخلصت الدراسة إلى بعض التوصيات التي من شأنها تفعيل دور الإدارة وجعلها اكثر ديناميكية من خلال إعادة النظر في صلاحيات المجالس ليتلاءم مع القدرة المالية، وإعادة النظر في الرقابة بحصرها بجهة واحدة منعا للازدواجية ومن خلال تفعيل دور إدارة منطقة البرنامج.

The Administrative Status of the Badia

AWNI HALASSA
RAJA' FAYYAD
(*General Administration Institution*)

A better understanding of the status of the local, public administration has been achieved, as has its role in community development. Also studied were the problems and constraints that prevent the region encouraging human and economic development. An administrative structure aimed at achieving social and economic development is proposed.

Town councils were visited including those at the Syrian and Iraqi borders.

A detailed description of the status of the administration and the authority of the governor is given in relation to the various laws in place: the municipal and national laws, maintenance and rural project laws and the laws of the joint services councils. The authority of the administrative councils of the governorate, the executive council, the advisory council and of the administrative municipalities and joint services councils are discussed.

The study also indicates the administrative problems of the area, within the limits of the legal framework and the specialist powers of the village and town councils, concerning land use, public services and health services, and the cultural authorities and the governorate on the economic and financial environment.

The study also deals with the supervision of village and town councils, the roles of the Ministry of Municipal, Village and Environmental Affairs and the staff and employees of the town council. It also deals with the supervision by the governor of the finance department and the Urban and Rural Development Bank. The financial resources of the councils are analysed, together with their related problems and weaknesses, and their dependence on loans as a result of low incomes and allocations.

Recommendations are made to increase the role and the effectiveness of the administration and make it more dynamic by reviewing the powers of the councils and their match with their financial powers, to revise the inspection procedure and to recommend only one authority so as to avoid duplication and increase efficacy.

الثروة الحيوانية في البادية الأردنية

نقيب كريم الشوشان

تناولت الدراسة نبذة تاريخية عن وضع الثروة الحيوانية والأسباب التي أدت لتقليل الاعتماد عليها، وهي أما أسباب اجتماعية أو اقتصادية.

وحددت الدراسة المشكلات التي تواجه قطاع الثروة الحيوانية في البادية مثل تقلص مساحة المراعي وتدني إنتاجها، وارتفاع تكلفة الإنتاج نتيجة لارتفاع تكاليف مدخلاتها كالعمالة والأعلاف، وتدني موارد المياه وضعف التسويق للمنتوجات الحيوانية.

كما تناولت الدراسة أساليب الإنتاج (الحليب، الصوف،...) وطرقه ومشكلاته، ومشكلات التسويق وأسباب زيادة أعداد الثروة الحيوانية في البادية خلال منتصف السبعينات وحتى الآن، نتيجة لتوفر المياه في أماكن واسعة ودعم الدولة للأعلاف وتسهيلات وسائل النقل وعودة بعض الأفراد من المدينة وتوفر رعاة وسهولة تسويق الحليب، وارتفاع أسعار اللحوم والمنتجات الحيوانية.

وانتهت الدراسة ببيان الأسباب التي أدت إلى غياب الإبل من البادية وهي تتلخص بقلة مردودها الاقتصادي والاعتماد على السيارات في النقل والارتحال وغيره وبطء تكاثر الإبل وضيق المراعي وانتشار المحميات وحاجتها الكبرى للأعلاف المدعومة وبطئها في التنقل.

Livestock Resources

NAQUIB KARIM SHOSHAN
(*Jordan Badia Research and Development Programme*)

The historical background of livestock resources is given together with the social and economic reasons that have led to diminishing dependence on them.

The problems facing the livestock sector are identified including the reduction of the range area, its low productivity and rising costs of production due to the higher cost of inputs such as labour and feed, and the limited water resources and the weakness of marketing animal products.

The means of production of milk and wool are considered together with their associated problems. Marketing issues and the reasons for the growth in livestock numbers since the mid-1970s to the present day are examined. These include the widespread introduction of more water pools, the government feed subsidies, transport, the return of some people from the urban centres to the Badia, the availability of more shepherds, a market for milk and the high prices of meat and animal products.

The study also lists the reasons for the absence of camels in the Badia, including their low economic returns, smaller range areas, the introduction of protected reserves, their high feed requirement, and their replacement by vehicles.

الأعلاف في البادية الأردنية

نقيب كريم الشوشان

بينت الدراسة الطرق التي كانت تتبع من قبل مربي المواشي لتوفير الأعلاف لها. كما بينت دور الحكومة في استيراد وتوفير ودعم الأعلاف للمربين. وبينت الدراسة كذلك الزيادة الحاصلة خلال الأعوام الأخيرة في أعداد المواشي على الرغم من الجفاف.

قدمت الدراسة بعض الإحصائيات حول أعداد المواشي في المحافظات المختلفة خلال الأعوام الماضية إذ بلغ مجموعها عام ١٩٩١ ٣,٣٤٦,٠٠٠ رأس.

اشتركت وزارة الزراعة مع وزارة التموين في وضع الإحصائيات عنها وتوفير الأعلاف لكل نوع حسب بطاقات علفية تصرف للمربين، وكمية الدعم بالدينار إذ بلغت حوالي ٤٣,٤٥٨,٠٠٠ دينار .

كما بينت الدراسة بعض الإيجابيات والعيوب لهذه السياسة، ويمكن تلخيص الإيجابيات بالاستقرار المعيشي للمربين وزيادة أعداد المواشي ومنتوجاتها والتصدير منها، أما السلبيات فهي عدم الاكتفاء الذاتي في الأردن فيما يتعلق باحتياجات اللحوم وزيادة الاستيراد للأعلاف وتزايد الدعم من قبل الحكومة نتيجة لزيادة العدد، وعدم وصول الدعم للمربين الحقيقيين في بعض الأحيان وسوء استغلال الدعم من قبل العديد من الحاصلين على بطاقات أعلاف.

وخلصت الدراسة إلى بعض التوصيات منها توحيد السعر واختيار مواقع مناسبة للتوزيع ومراقبة المخالفين وتحقيق العدل في توزيع بطاقات الأعلاف.

Feeds for Livestock

NAQUIB KARIM SHOSHAN
(*Jordan Badia Research and Development Programme*)

The means by which farmers provide feeds to their livestock are indicated as is the role of the government in providing subsidized feeds to the farmers. The study also estimates the increase in livestock in recent years in spite of drought. Statistics on livestock numbers are given by governorate up to 1991 when the total was almost 3.4 million.

The Ministry of Agriculture and the Ministry of Food Supply jointly surveyed livestock numbers, provided feed quota cards to the owners and issued feed according to the number of animals registered. The subsidy totalled JD43.5 million to mid-1996 when it was discontinued. The pros and cons of the feed subsidy policy are listed.

الحفائر

محمد السلاق و ابراهيم فرح

تتراوح معدلات كميات هطول الأمطار في منطقة المشروع بين ٦٩ ملم في محطة الوساد و ١٩٢ ملم في محطة جلعد. ويتراوح معدل جريان الفيضان السنوي طويل الأمد بين ٠,٠٣مليون متر مكعب في الحوض الصباب لحفيرة قطية و ٧,٧ مليون متر مكعب في الحوض الصباب التابع لحفيرة القاعدة الجوية. أما التبخر في الحفرتين فبلغ ٠,٨٥ مليون متر مكعب و ٢٣٥,٦ مليون متر مكعب على التوالي. وبذلك يكون معدل التبخر في الأحواض الصبابة للحفائر بين ٩٤ و ٩٥٪ ، ونسبة الفيضانات بين ٣,١ و ٣,٥.%

ونظرا لتدني الموارد المائية في المنطقة وارتفاع التبخر فمن الضروري عمل حفائر وسدود ترابية صغيرة لتجميع المياه فيها.

وقد أجريت هذه الدراسة لجمع وتقييم نتائج الفيضانات في منطقة المشروع وبالتحديد في مناطق الحفائر ووضع تصاميمها، إضافة لاقتراح مشاريع بحوث ودراسات للمنطقة، بالاستناد على المعلومات المتوفرة عنها التي تم تلخيصها في الدراسة مثل الأمطار والفيضانات والتبخر وخصوصية الأحواض.

كما قدمت هذه الدراسة شرحا لجيولوجية المنطقة وهيدرولوجيتها ومناخها وتربتها. كما وقد تم تحديد ١٥ موقعا لحفائر معظمها في وادي راجل، مع الأخذ بعين الاعتبار التجمعات البدوية والطرق في المنطقة. كما تم وضع مواصفات لهذه الحفائر من حيث التسليح والدحل وتبطين الاقنية ومواصفات المواسير والحاجز التحويلي والقناة التحويلية والحوض الترسيبي والحفرة الرئيسية وبئر التصريف والخط الناقل بين الحفرة الرئيسية وبئر التصريف، ومهرب المياه.

Water Ponds

MOHAMMAD SALLAQ
IBRAHIM FARAH
(*Ministry of Water and Irrigation*)

The rainfall range is estimated at 69mm (in Wisad) to 192mm (in Jalad). The long-term surface water flow is estimated at 0.03 million m^3 in the Sabbab catchment associated with the Qatya water pond and 7.7 million m^3 in the Sabbab catchment associated with the water pond at the air force base. The evaporation in the two water ponds reaches 0.85 and 235.6 million m^3 respectively. So, average evaporation from the water ponds in the Sabbab catchments is 94-95 per cent. Floods constitute 3.1-3.5 per cent of the rainfall.

Due to the low water resources in the area and the high evaporation, it is necessary to construct water ponds and small earth dams for water harvesting.

The study was carried out to collect and interpret flood flow data for the area, with the objective of designing and locating water ponds in addition to proposing research projects summarized in the study such as those on rainfall, floods, evaporation and catchment characteristics.

A description of the geology, hydrology, climate and soils of the area is presented. Fifteen locations for water ponds are identified, mostly in Wadi Rajil. The locations were selected taking into consideration groupings of Bedouin and the network of roads in the area. Specifications of the ponds are also given with respect to reinforcement, compaction, canal pavement, pipe sizes, conveyance dams, settlement pond, main pond, drainage well, the line from the main pond to the drainage well, and the main water drain.

مشروع جمع وتوثيق نقوش جديدة من منطقة الأزرق وآثارها

(1)المنطقة الشمالية : وادي سلمى

محمود محمد الروسان

يقع وادي سلمى في المنطقة الشمالية من الأزرق وتشمل هذه المنطقة أيضا الصفاوي والشبيكه. ويمتاز هذا الوادي بكثرة أعشابه. ويوجد على حافتيه عدد من الرجوم فيها مجموعة كبيرة من النقوش المختلفة وكذلك الرسوم ويحاذي الوادي مجموعة من الجبال العالية ذات صخور بركانية.

هدفت الدراسة إلى جمع نقوش وأدوات واوان فخارية لرسم صورة الإنسان الذي عاش في تلك المنطقة.

قسمت المنطقة إلى عشرة مواقع وقد تم وصف كل موقع من حيث ارتفاعه ومدافنه ونقوشه وحجارته وفخاره، وتم تحديد تاريخ المدافن فيها وتحليل النقوش والرسم عليها من حيث كونها يونانية أو صفوية أو عربية أو بيزنطية أو إسلامية.

Collection and Documentation of Newly Found Inscriptions and Ruins in the Azraq Area: North Region - Wadi Salma

MAHMOUD MOHAMMAD RUSAN
(*Yarmouk University*)

Wadi Salma is located in the northern part of the Azraq Basin, an area which also includes Safawi and Shubayka. The wadi is relatively well vegetated. On both sides of it are a series of artificial stony mounds on which are found a large collection of inscriptions and carvings. Surrounding both sides of the wadi are high mountains of volcanic origin.

Inscriptions, tools and ceramics were collected in the hope that they would yield information about the people who lived in the area.

The study area was divided into ten locations. Each is described with respect to height, graves, inscriptions, stones and ceramics. The ages of the graves were identified and the descriptions and drawings on them categorized as being ancient Greek, Sufi, Arab, Byzantine or Islamic.

الوضع الحالي للبرك والسدود والحفائر في منطقة الصفاوي

عزم الحمود، محمد السلاق، عمر ملكاوي، سعد العياش

تحتوي الدراسة على وصف للوضع الحالي لبرك مثناة راجل، الجدعاء، دير القن(١)و(٢)، تل الرماح، دير الكهف (شمال ووسط البلدة)، الجبية، ارينبة النعيمات، الرفاعيات، قاسم، الثلاج، المنيصة، منشية القنو، أبو الفرث، المكيفته، وأم القطين (شمال ووسط البلدة). وقد تضمنت الدراسة شرحا عن مقياس البركة وتبطينها وعمقها، وارتفاع جدارها ووجود حوض ترسيب فيها وطول المهرب وحالته بشكل مفصل والإصلاحات اللازمة لها.

وتناولت الدراسة بالتفصيل ماهية الأعمال الإنشائية اللازمة لاصلاح الحفائر والبرك والسدود عليها، والكميات اللازمة من الحصمة والباطون والرمل والاسمنت والماء لاصلاح هذه المنشآت.

The Present Condition of Pools, Dams and Water Ponds in the Safawi Region

AZM HOMOUD
(*Jordan University for Science and Technology*)

MOHAMMAD SALLAQ
(*Ministry of Water and Irrigation*)

OMAR MILKAWI
(*Royal Jordanian Geographic Centre*)

SAAD AYYASH
(*Jordan Badia Research and Development Programme*)

The existing pools of Mathnat Rajil, Jad'a, Dayr al-Qinn, Tall ar-Rimah, Dayr al-Kahf (the north and centre of the town), Jubaya, Urayniba Nuaymat, Rifa'iyyat, Qasim, Thilaj, Munaysa, Manshiyyat al-Qinu, Abu al-Farth, Mukayfita, and Umm al-Quttayn (the north and centre of the town) were studied. The study includes the dimension of each pool, its pavement, depth, surrounding wall height, presence of a settlement pond, length of the drain, detailed shape and required restoration.

The study describes in detail the work required to restore the water ponds, pools and earth dams and the quantities of gravel, cement, sand, reinforcing and water needed to repair these constructions.

الواقع التربوي في منطقة مشروع بحث وتطوير البادية الأردنية

موسى النبهان و ماجد القرعان

تناولت الدراسة واقع المدارس في منطقة المشروع من حيث كفاءات المعلمين وأعداد الطلبة وتوزيعهم وتحصيلهم العلمي.

بينت النتائج عدم توفر طاقم إداري متكامل إضافة لغياب المرافق التعليمية وعدم توفر مدارس مهنية. كما بينت الدراسة ضعف مستوى التحصيل العلمي عند الطلبة، وبينت أيضا عدم كفاءة المعلمين بسبب حداثتهم وقلة خبرتهم وشعورهم بعدم الاستقرار.

وانتهت الدراسة بتوصيات منها: زيادة عدد الغرف الصفية وإيجاد مرافق وبناء مدارس جديدة وتوفر المواصلات للطلبة وإقامة مدارس مهنية، أما بالنسبة للمعلمين فهناك حاجة لتأمين السكن والخدمات الضرورية لهم، إضافة لعقد دورات تدريبية لهم أيضا.

The Status of Education

MUSA NABHAN AND MAJED QURA'AN
(*National Centre for Human Resource Development*)

The status of schools is given with respect to staff standards and student numbers and the distribution and qualifications of both staff and students.

Results show the lack of fully qualified administrators in addition to the absence of educational infrastructure and vocational schools. The findings also indicate the poor academic standards of the students and low staff qualifications, the latter associated with the staff being young, inexperienced and feeling insecure because of lack of accommodation.

Recommendations are made for increasing classroom numbers, with some infrastructure, and new schools. Also recommended are: providing the means of transportation for the students, constructing vocational schools, also, for the teachers, providing lodging and essential services and organizing teacher training courses.

The study deals with the status of schools in the project region, with respect to their staff standards, student numbers, their distribution and academic qualifications.

The results show the lack of administrators, educational infrastructure and professional (polytechnic) schools. The results also indicate the poor academic standards of students, and of staff qualifications due to the staff being young, inexperienced and insecure.

The study concludes with some recommendations for increasing classroom numbers including some infrastructure, and new schools. It also recommends the following: providing transportation for both staff and students; constructing polytechnic schools; providing lodging and essential services for staff; and organizing staff training courses.